Advances in Experimental Medicine and Biology

Volume 1010

More information about this series at http://www.springer.com/series/5584

Xiaochu Zhang • Jie Shi • Ran Tao
Editors

Substance and Non-substance Addiction

Springer

Editors
Xiaochu Zhang
Key Laboratory of Brain Function and Disease, Chinese Academy of Sciences, and School of Life Sciences
University of Science & Technology of China
Hefei, Anhui, China

School of Humanities & Social Science
University of Science & Technology of China
Hefei, Anhui, China

Centers for Biomedical Engineering
University of Science & Technology of China
Hefei, Anhui, China

Ran Tao
Department of Psychological Medicine
PLA Army General Hospital
Beijing, China

Jie Shi
National Institute on Drug Dependence
Peking University
Beijing, China

ISSN 0065-2598 ISSN 2214-8019 (electronic)
Advances in Experimental Medicine and Biology
ISBN 978-981-10-5561-4 ISBN 978-981-10-5562-1 (eBook)
DOI 10.1007/978-981-10-5562-1

Library of Congress Control Number: 2017954505

Printed on acid-free paper

This Springer imprint is published by Springer Nature
The registered company is Springer Nature Singapore Pte Ltd.
The registered company address is: 152 Beach Road, #21-01/04 Gateway East, Singapore 189721, Singapore

Foreword

Like most other neuropsychiatric diseases, addiction remains stubbornly difficult to treat, with recidivism rates that may often exceed 90%. Indeed, the chronic relapsing nature of the disease is supported by the multiple times an individual presents for treatment before some are able to successfully develop long-term abstinence. The absence of a clinically verified biomarker for disease severity, and thus the ability to assess treatment efficacy, precludes declaring that sustained abstinence, in fact, indicates the absence of disease, rather than the absence of the behavior. This is aptly demonstrated by the high relapse rate seen after an individual returns to society following prolonged inpatient treatment or penal incarceration. The best current measurement of treatment success for drug dependence is urine testing, which is doubly unfortunate since it merely signals the presence or absence of a substance in the body, usually within a limited time window. However, the absence of drug does not signal absence of disease, nor does the presence of drug signal the presence of disease (e.g., the ability of most to partake in social alcohol consumption).

Over the past 40 plus years, tremendous new insights into the molecular, cellular, and neurobiological underpinnings of drug dependence have been gained, and the introduction of preclinical models that attempt to mirror addictive behaviors, e.g., drug self-administration, conditioned place preference and aversion, and cue- and stress-induced drug reinstatement, has led to a better understanding of the behavioral consequences of acute and chronic drug administration. Additionally, the explosion of research that has followed the introduction of truly breakthrough neurobiological and genetic tools, such as optogenetics and DREDD, has led to greater insight into the underlying neuronal cell types, circuits, and networks that demonstrate addiction-dependent plasticity. However, few new therapeutic agents have been discovered, and novel drug targets are still being sought. Indeed, only tobacco, alcohol, and opiate dependence even has approved medications—and these all have limited efficacy. And there are currently no efficacious pharmacotherapies for stimulants, i.e., cocaine and amphetamine or marijuana. Moreover, the so-called behavioral addiction, including pathological gambling, internet addiction, and binge eating, also is without specific pharmacological interventions. That said, various behavioral interventions, including cognitive behavioral therapy, contingency management, and motivational enhancement therapy, have been shown efficacious in a subset of both substance- and non-substance-dependent individuals.

Why has this veritable explosion of neuroscience knowledge not translated to better outcomes for our patients, whose lives and those of their families have been so devastated by addiction? Many hypotheses have been offered including the inability of animal models to fully capture the totality of addiction and thus do not provide a useful platform upon which to test novel pharmacotherapies. For example, while dopaminergic transmission and specific receptor subtypes have long been known to be dramatically altered following both acute drug administration and chronic drug administration in both preclinical and human studies, neither dopamine receptor agonist nor antagonists have been found to be clinically useful. Nevertheless, medicinal chemists continue the search for better, more specific dopaminergic ligands, with much emphasis today on the D3 subtype.

An alternative hypothesis is that addiction is a uniquely human disease that has very complex and intersecting causes including psychosocial, affective, socioeconomic, genetic, and neurobiological substrates. It is inherently a disease of pathological overlearning of a series of interlocking stimuli (environmental, social, pharmacological) with specific affective and cognitive outcomes. Indeed, the human literature has posited the dysregulation of complex cognitive constructs including reward sensitivity, impulsivity and response inhibition, decision-making and value determination, acute and long-term affective regulation, and ability to foresee and plan for the future. It is not clear how one captures these constructs with one or more preclinical models, some of which may have faced validity although for the most part, none have been shown to have predictive validity. Moreover, multiple constructs are likely to be dysregulated to comprise underlying behavioral regulation. For example, both impulsivity and cognitive control could both be compromised leading to compulsive behaviors, rather than simply too much of the former or not enough of the latter.

To make matters even more complex, it has long been appreciated that those who suffer from addiction also present clinically with other neuropsychiatric disorders, the most prevalent of which include depression, anxiety, and psychosis. Moreover, in most cases, the individual is not simply addicted to a single substance, but rather presents with dependence and/or abuse to multiple drugs. Thus, it would seem difficult with our current knowledge to propose a single pharmacological or even behavioral intervention to capture both the dual dependence and dual diagnoses of most of our patients. Moreover, treatment and research advances in the field of behavioral addictions have significantly lagged that of the pharmacological addictions, even with the growing acceptance that such compulsive behaviors as gambling follow many of the principles learned from substance dependence.

Despite the abovementioned challenges, significant advances have been made in recent years that leave one more optimistic. The field of noninvasive brain imaging, mostly MRI based, has provided for the first time the ability to directly observe changes in brain chemistry, structure, and function in the behaving human. Positron emission tomography has provided exciting insights into brain transmitter and receptor alterations in disease, and the explosion in human genetics and epigenetics has revealed a number of polymorphisms that may give insights into the risk levels and treatment options (i.e., personalized medicine) for our patients, all of which bring me to the current volume.

Zhang and his colleagues provide an extensive review of the current state of the art in human addictions. They powerfully demonstrate the behavioral and cognitive parallels between substance and non-substance dependence and argue compellingly in several of the chapters how further knowledge of the latter may profitably inform the former. That is, from a neurobiological perspective, the behavioral addictions may have some research advantages in that detected alterations in brain structure and function are not likely the result of an exogenously administered drug, which has the ability to engage multiple brain and peripheral (e.g., hormonal) systems both directly and indirectly. Indeed it is often difficult to disambiguate the pharmacological effects of a drug (e.g., attentional and working memory improvements following nicotine administration) from the dependence-producing properties of the agent. This is not the case for behavioral addictions where brain alterations are most likely the result of (or directly cause) the compulsive and destructive behaviors underlying the disease. Thus, what is importantly argued by many of the chapters is that much of what can be learned by studying these behavioral addictions may be profitably applied to all addictive disorders.

An important strength of the chapters in this book is the continued attempt to link factors that relate the behavioral to drug addictions; similarities and differences in the cognitive aspects of sensation seeking, intertemporal choice behavior, attentional bias, or inhibitory control are compared and contrasted. An important discussion of the difficulties of developing predictive preclinical models of compulsive behavioral addictions is also presented. Another strength of this book is the inclusion of somewhat less studied and less appreciated factors in dependence including neurotrophic factors, inflammatory factors (which notably has become much more appreciated of late in the field), neurovascular injury, as well as potential genetic and epigenetic biomarkers such as peripheral microRNAs. Finally, various potential therapeutic interventions are reviewed from the more traditional pharmacological and behavioral treatments to those less well studied including, nutritional, physical therapy and traditional Chinese medicine approaches.

What may be missing from the data presented and arguments made in these chapters is the appreciation that it may take more than understanding *differences* between the addicted and non-addicted brain and differences in substance and non-substance abuse to ultimately provide better treatments. While of course such basic knowledge is extremely important, what is needed is a better understanding of which one or more of these differences are, in fact, *predictive* of disease severity and treatment outcome success. This will require significant additional and logistically difficult and financially costly longitudinal studies. Much of the literature is comprised of cross-sectional research where a particular dependent variable is often shown to be different between populations. But two things that are different from each other are not necessarily predictive of the future. They could merely be different, and while such differences could be the result of the addiction, they could also represent premorbid, inherited differences and not be related to disease trajectory. This can only be determined from longitudinal studies.

Another missing area of research is that of risk factors for developing addiction. Once again, this is both vitally important and very difficult to obtain data. Moreover, such risk factors are often confounded by various normal neurodevelopmental changes to the individual. It is well known that most addictions begin during the early to late adolescent years—a time when the brain is undergoing rapid changes and thus very susceptible to environmental and pharmacological challenges.

That said, it is almost impossible for any single volume to cover the totality of addiction, and some topics are inevitably given less attention than others. This volume has taken a different approach and, rather than tackling the totality of the field, has chosen and successfully accomplishes the more novel and socially and medically important aspects of comparing the behavioral and drug addictions, with the premise that they share common neurobiological mechanisms and thus knowledge of one can inform the other. This book is recommended for students just starting out in the field, experienced treatment providers, and others who are interested in better understanding the complexities of the addictive disorders that so devastate communities around the globe. It provides a timely and up-to-date review of many of the current clinical and basic research issues and points out important gaps in our knowledge that need to be filled to improve the outlook of our patients.

Neuroimaging Research Branch, National Institute on Drug Abuse, Intramural Research Program
National Institutes of Health,
Bethesda, MD, USA

Elliot A. Stein

Contents

Contributors

Si-Zhi Ai National Institute on Drug Dependence, Peking University, Beijing, China

Hong An English Department, Tianjin University of Technology and Education, Tianjin, China

Hong Chen Faculty of Psychology, Southwest University, Chongqing, China

Manli Chen Department of Pharmacology, School of Basic Medical Sciences, Peking University Health Science Center, Beijing, China

National Institute on Drug Dependence, Peking University, Beijing, China

Xueli Chen Key Laboratory of Brain Function and Disease, Chinese Academy of Sciences, School of Life Sciences, University of Science & Technology of China, Hefei, Anhui, China

Yu Chen Department of Pharmacology, School of Basic Medical Sciences, Peking University Health Science Center, Beijing, China

National Institute on Drug Dependence, Peking University, Beijing, China

Federico d'Oleire Uquillas Department of Neurology, Massachusetts General Hospital, Harvard Medical School, Boston, MA, USA

Jianrui Ding Faculty of Psychology, Southwest University, Chongqing, China

Ri-Hui He RiHuiAddiction and Mental Disorders Medical Center, Guangzhou, China

Jason J. Li Program in Human Biology, Stanford University, Stanford, CA, USA

Mingzhu Li School of Educational Science, Anhui Normal University, Wuhu, China

Peng Li School of Educational Science, Anhui Normal University, Wuhu, China

Jie Liang Department of Pharmacology, School of Basic Medical Sciences, Peking University Health Science Center, Beijing, China

National Institute on Drug Dependence, Peking University, Beijing, China

Xiao Lin Peking-Tsinghua Center for Life Sciences and PKU-IDG/McGovern Institute for Brain Research, Peking University, Beijing, China

Haixiong Liu Zhejiang Provincial Key Laboratory of Addiction Research, Medical School of Ningbo University, Ningbo, Zhejiang Province, China

Huifeng Liu Zhejiang Provincial Key Laboratory of Addiction Research, Medical School of Ningbo University, Ningbo, Zhejiang Province, China

Lin Lu Institute of Mental Health/Peking University Sixth Hospital and National Clinical Research Center for Mental Disorders & Key Laboratory of Mental Health, Peking University, Beijing, China

Yamikani Ndasauka Chancellor College, University of Malawi, Zomba, Malawi

School of Humanities & Social Science, University of Science & Technology of China, Hefei, Anhui, China

Wenwen Shen Zhejiang Provincial Key Laboratory of Addiction Research, Medical School of Ningbo University, Ningbo, Zhejiang Province, China

Jie Shi National Institute on Drug Dependence, Peking University, Beijing, China

Delin Sun Duke-UNC Brain Imaging and Analysis Center, Duke University Medical Center, Durham, NC, USA

Yan Sun National Institute on Drug Dependence, Peking University, Beijing, China

Yan-Kun Sun Department of Pharmacology, School of Basic Medical Sciences, Peking University Health Science Center, Beijing, China

National Institute on Drug Dependence, Peking University, Beijing, China

Ran Tao Department of Psychological Medicine, PLA Army General Hospital, Beijing, China

Bin Xuan School of Educational Science, Anhui Normal University, Wuhu, China

Huijun Wang Faculty of Psychology, Southwest University, Chongqing, China

Xiaomei Wang Faculty of Psychology, Southwest University, Chongqing, China

Zhengde Wei Key Laboratory of Brain Function and Disease, Chinese Academy of Sciences, School of Life Sciences, University of Science & Technology of China, Hefei, Anhui, China

Li-Jun Xiao Department of Psychological Medicine, PLA Army General Hospital, Beijing, China

Xiaohu Xie Zhejiang Provincial Key Laboratory of Addiction Research, Medical School of Ningbo University, Ningbo, Zhejiang Province, China

Liping Yang School of Educational Science, Anhui Normal University, Wuhu, China

Xiaochu Zhang Key Laboratory of Brain Function and Disease, Chinese Academy of Sciences, and School of Life Sciences, University of Science & Technology of China, Hefei, Anhui, China

School of Humanities & Social Science, University of Science & Technology of China, Hefei, Anhui, China

Centers for Biomedical Engineering, University of Science & Technology of China, Hefei, Anhui, China

Yang Zhang Department of Pharmacology, School of Basic Medical Sciences, Peking University Health Science Center, Beijing, China

National Institute on Drug Dependence, Peking University, Beijing, China

Yun-Rong Zheng English Department, Tianjin University of Technology and Education, Tianjin, China

Jing Zhou School of Educational Science, Anhui Normal University, Wuhu, China

Wenhua Zhou Zhejiang Provincial Key Laboratory of Addiction Research, Medical School of Ningbo University, Ningbo, Zhejiang Province, China

Zhiling Zou Faculty of Psychology, Southwest University, Chongqing, China

Part I
Overview of Substance and Non-substance Addictions

Chapter 1
Received View of Addiction, Relapse and Treatment

Yamikani Ndasauka, Zhengde Wei, and Xiaochu Zhang

Abstract It is important to highlight that attempts at understanding and explaining addiction have been made for centuries. It is, however, just five decades ago, with the growth of science and technology that more interest has been observed in this field. This chapter examines different views and theories that have been posited to understand and explain addiction. More attention will be given to prominent views that seem to draw consensus among researchers and medical practitioners. The first section of the chapter introduces the addiction debate, the different theories that have been provided to explain it from different perspectives and disciplines such as neurosciences, philosophy and psychology. Then, the chapter discusses different views on the role of relapse and what it entails in understanding addiction. The second section discusses different proposed and used forms of treating addiction. Thus, the chapter discusses the received view of addiction, the understanding of relapse as a critical element in addiction and treatments.

Keywords Addiction • Relapse • Treatment

Y. Ndasauka
Chancellor College, University of Malawi, Zomba, Malawi

School of Humanities & Social Science, University of Science & Technology of China, Hefei, Anhui 230026, China

Z. Wei
Key Laboratory of Brain Function and Disease, Chinese Academy of Sciences, School of Life Sciences, University of Science & Technology of China, Hefei, Anhui 230027, China

X. Zhang (✉)
Key Laboratory of Brain Function and Disease, Chinese Academy of Sciences, and School of Life Sciences, University of Science & Technology of China, Hefei, Anhui 230027, China

School of Humanities & Social Science, University of Science & Technology of China, Hefei, Anhui 230026, China

Centers for Biomedical Engineering, University of Science & Technology of China, Hefei, Anhui 230027, China
e-mail: zxcustc@ustc.edu.cn

X. Zhang et al. (eds.), *Substance and Non-substance Addiction*, Advances in Experimental Medicine and Biology 1010, DOI 10.1007/978-981-10-5562-1_1

1.1 Understanding Addiction

What is addiction? In this section, we discuss this question as tackled from three different perspectives (psychological, biological and social-cultural perspectives) and demonstrate how each perspective, if independently conceived and propounded falls short in adequately addressing the questions. We will consequently present a nuanced view of addiction, largely accepted among scholars termed biopsychosocial model of addiction, taking into account the three perspectives whilst avoiding the weaknesses of independently holding either. Although this model has been discussed in key literature on addiction, especially substance related addiction, there is minimal discussion of the sociological/cultural perspective in behavioural addictions. This chapter seeks to add to this knowledge gap and will pay much attention to the social-cultural connotations of addiction.

1.1.1 Biological/Medicinal Perspective of Addiction

American Society of Addictive Medicine defines addiction as a primary, chronic disease of brain reward, motivation, memory and related circuitry [4]. Addiction is thus characterized by inability to constantly abstain, diminished behavioural control, craving, diminished recognition of significant problems with one's behaviour and dysfunctional emotional response. Addiction affects neurotransmission and exchanges within reward structures of the brain, including the nucleus accumbens, anterior cingulate cortex, basal forebrain and amygdala, such that motivational structures are altered. Addictive behaviours supersede healthy and self-care related behaviours. Addiction also affects neurotransmission and interactions between cortical and hippocampal circuits and brain reward structures, such that the memory of previous exposures to rewards (such as food, sex, alcohol, drugs and the internet) leads to a biological and behavioural response to external cues, in turn triggering craving and/or engagement in addictive behaviours [4].

It is widely accepted that the initial reinforcing effects of most addictive substances and behavior rely heavily upon the induction of large and rapid increases in the level of DA in the nucleus accumbens. DA, a multifaceted neurotransmitter, is involved in the fine-tuning of motor and cognitive function, modulation of salience attribution and attention, and regulation of reward and motivation. For instance, imaging studies have shown that in drug-addicted individuals, supra physiological levels of DA in the nucleus accumbens are followed by marked decreases in dopamine function [5, 20, 39–41, 77].

Further, in trying to understand and treat addiction, the biological/medicinal perspective relies on the view of addiction as being driven by reward. Reward is generally identified with pleasure, which reinforces the addictive behaviour. "The mesolimbic dopamine circuit is a hard-wired system in the brain … that provides pleasure in the process of rewarding certain behaviour" ([8]:132). In addicted individuals, natural rewards as well as the entire reward system are compromised. Addictive substance and behaviors are simply effective at temporarily blocking the negative reinforcement that addicted individuals experience during abstinence, resulting in conditioned learning towards the substance or behavior and previously neutral stimuli associated with it expounds [2, 31]. This increases the risk for relapse when an addicted person is exposed to the addictive substance or behavior or their cues [78].

In this view, both seeking and use or engagement in behaviour are motivated by innate pleasures caused by the addictive substance/behaviour. When the substance/behaviour no longer produces the same positive effect, people tend to increase use or engagement thereby leading to abuse. Addicts enter into withdrawal, taking or engaging more and more of the substance or the behaviour to produce pleasurable effects to counter the negative experiences of withdrawal [34].

Some neurological studies have proposed a different approach on the role of brain systems in addiction that emphasizes how addictive substances and behaviours affect motivations and incentives that individuals experience [12, 54, 62]. The emphasis of these theories is on seeking over using or engagement, proposing that "wanting" and seeking are central components of addictive behaviour.

Based totally on analysis with animal models, Robinson and Berridge [60, 61] developed a psychological model of brain performance and abuse. Rather than reinforcement, Robinson and Berridge projected that the central brain system concerned in substance abuse- the mesolimbic Dopastat system mediates incentive saliency. They "suggested that it's the method of incentive saliency attribution that transforms … the neural and psychological representations of stimuli, in order that they become particularly salient stimuli, stimuli that attract attention and become particularly engaging and wished, thereby eliciting approach and guiding behaviour to a particular goal" ([61]:104). Robinson and Berridge highlight 'wanting' as the individual feeling of incentive saliency, providing the need to pursue and use a substance or have interaction during an explicit behaviour.

In addition, Robinson and Berridge's theory emphasised the importance of associative learning and context in shaping the attribution of saliency, instead of substances/behaviour making an interior feeling of enjoyment. This view, wherever saliency is an element of larger activity interactions with the surroundings, helps open the door for biopsychosocial analysis through the thought of psychological and cognitive content processes. Robinson and Berridge did not deny the importance of positive reactions to addictive behaviour and substances, positing that "liking" plays a central role within the initiation of substance use and engagement in additive behaviour. However, they projected that association in nursing individual's sensitization to wanting drives problematic use. Through sensitization or associated accumulated reaction to substance/behaviour, "substance cues trigger excessive incentive motivation for substances, resulting in compulsive seeking" ([63]:36). Thus, Robinson and Berridge's theory projected that with increasing levels of use, the affected brain cells manufacture a greater-than-normal saliency signal [38]. Supported, this increase within the quantity of saliency signalled the drug-addicted individual experiences a strong need for substance/behaviour.

In clinical terms, this heightened prominence of substance and activity cues and connected behaviours corresponds to the compulsive seeking seen in drug abuse. In different words, the excessive prominence drives the compulsion to use, from seeking out medicine that now do not offer an equivalent enjoyable result to issues with relapse once addicts are trying to keep up abstinence. Thus, as compared to the classic reward approach, this theory helps make a case for sure problematic aspect of substance abuse. However, the analysis that led to the creation of this theory was primarily based totally on work with rats and needed some remodelling to be utilized in a social science project [38].

Robinson and Berridge [60–62] have conjointly emphasised wanting because of the subjective expertise associated with incentive prominence attribution. This wanting- an acutely awareness need for substance/behaviour drives each seeking and relapse. Thus, the excessive wanting of incentive prominence is seen because of the proximate mechanism driving cravings, the compulsive urge and/or need to use substance or interact in habit-forming behaviour [25]. Anthropology analysis has confirmed that "wanting" is a typical means that drug abusers represented the expertise of desire [45]. Thus, wanting received support as a relevant domain for exploration.

1.1.2 Psychological Perspective of Addiction

Addiction and Recovery, a popular site for information on addiction and people seeking help for addiction in the USA, defines addiction as a relationship with drugs or alcohol [behaviour] in which you use more than you would like to use, and you continue to use despite negative consequences [1]. People use drugs or alcohol and engage in some behaviour to escape, relax, or to reward themselves. But over time, drugs/alcohol and even some behaviour make people believe that they cannot live without them, or that they cannot enjoy life without using or engaging in them.

Psychology Today [55] defines addiction as a condition that results when a person ingests a substance (e.g., alcohol, cocaine, nicotine) or engages in an activity (e.g., gambling, sex, shopping) that can be pleasurable but the continued use/act of which becomes compulsive and interferes with ordinary life responsibilities, such as work, relationships, or health. Users may sometimes not be aware that their behaviour is out of control and causing problems for themselves and others.

According to the criteria of the American Psychiatric Association (DSM-IV) and World Health Organization (ICD-10), addiction should meet three of the following; (1) **Tolerance;** using more and more drugs/alcohol or engaging more and more in a particular behaviour over time. (2) **Withdrawal**; experiencing physical or emotional withdrawal when you have stopped using or engaging in a particular behaviour. Some signs of withdrawal include anxiety, irritability, shakes, sweats, nausea, or vomiting when abstaining from the particular addictive drug or behaviour. (3) **Limited control;** using a substance or engaging in behaviour more than you would like. This often times leads to regret after the activity but you still feel the need to continue using the substance or engaging in the behaviour. (4) **Negative consequences;** continued use of substance or engagement in a behaviour even after experiencing negative consequences to mood, self-esteem, health, job, education or family. (5) **Neglected or postponed activities;** putting off or reducing social, recreational, work, educational or household activities because of substance use or engagement in some behaviour. (6) **Significant time or energy spent;** spending a significant amount of time obtaining, using, concealing, planning, or recovering from use of substance or engagement in behaviour. This also involves thinking about the substance or behaviour, concealing and minimizing usage and engagement but failing to sustain it. (7) **The desire to cut down;** thinking about cutting down or controlling usage of substance or engagement in behaviour and unsuccessfully attempting to cut down or control usage or behaviour [55].

Although these criteria are largely applied to substance addiction, they have recently been adopted in some behavioural addictions like gambling addiction, Internet addiction and video and computer game addiction. One key question that arises when discussing addiction, especially behavioural addiction, is the fact that most behaviours that society may consider improper may well meet the above criteria. So, what really distinguishes addiction from other 'improper' behaviours or substances? This question is what distinguishes the biological, psychological cum philosophical and social cum cultural perspectives.

The above descriptions of addiction make mention or two critical notions that merit further discussion, thus, addiction as loss of control and thus involuntary and addiction as an impulsive disorder.

1.1.2.1 Addiction and Voluntary Action

Does an addicted person act freely and is the engagement in addictive behaviour a voluntary act? As noted earlier, addicted persons feel a strong urge to engage a particular behaviour. These persons find it difficult to resist the urge and consequently feel obliged to fulfil it in order to curb the pain that follows from not fulfilling it. For the medicinal perspective of addiction, this struggle and failure to resist indicates a form of compulsive disorder that ultimately is some sort of dysfunction in the brain pattern. As a result, an individual is not herself but is compelled to act in a particular manner by the 'disease'. On the contrary, the psychological perspective holds that addicted persons act voluntarily. In this section, we argue that addicted persons act voluntarily in the minimalist sense. That is to say, the disruption of the reward system in the brain acts as a major hindrance for the psyche to put into action decisions produced under the normal deliberative process. Aristotle's conception of *akrasia*, or weakness of the will, gives us an intermediary interpretation on the cause and process of addiction. This compromising position between the medicinal perspective and psychological perspective assumes both physical and mental impairment of an individual as the cause of addiction.

One characteristic of *akrasia,* which seems to be the dividing point of the two perspectives, concerns freewill and intentionality. This characteristic, as noted by Mele [43], is that incontinent action is "free, intentional action contrary to the agent's better judgment." However, not all intentional actions against one's better judgement may be considered as *akratic*. For Mele [44], some actions are compelled. This is the main point employed by the medicinal perspective in explaining drug addiction. The medicinal perspective claims that addicts are compelled. Though they act intentionally against their better judgement, they are compelled to act such by the disease in them. For the psychological model, this compulsion in addiction does not determine action; thus, it is not a necessary cause of action. Now, how is it possible to act voluntarily against one's own better judgment?

For Aristotle, the weak-willed know in so far as the relevant facts are available to them. As Sarah Broadie and Christopher Rowe [9] note, these individuals "are not unconscious or hypnotized," and they see no need to check if they have made the right decision. On the other hand, their knowledge seems not to be making any difference to their choices. Thus, "it is not on active duty when it ought to be, or not fully so (for it might be making them ashamed even as they act)" [9]. The knowledge that these individuals have is not practically realised because it is not impacting them or making any difference in them. Aristotle is a man of action. For Aristotle, the actual point of knowledge or being aware of what one should be doing is to do it, and not to contrast it with what one thought would have been done and feeling ashamed. The weak-willed demonstrate a failure to translate universals into particulars and use them in their present situation. As pointed out earlier, this failure is due to lack of discipline. For Aristotle, proper training and character building is fundamental in making the mind the right motivator of human action.

Now, are addicts free or do they engage in addictive behaviour voluntarily? Addiction may be understood as a case of weak-will. As noted in Aristotle, it may well be categorised as a battle of the mind and body/desires. What is central at this point is the power of deliberation and its ability to motivate action. The soul, despite being influenced by physical processes of the body, retains its agency and ability to influence the body. An influential approach on motivation of action is what Wallace calls the hydraulic conception of desire [79]. This position is inclined towards Hume's conception of passion as the motivating factor of action. Desires are thought as vectors of causal force to which we as agents are subject and which determine the actions we end up performing. This approach seems inadequate because it deprives an individual of her self-determination and the agent is depicted as subject to forces which are irresistible in that situation. This conception goes against phenomenological evidence of human agency and self-determination.

1.1.2.2 Addiction, Compulsivity and Impulsivity

To clearly discuss the concept of compulsivity, it is essential to discuss its relation to impulsivity, a multifaceted construct bearing numerous possible definitions [59]. Key elements of impulsivity include a predisposition toward rapid reactions, automatic and quick response to desires, insensitivity to delayed rewards and lack of reflection when making decisions [14]. Although healthy individuals may possess impulsive personality traits, excessive impulsivity is a key defining feature of many neuropsychiatric conditions, including attention deficit/hyperactivity disorder, substance use disorder, antisocial behaviour, as well as many behavioural addictions [11, 58].

Compulsivity refers to persistent or perseverative behaviour that's inappropriate to associated things, which have no obvious relationship to an overall goal. Compulsive behaviours are typically unpleasantly monotonous and done as habits [13]. Compulsive behaviours show hanging persistence, generally enduring for long periods despite being harmful to the individual. Compulsiveness may be a hallmark of many psychiatric conditions and is determined most notably in neurotic disorder (OCD). Alternative conditions that are thought to exhibit compulsive options embody uptake disorders, substance dependence and behavioural addictions like compulsive Internet use. The same as impulsivity, compulsiveness might arise from failures in response inhibition or "top-down" reasoning, associated over-stimulated drive state, or a mixture of those factors [23, 65].

Clinical observation has shown that patients with primary impulsive psychopathology (addictive gambling or addictive gaming) might demonstrate neurotic options, whereas patients with primary compulsive psychopathology (e.g., OCD) might score high on impulsivity ratings, and/or have impulsive-aggressive symptoms or comorbid impulse management disorders [66]. Additionally worth noting is that overlapping impulsive and compulsive options might develop in otherwise equivalent disorder. Totally different trajectories of impulsivity and compulsiveness among single disorders complicate the manner in which these disorders are understood by clinicians, also on how they're treated [23]. Rather than considering the two constructs as polar opposites, an alternate and additional realistic position is to contemplate impulsivity and compulsiveness as orthogonal factors across a variety of disorders, wherever either construct could be identified, there is a likelihood, higher or lower for the availability of the opposite construct.

Now, addiction provides a very important opening into debates regarding compulsiveness and impulsivity. Current psychobiological models perceive addiction as transition between impulsive and compulsive behaviour [11, 35]. Impulsivity may be at the foundation of an inclination to pursue short-run rewards led to by addictive behaviour and substances and is a powerful mechanism within the early stages of addiction (thus, problematic use or abuse) [18, 76]. With regular engagement in addictive behaviour or use of substance over extended amount of your time, impulsivity as a learning mechanism becomes over-trained and then develops into compulsive habits.

Two vital options of habitual behaviours and substance taking are: (1) the behaviour is driven by associations triggered by stimuli and not by an evident goal/reward; and (2) the individual is unable to reverse the repetitive activity pattern, therefore resulting in compulsive behaviour. In biological science terms, the transition from voluntary action in behaviour or substance use to a lot of habitual or compulsive modes looks to represent a shift in brain systems. This shift on top of things correlates clinically with a move aloof from absolutely strengthened behaviour actuated by reward seeking in problematic behaviour or drug abuse toward negatively strengthened behaviour actuated by the rejection of withdrawal symptoms in full-blown dependence and addiction [15].

Impulsivity plays a significant role in the early stage of addiction while maladaptive learning processes and habit formation lead to the development of compulsive behaviour in the later stage of addiction. The two should not hence be taken as parallels, but rather as one that builds on another. Dealing with impulsivity may help prevent addiction, but dealing with compulsivity may help in treating addiction.

In summary, a psychological cum philosophical perspective of addiction views addiction as a product of psychological maladjustment and functional impairment.

1.1.3 Social/Cultural Perspective of Addiction

We have been discussing the Biopsychosocial model of addiction. We have used this model to answer the questions: "what is addiction?" "What leads to addition?" and "what sustains addiction?" So far, we have reviewed the biological reasons people can get addicted. The biological portion of the BPS model considers addiction a brain disease with biological, chemical, and genetic roots. We have also reviewed the psychological reasons people can get addicted. The psychological portion of the model views addiction as a learned behaviour, a problem of faulty thinking, or of developmental delay. Other psychological disorders also contribute directly or indirectly to the development of an addiction. However, as we have attempted to demonstrate, holding polar views of either psychological or biological/medicinal perspectives does not fully account for the problem of addiction. Rather, a more nuance position, emanating from the discussion of voluntary action in Aristotle avoids weaknesses of the independent perspectives whilst retaining their strengths. This position acknowledges the role of impulsivity and compulsivity as a learning process and takes into account vulnerabilities that affect the biological brain processes or reward.

As psychology is concerned with understanding individual human behaviour, sociology is concerned with understanding the behaviour of larger groups (families, organizations, societies, cultures). Sociologists and psychologists both study the influence of these groups on individual behaviour. From a sociological perspective, addiction is a harmful behaviour that affects both individuals and groups. As such, we can only understand and correct addiction within the context of the society in which it occurs.

In this section, apart from referring to published texts and discourse on sociology of addiction, we take a steep turn to refer to text from literally writer, who attempts to elucidate the problem of addiction in a literal manner. This text is written by an individual who has had experience living with addicted persons and done informal research and talked to people from different backgrounds and cultures on this topic. Before getting into the text, we briefly discuss the social-cultural perspective of addiction.

Milkman and Sunderwirth [46] view addiction as a learned behaviour. The authors also note the consistently strong correlation between one's addiction and concurrent engagement or substance use by friends—a finding that suggests greater support for a sociological understanding of addiction. Social construction explanation posits that addiction is meaningful only within the conceptual categories available within culture and framed by social context [6]; therefore, the "particular features of and the meanings attributed to addiction experiences, as well as the behaviour thought to follow from them, are culturally specific" ([57], p. 316).

Culture may simply be defined as a group's learned and shared pattern of values and beliefs. These values and beliefs guide group members' behaviour and their social interactions. Cultural norms, practices and conceptions are transmitted from one generation to another through families. For instance, if one culture experienced oppression in the past; through learning of family history and imitation, feelings developed as a result of the oppression such as feelings of hopelessness, fear and loss are passed on to next generation who pass on to the next generation and so forth.

Such understanding of social and cultural forces helps in answering the question of how people get addicted. Three primary socio-cultural influences are important in responding to this question, namely, culture, families, and social support. Below, I will briefly discuss an excerpt from a TED TALK in order to highlight the role of family and social support in the development process of addiction. Johann Hari speaking in 2013 at a TED Talk Show- titled *Everything You Think You Know About Addiction Is Wrong*, said:

> Professor Alexander built a cage that he called "Rat Park," which is basically heaven for rats. They've got loads of cheese, they've got loads of coloured balls, and they've got loads of tunnels. Crucially, they've got loads of friends. They can have loads of sex. And they've got water bottles, the normal water and the drugged water. But here's the fascinating thing: In Rat Park, they don't like the drug water. They almost never use it. None of them ever use it compulsively. None of them ever overdose. You go from almost 100 percent overdose when they're isolated to zero percent overdose when they have happy and connected lives… Now, when he first saw this, Professor Alexander thought, maybe this is just a thing about rats, they're quite different to us. Maybe not as different as we'd like, but, you know—but fortunately, there was a human experiment into the exact same principle happening at the exact same time. It was called the Vietnam War. In Vietnam, 20 percent of all American troops were using loads of heroin, and if you look at the news reports from the time, they were really worried, because they thought, my God, we're going to have hundreds of thousands of junkies on the streets of the United States when the war ends; it made total sense. Now, those soldiers who were using loads of heroin were followed home. The Archives of General Psychiatry did a really detailed study, and what happened to them? It turns out they didn't go to rehab. They didn't go into withdrawal. Ninety-five percent of them just stopped… Professor Alexander began to think there might be a different story about addiction. He said, what if addiction isn't about your chemical hooks? What if addiction is about your cage? What if addiction is an adaptation to your environment?...Looking at this, there was another professor called Peter Cohen in the Netherlands who said…, maybe we shouldn't even call it addiction. Maybe we should call it bonding. Human beings have a natural and innate need to bond, and when we're happy and healthy, we'll bond and connect with each other, but if you can't do that, because you're traumatized or isolated or beaten down by life, you will bond with something that will give you some sense of relief. Now, that might be gambling, that might be pornography, that might be cocaine, that might be cannabis, but you will bond and connect with something because that's our nature. That's what we want as human beings. [26]

The series of experiments quoted in this excerpt highlight the role of the environment; of society and family; and their role in addiction. The history of a culture and some cultural values may advertently lead to excessive engagement in some behaviour. For instance, in "cultures where drinking is integrated into religious rites and social customs, where the place and manner of consumption are regulated by tradition and where, moreover, self-control, sociability, and 'knowing how to hold one's liquor' are matters of manly pride, alcoholism problems are at a minimum, provided no other variables are overriding. On the other hand, in those cultures where alcohol has been but recently introduced and has not become a part of pre-existing institutions, where no prescribed patterns of behaviour exist when 'under the influence,' where alcohol has been used by a dominant group the better to exploit a subject group, and where controls are new, legal, and prohibitionist, superseding traditional social regulation of an activity which previously has been accepted practice, one finds deviant, unacceptable and asocial behaviour, as well as chronic disabling alcoholism. In cultures where ambivalent attitudes toward drinking prevail, the incidence of alcoholism is also high," [7]). With such cultural differences in perspectives towards alcohol, prevalence of addiction to alcohol will also differ between cultures. This same understanding of cultural differences can also be applied to problematic engagement in other behaviours as predicted by culture. However, there are minimal studies to this effect, an area requiring further research, hence part of this thesis. In addition, as we discuss in the next chapter, studies have also shown that negative real life events, lack of social support and loneliness are some of key predictors of behavioural addictions. So, society and environment play a critical role not only in drug addiction but also in behavioural addictions.

From the reviews and deductions made, addiction is a crosscutting phenomenon, thus, it should be explained from different perspectives to present a holistic picture of what it is and its developmental process. Each perspective, separately, has limitations. The biological/medical perspective seems not to consider some phenomenological element of tendencies expressed by addicted persons. Whilst some people take a long road to recovery, often characterised by relapse and withdrawal symptoms, some people seem to find it easy to deal with addiction. Such evidence shows the limit of the biological perspective of addiction.

From this limitation, I note the importance of another perspective of addiction, thus, the socio-cultural perspective. Studies have also shown that addicted persons that receive proper support; are not estranged by family members but are accepted and supported to deal with addiction, find it less hard to recover from addiction than those who lack social and family support. On its own, the social cultural perspective tends to undermine the compulsive power of addiction, its compulsive nature that cannot be resolved by social support alone, but by directed and deliberate therapeutic intervention. For instance in behavioural addictions, people begin to engage in a particular activity for right reasons. However, due to excessive use and psychological factors not related to social factors, some people end up compulsively engaging in the activity. Though in its infancy, recent neuroscience evidence has shown that some people are genetically more vulnerable to addiction than others whilst some are vulnerable as a result of non-social psychological traits.

These studies reveal the complex nature of addiction. The non-social psychological traits that people attain in childhood development also play a critical role in explaining addiction. The interplay of these factors may lead to different results in different people. From the foregoing, it is hence imperative and essential to promote the biopsychosocial perspective of addiction, which will help develop and encourage diverse interventions to deal with the problem by addressing important factors.

1.2 Treating Addiction and Relapse: Biopsychosocial Perspective

As a clinical disorder, addiction requires long-term treatment that should and can only be measured in months and years. Is it recommended to individualize the treatment process of addiction [50]. Further, a complete evaluation is required in order to trace any co-existing medical, psychiatric and social problem that require redress together with the addiction treatment. Key to successful treatment of addiction is long-term prevention of relapse by pharmacological and behavioural means.

A biopsychosocial model of addiction entails treating addiction from all three fronts, thus, medicinal, psychological and social. It should be noted that addiction treatments vary depending on the form and level of addiction. Traditionally, strategies for preventing relapse have involved counselling and/or psychotherapy. However, more recently, pharmacotherapies and technologies combined with cognitive psychotherapies have been adopted and employed in treating addiction and preventing relapse.

1.2.1 Pharmacological Perspective

Some of the effective medications for treating opiate dependence include buprenorphine (commonly known as Suboxone®), Vivitrol® (extended-release naltrexone), and methadone [71–75]. Each of these three medications has been proven significantly more effective at preventing drug use relapse than a placebo in rigorous, double-blind experimental studies [3, 32, 33, 67].

The types of medication that have been found to be effective when combined with behavioural treatment in preventing relapse (like those stated above) can be classified as agonists, antagonists and anti-craving medications. These medications work through a variety of mechanisms. For instance, methadone is a full agonist and works by activating the opiate receptor, diminishing cravings for opiates and preventing euphoria if the patient abuses opiates [68]. On the other hand Vivitrol® contains extended-release naltrexone, which is a complete mu-receptor antagonist, meaning it completely blocks the mu-receptor. As a result, Vivitrol® prevents an individual from experiencing euphoria if he or she abuses any opiate, helping to prevent relapse [36], whilst Buprenorphine is a partial mu-agonist [72]. It prevents the patient from going into withdrawals or experiencing cravings, while preventing euphoria from any opiate used (including too much buprenorphine).

Aside pharmacological interventions, recently, researchers have explored the use of non-invasive brain stimulation techniques to treat addiction. From a symptomatic approach, it is tempting to think that non-invasive brain stimulation (NIBS) techniques, such as rTMS and transcranial Direct Current Stimulation (tDCS), may be of interest for individuals suffering from IA, as is the case in SUD. Indeed, the PubMed/Medline database contains more than thirty studies on the use of NIBS techniques to treat substance use disorder, including alcohol, tobacco, cocaine, cannabis, and methamphetamine. In most of these studies, brain stimulation seemed to lead to a significant decrease in craving, both in baseline and cue-induced craving, and may have led to an improvement in decision-making by reducing both impulsivity and risk-taking behaviour [16, 17, 21, 22].

1.2.2 Psychological Perspective

There are a number of behavioural interventions that have been found to effectively treat addictions. These interventions mainly involve behavioural therapy including motivation interviewing, contingency management therapy and the most adopted interventions- cognitive behavioral therapy. Motivational Interviewing is a counseling approach used to explore and resolve ambivalence about behavior change. There is a strong evidence base that it reduces substance use problems and a growing evidence base for other problems [30]. It has been defined as "a client-centered, directive method for enhancing intrinsic motivation to change by exploring and resolving ambivalence" ([47], p. 25). Contingency management provides tangible reinforcers for achieving target behaviors to increase the likelihood of those behaviors reoccurring. Typically, contingency management interventions identify an appropriate target behavior (e.g., abstinence as verified by a negative urine toxicology test) and provide tangible reinforcers each time the target behavior occurs [52]. The reinforcers are most often monetary- based vouchers exchangeable for retail goods and services or the chance to win prizes of varying magnitudes. If the target behavior does not occur, the reinforcers are removed [28, 53].

Cognitive behavioral therapy is an individualized, collaborative approach to psychotherapy that emphasizes the importance of thoughts, feelings, and expectancies and also incorporates more traditional behavioral approaches that utilize counter-conditioning and contingency management in addressing the problem of addiction [51]. It combines two very effective kinds of psychotherapy—cognitive therapy and behavioral therapy. Cognitive behavioral therapy is based on a number of theories including, social learning theory, stress theory and coping theory. It underlines that the learning processes play an important role in the development and continuation of addiction as well as reducing and treating addiction. Further, this intervention is cognizant of the view that stressors are likely to trigger addictive behavior as a coping strategy to avoid experiencing distress. As such, cognitive behavioral therapy focuses on challenging individuals' positive expectancies about substance use, enhancing their self-confidence and self-efficacy to resist addictive behavior and tendencies.

Mainly, cognitive behavioral therapy helps clients in two major behavioral ways. The first is to help reduce the intensity and frequency of their urges to use or engage in addictive substance or behavior, by undermining their underlying beliefs or cognitions about the substance or behavior. The second is to teach the clients specific techniques for controlling or managing their urges to use or engage in addictive substance or behavior. Cognitive behavioral therapy has been demonstrated to facilitate effectively improvement for a number of mainstream addictions. Reductions in drinking and drug use were seen mostly when clients were motivated to change and possessed at least a low average intelligence level needed to process and relate thought patterns with behavioral reactions [42]. Treatment gains with respect to stimulant use have been well established, with evidence that gains persist and grow over periods of 6–12 months [10, 56].

1.2.3 Social Perspective

In treating addiction, a biopsychosocial perspective highlights the role of the society/community in prevention of relapse for addicted persons. As highlighted earlier, parents, siblings, friends and the community at large play a critical role in safeguarding treatments of addiction and encouraging addicted persons to stay on course of their recovery. One of the systematic community approaches that has been proposed and found to be effective is the community reinforcement approach therapy. Hunt and Azrin [29] developed this intervention and tested it on persons with alcohol dependence. The community reinforcement approach is based on the theoretical view that individuals use substances or engage in some behavior for their positive, reinforcing effects and that the relative lack of alternative, non-drug and non-addictive reinforcers maintains dependence. The development of alternative reinforcing activities that are incompatible with drug use is therefore central to the community reinforcement approach [52].

In the community reinforcement approach, the therapist places a great deal of emphasis on changing environmental contingencies in the client's life. Employment, recreation, and family systems are all addressed to promote a lifestyle that is more reinforcing than substance use and/or additive behaviour. Rather than being entirely office-based, the community reinforcement approach is typically performed, at least in part, in the community. If clients do not attend treatment or do not follow through with an employment or recreational goal, the therapist may go to their homes, take them to job interviews, or help them try a new recreational activity. The purpose of expanding the treatment beyond the office setting is to increase the positive reinforcing effects of non-substance-using activities by direct exposure.

Studies have found the community reinforcement approach to be of therapeutic benefit to alcohol-dependent individuals [49, 64]. Further, several reviews and meta- analyses have concluded that the community reinforcement approach is an important, established, and effective treatment for alcohol use disorders [19, 48].

1.3 Conclusion

This chapter has presented a biopsychosocial perspective of understanding and treating addiction. It has stressed the role of society and community in treating addiction, underscoring the social nature of addiction. The assumptions, studies and recommendations presented herein are general, thus, covering both substance and behavioural addictions. This is based on the understanding that there are striking similarities between substance and behavioural addictions at the clinical, neurobiological and neurofunctional level [24, 37]. This understanding suggests therefore that interventions in treating addiction such as neurostimulation techniques, which are effective in treating substance addiction, could also be effective in treating similar symptoms in behavioural addictions.

As we conclude, it is important to stress that treatment of addiction is complex and should require a combination of techniques in order to adequately treat it and prevent relapse, hence a biopsychosocial approach to treating addiction. Experimental studies have found that the combination of medication and counselling is more effective than counselling alone at preventing relapse [27, 73]. Also, combining pharmacological and behavioural treatment leads to greater rate of retention than for either counselling or 12-step groups [69, 80]. According to the World Health Organization, the most effective treatment for opiate dependence is medication combined with counselling [70].

Acknowledgments This work was supported by the National Natural Science Foundation of China (31171083, 31230032, 31471071, 31771221), and the Fundamental Research Funds for the Central Universities of China, the National Key Basic Research Program (2016YFA0400900).

References

1. Addiction and Recovery (2015) http://www.addictionsandrecovery.org/what-is-addiction.html
2. Ahmed SH, Koob GF (2005) Transition to drug addiction: a negative reinforcement model based on an allostatic decrease in reward function. Psychopharmacology (Berl.) 180:473–490
3. Amato L, Davoli M, Perucci C, Ferri M, Faggiano F et al (2005) An overview of systematic reviews of the effectiveness of opiate maintenance therapies: available evidence to inform clinical practice and research. J Subst Abus Treat 28:321–329
4. ASAM (2016) http://www.asam.org/for-the-public/definition-of-addiction
5. Baler RD, Volkow ND (2006) Drug addiction: the neurobiology of disrupted self-control. Trends Mol Med 12(12):559–566
6. Becker HS (1953) Becoming a marijuana user. Am J Sociol 59:235–242
7. Blum RH, Blum EM (1969) A cultural case study., pp. 188–227. In: Blum RH et al (eds) Drugs I: society and drugs. Jossey-Bass, San Francisco, pp 226–227
8. Blum K, Cull JG, Braverman ER, Comings DE (1996) Reward deficiency syndrome. Am Sci 84:132–145
9. Broadie S, Rowe C (2002) Aristotle: Nicomachean ethics. Oxford University Press, Oxford
10. Carroll KM, Rounsaville BJ, Gordon LT et al (1994) Psychotherapy and pharmacotherapy for ambulatory cocaine abusers. Arch Gen Psychiatry 51:177–187
11. Dalley JW, Everitt BJ, Robbins TW (2011) Impulsivity, compulsivity, and top-down control. Neuron 69:680–694
12. Di Chiara G (1998) A motivational learning hypothesis of the role of mesolimbic dopamine in compulsive drug use. J Psychopharmacol 12:54–67
13. Ersche KD, Clark L, London M, Robbins TW, Sahakian BJ (2006) Profile of executive and memory function associated with amphetamine and opiate dependence. Neuropsychopharmacology 31(5):1036–1047
14. Evenden JL (1999) Varieties of impulsivity. Psychopharmacology 146:348–361
15. Everitt BJ, Robbins TW (2005) Neural systems of reinforcement for drug addiction: from actions to habits to compulsion. Nat Neurosci 8:1481–1489
16. Fecteau S, Agosta S, Hone-Blanchet A, Fregni F, Boggio P, Ciraulo D et al (2014) Modulation of smoking and decision-making behaviors with transcranial direct current stimulation in tobacco smokers: a preliminary study. Drug Alcohol Depend 140:78–84
17. Feil J, Zangen A (2010) Brain stimulation in the study and treatment of addiction. Neurosci Biobehav Rev 34:559–574

18. Fernández-Serrano MJ, Perales JC, Moreno-López L, Pérez-García M, Verdejo- García A (2012) Neuropsychological profiling of impulsivity and compulsivity in cocaine dependent individuals. Psychopharmacology 219:673–683
19. Finney JW, Monahan SC (1996) The cost- effectiveness of treatment for alcoholism: a second approximation. J Stud Alcohol 57:229–243
20. Garavan H et al (2000) Cue-induced cocaine craving: neuroanatomical specificity for drug users and drug stimuli. Am J Psychiatry 157:1789–1798
21. Gorelick DA, Zangen A, George MS (2014) Transcranial magnetic stimulation in the treatment of substance addiction: TMS as addiction treatment. Ann N Y Acad Sci 1327:79–93
22. Grall-Bronnec M, Sauvaget A (2014) The use of repetitive transcranial magnetic stimulation for modulating craving and addictive behaviours: a critical literature review of efficacy, technical and methodological considerations. Neurosci Biobehav Rev 47:592–613
23. Grant JE, Brewer JA, Potenza MN (2006) The neurobiology of substance and behavioral addictions. CNS Spectr 11:924–930
24. Griffiths MD (2005) A "components" model of addiction within a biopsychosocial framework. J Subst Use 10:191–197
25. Halikas JA (1997) Craving in substance abuse: a comprehensive textbook. In: Lowinson JH, Ruiz P, Millman RB, Langrod JG (eds) . Williams and Wilkins, Baltimore, pp 85–89
26. Hari J (2013) Everything you think you know about addiction is wrong. TED TALK
27. Hesse M, Pederson M (2008) Easy-access services in low-threshold opiate agonist maintenance. Int J Ment Heal Addict 6:316
28. Higgins ST, Budney AJ, Bickel WK, Foerg FE, Donham R, Badger GJ (1994) Incentives improve outcome in outpatient behavioral treatment of cocaine dependence. Arch Gen Psychiatry 51:568–576
29. Hunt GM, Azrin NH (1973) A community reinforcement approach to alcoholism. Behav Res Ther 11:91–104
30. Ingersoll KS, Wagner CC (2011) Motivational interviewing: emerging theory, research, and practice. In: Johnson BA (ed) Addiction medicine. Springer, New York, pp 705–726
31. Ito R et al (2000) Dissociation in conditioned dopamine release in the nucleus accumbens core and shell in response to cocaine cues and during cocaine-seeking behavior in rats. J Neurosci 20:7489–7495
32. Jones CM, Campopiano M, Baldwin G, McCance-Katz E (2015) National and state treatment need and capacity for opioid agonist medication-assisted treatment. Am J Public Health 105(8):e55–e63
33. Kjome K, Moeller F (2011) Long-acting injectable naltrexone for the management of patients with opioid dependence. Subst Abuse: Res Treat 5:1–9
34. Koob GF, Le Moal M (1997) Drug abuse: hedonic homeostatic dysregulation. Science 278:52–58
35. Koob GF, Volkow ND (2009) Neurocircuitry of addiction. Neuropsychopharmacology 35(1):217–238
36. Krupitsky E, Nunes EV, Ling W, Illeperuma A, Gastfriend DR et al (2001) Inject- able extended-release naltrexone for opioid dependence: a double-blind, placebo-controlled, multicentre randomized trial. Lancet 377(9776):1506–1513
37. Kuss DJ, Griffiths MD (2012) Internet and gaming addiction: a systematic literature review of neuroimaging studies. Brain Sci 2:347–374
38. Lende DH (2005) Wanting and drug use : a biocultural approach to the analysis of addiction. Ethos 33(1):100–124
39. López-Moreno JA, González-Cuevas G, Moreno G, Navarro M (2008) The pharmacology of the endocannabinoid system: functional and structural interactions with other neurotransmitter systems and their repercussions in behavioural addiction. Addict Biol 13(2):160–187
40. Martin-Soelch C et al (2001) Changes in reward-induced brain activation in opiate addicts. Eur J Neurosci 14:1360–1368

41. Martin-Solch C et al (2001) Changes in brain activation associated with reward processing in smokers and nonsmokers. A positron emission tomography study. Exp Brain Res 139:278–286
42. Mattson ME, Allen JP, Longabaugh R et al (1994) A chronological review of empirical studies matching alcoholic clients to treatment. J Stud Alcohol Suppl 12:16–29
43. Mele AR (1987) Irrationality: an essay on akrasia, self-deception, and self-control. Oxford University Press, New York
44. Mele AR (2006) Free will and luck. Oxford University Press, New York
45. Merikle EP (1999) The subjective experience of craving: an exploratory analysis. Subst Use Misuse 34:1101–1115
46. Milkman H, Sunderwirth S (1995) Doorway to excess. In: Inciardi JA, McElrath K (eds) The American drug scene: an anthology. Roxbury, Los Angeles, pp 12–22
47. Miller WR, Rollnick S (2002) Motivational interviewing: preparing people for change, 2nd edn. Guilford, New York
48. Miller WR, Wilbourne PL (2002) Mesa Grande: a methodological analysis of clinical trials of treatments for alcohol use disorders. Addiction 97:265–277
49. Miller WR, Meyers RJ, Tonigan JS (2001) A comparison of CRA and traditional approaches. In: Meyers RJ, Miller WR (eds) A community reinforcement approach to addiction treatment. Cambridge University Press, Cambridge, UK
50. O'Brien CP (2008) Evidence-based treatments of addiction. In: The neurobiology of addiction: new vistas, Philosophical Transactions: Biological Sciences, Vol. 363, No. 1507. Oxford University Press, Oxford, pp 3277–3286
51. Penberthy JK, Wartella JA, Vaughan M (2011) Cognitive behavioral therapy for addiction. In: Johnson BA (ed) Addiction medicine. Springer, New York, pp 729–750
52. Petry NM, Barry D (2011) Community reinforcement approach and contingency management therapies. In: Johnson BA (ed) Addiction medicine. Springer, New York, pp 751–773
53. Petry N, M. (2000) A comprehensive guide to the application of contingency management procedures in clinical settings. Drug Alcohol Depend 58:9–25
54. Pihl R, Peterson JB (1995) Alcoholism: the role of different motivational systems. J Psychiatry Neurosci 20:372–396
55. Psychology Today (2015) https://www.psychologytoday.com/basics/addiction
56. Rawson RA, Huber A, McCann M et al (2002) A comparison of contingency management and cognitive-behavioral approaches during methadone maintenance treatment for cocaine dependence. Arch Gen Psychiatry 59:817–824
57. Reinarman C (2005) Addiction as accomplishment: the discursive construction of disease. Addict Res Theory 13:307–320
58. Robbins TW, Gillan CM, Smith DG, de Wit S, Ersche KD (2011) Neurocognitive endophenotypes of impulsivity and compulsivity: towards dimensional psychiatry. Trends Cogn Sci 16:81–91
59. Robbins TW, Curran HV, de Wit H (2012) Special issue on impulsivity and compulsivity. Psychopharmacology 219:251–252
60. Robinson TE, Berridge KC (1993) The neural basis of drug craving: an incentive-sensitization theory of addiction. Brain Res-Brain Res Rev 18:247–291
61. Robinson TE, Berridge KC (2000) The psychology and neurobiology of addiction: an incentive-sensitization view. Addiction 95(Suppl 2):S91–S117
62. Robinson TE, Berridge KC (2001) Incentive-sensitization and addiction. Addiction 96:103–114
63. Robinson TE, Berridge KC (2003) Addiction. Annu Rev Psychol 54:25–53
64. Smith JE, Meyers RJ, Delaney HD (1998) The community reinforcement approach with homeless alcohol-dependent individuals. J Consult Clin Psychol 66:541–548
65. Stein DJ, Hollander E (1995) Obsessive-compulsive spectrum disorders. J Clin Psychiatry 56:265–266
66. Stein DJ, Lochner C (2006) Obsessive-compulsive spectrum disorders: a multidimensional approach. Psychiatr Clin N Am 29:343–351

67. Stotts A, Carrie Dodrill C, Kosten T (2009) Opioid dependence treatment: options in pharmacotherapy. Expert Opin Pharmacother 10(11):1727
68. Substance Abuse and Mental Health Services Administration (SAMHSA) (2014) Adult drug courts and medication-assisted treatment for opioid dependence. In Brief 8(1)
69. Trautmann S, Wittchen HU (2012) An analysis of German settings providing opioid maintenance therapy. Subst Use Misuse 47:2223
70. United Nations, Office on Drugs and Crime, World Health Organization (2009) Discussion paper: principles of drug dependence treatment. https://www.unodc.org/documents/drug-treatment/UNODC-WHO-Principles-of-Drug-Dependence-Treatment-March08.pdf
71. United States Department of Health and Human Services (USDHHS) (2015) About buprenorphine treatment. Retrieved from, http://buprenorphine.samhsa.gov/about.html
72. United States Department of Health and Human Services (USDHHS) (2015) Executive summary: opioid abuse in the U.S. and HHS actions to address opioid-drug related overdoses and deaths. Retrieved from, http://aspe.hhs.gov/sp/reports/2015/OpioidInitiative/es_OpioidInitiative.pdf
73. United States Department of Health and Human Services, National Institute on Drug Abuse (USDHHS) (2015) Principles of drug addiction: a research-based guide. What is drug addiction treatment? Retrieved from, http://www.drugabuse.gov/publications/principles-drug-addiction-treatment-research-based-guide-third-edition/frequently-asked-questions/what-drug-addiction-treatment
74. United States Department of Health and Human Services, National Institutes of Health (USDHHS) (2015) About NCATS, budget. Retrieved from, http://www.ncats.nih.gov/about/budget/budget.html
75. United States Department of Health and Human Services, National Institutes of Health (USDHHS) (2015) Trends & statistics. Retrieved from, http://www.drugabuse.gov/related-topics/trends-statistics
76. Verdejo-García A, Lawrence AJ, Clark L (2008) Impulsivity as a vulnerability marker for substance-use disorders: review of findings from high-risk research, problem gamblers and genetic association studies. Neurosci Biobehav Rev 32:777–810
77. Volkow ND et al (1997) Decreased striatal dopaminergic responsiveness in detoxified cocaine-dependent subjects. Nature 386:830–833
78. Volkow ND et al (2006) Cocaine cues and dopamine in dorsal striatum: mechanism of craving in cocaine addiction. J Neurosci 26:6583–6588
79. Wallace J (1999) Addiction as defect of the will: some philosophical reflection. Law Philos 18(6):621–654
80. Willenbring M, Hagedorn H, Postier A (2004) Variations in evidence-based clinical practices in nine United States veterans administration opioid agonist therapy clinics. Drug Alcohol Depend 75(1):97–106

Chapter 2
Definition of Substance and Non-substance Addiction

Zhiling Zou, Huijun Wang, Federico d'Oleire Uquillas, Xiaomei Wang, Jianrui Ding, and Hong Chen

Abstract Substance addiction (or drug addiction) is a neuropsychiatric disorder characterized by a recurring desire to continue taking the drug despite harmful consequences. Non-substance addiction (or behavioral addiction) covers pathological gambling, food addiction, internet addiction, and mobile phone addiction. Their definition is similar to drug addiction but they differ from each other in specific domains. This review aims to provide a brief overview of past and current definitions of substance and non-substance addiction, and also touches on the topic of diagnosing drug addiction and non-drug addiction, ultimately aiming to further the understanding of the key concepts needed for a foundation to study the biological and psychological underpinnings of addiction disorders.

Keywords Substance addiction • Drug addiction • Behavioral addiction • Non-substance addiction • Pathological gambling • Food addiction • Internet addiction • Mobile phone addiction

2.1 Introduction

Alcohol, tobacco, heroin and many other drugs can be found in our society. While illness, death, low productivity, and crime are all associated with drug addiction, overall it has an immeasurable emotional and social cost. Psychologists and psychiatrists have defined addiction as a neuropsychiatric disorder characterized by a recurring desire to continue taking the drug despite harmful consequences [36].

Zhiling Zou and Huijun Wang contributed equally to this work and share first authorship.

Z. Zou (✉) • H. Wang • X. Wang • J. Ding • H. Chen
Faculty of Psychology, Southwest University, Chongqing, China
e-mail: zouzl@swu.edu.cn

F. d'Oleire Uquillas
Department of Neurology, Massachusetts General Hospital, Harvard Medical School, Boston, MA, USA

X. Zhang et al. (eds.), *Substance and Non-substance Addiction*, Advances in Experimental Medicine and Biology 1010, DOI 10.1007/978-981-10-5562-1_2

Concrete diagnostic criteria for substance addiction (or drug addiction [16]), is set in the Diagnostic and Statistical Manual of Mental Disorders (DSM-V), or International Classification of Diseases and Related Health Problems (ICD-10), and have been widely used to diagnose addiction and evaluate its treatment.

While we may be more-or-less familiar with drug addiction, non-substance addiction has become a new rising problem in modern society. Non-substance addiction involves similar reward system circuitry as substance addiction. Though all of them are "addiction", the diagnosis of different kinds of addiction disorders is often complex as they differ from each other. In this review, we discuss the definition of gambling disorder, internet addiction, food addiction and phone addiction, and also summarize the diagnosis of each. Gambling disorder is the only non-substance addiction that was included in the DSM-V, indicating that the understanding of non-substance addiction remains sparse.

2.2 Substance Addiction

2.2.1 What Is Substance Addiction?

Addiction can be defined as the loss of control over drug use, or the compulsive seeking and taking of drugs despite adverse consequences [56]. Substance addiction (or drug addiction) is a neuropsychiatric disorder characterized by a recurring desire to continue taking the drug despite harmful consequences [36]. This drug-seeking behavior is associated with craving and loss of control [66]. Addiction is caused by the actions of drug abuse and generally requires repeated drug exposure. This process is strongly influenced both by the genetic makeup of the person and by the psychological and social context in which drug use occurs.

However, addiction was largely seen in the past as a moral failure in will-power. In the late eighteenth century, Benjamin Rush held the idea that addiction was 'a disease of the will'. Addicts were seen as subject to opposing forces, motivations, and other sorts of processes that both impelled them towards and away from a drug [41]. In this view, drug addiction was regarded as a moral condition induced by an addicts' weakness in will [43].

As seen from Table 2.1 below, there was a significant change in the way addiction was perceived from the DSM-I [1] to the DSM-V. Whereas in the DSM-I addiction was seen as a product of aberrances in personality, in the DSM-II [2] the wording changed to 'dependence' while focusing more on psychobiological constructs (e.g., 'evidence of habitual use, or a clear sense of need for the drug'). By the DSM-III, a distinction was made between 'substance dependence' and substance 'abuse', with the former characterized by physiological dependence (e.g., tolerance and withdrawal) and again rooted in the framework that it is a psychobiological disorder rather than a problem of personality or the mind. In the DSM-IV, factors contributing to addiction were identified as including not only psychophysiology (tolerance and withdrawal), but also cognition – a definition that would carry over to future editions of the DSM (V and VI). By the DSM-V, we see a much more

Table 2.1 Changes in the definition of substance dependence from DSM-I to DSM-V

	Definition and diagnosis criteria
DSM-I – 1952	Drug addiction is usually symptomatic of a personality disorder, and will be classified here while the individual is actually addicted; the proper personality classification is to be made as an additional diagnosis. Drug addiction is symptomatic of organic brain disorders, psychotic disorders, psychophysiologic disorders, and psychoneurotic disorders are classified here as a secondary diagnosis.
DSM-II – 1968	Drug dependence is a category for patients who are addicted to or dependent on drugs other than alcohol, tobacco, and ordinary caffeine-containing beverages. Dependence on medically prescribed drugs is also excluded so long as the drug is medically indicated and the intake is proportionate to the medical need. The diagnosis requires evidence of habitual use or a clear sense of need for the drug. Withdrawal symptoms are not the only evidence of dependence; while always present when opium derivatives are withdrawn, they may be entirely absent when cocaine or marihuana are withdrawn. The diagnosis may stand alone, or be coupled with any other diagnosis.
DSM-III – 1980	Substance Dependence generally is a more severe form of Substance Use Disorder than Substance Abuse, and requires physiological dependence, evidenced by either tolerance or withdrawal. Almost invariably there is also a pattern of pathological use that causes impairment in social or occupational functioning, although in rare cases the manifestations of the disorder are limited to physiological dependence.
DSM-IV – 1994	The essential feature of Substance Dependence is a cluster of cognitive, behavioral, and physiological symptoms indicating that the individual continues use of the substance despite significant substance-related problems. There is a pattern of repeated self-administration that usually results in tolerance, withdrawal, and compulsive drug-taking behavior. A diagnosis of Substance Dependence can be applied to every class of substances except caffeine.
	Although not specifically listed as a criterion item, "craving" (a strong subjective drive to use the substance) is likely to be experienced by most (if not all) individuals with Substance Dependence.
DSM-V – 2013	Overall, the diagnosis of a substance use disorder is based on a pathological pattern of behaviors related to use of the substance.
	Criterion A: Development of a substance-specific syndrome due to the recent ingestion of a substance. Criterion B: Changes are attributable to the physiological effects of the substance on the central nervous system. Criterion C: The substance-specific syndrome causes clinically significant distress or impairment in social, occupational, or other important areas of functioning. Criterion D: The symptoms are not attributable to another medical condition and are not better explained by another mental disorder.

holistic definition of substance dependence, emphasizing the psychobiological changes that occur from drug abuse which promote a lack of cognitive control over the use of the drug.

With the continuous development of advanced research techniques, various approaches have been applied to the field, and these have produced comprehensive insights into the processes underlying drug addiction. Via neuroimaging technology, experts have observed that chronic drug exposure causes stable changes in the brain at the molecular and cellular level, and that these changes may perhaps underlie behavioral abnormalities [56]. Gene knockout technology and genomic scanning enable us to identify both genes that contribute to individual risk for addiction and those through which drugs may cause addiction [56]. Based on this empirical evidence, experts tend to consider drug addiction as a kind of brain disease [43]. While early use of a drug may indeed be by choice, the neurobiological changes that occur with continued use, particularly to the prefrontal cortex among other regions related with executive function, compromise inhibitory control which when coupled with physiological and psychological craving for the drug lead to uncontrolled drug use [36, 37]. Thus, it is the mechanisms that occur as a result of taking the drug that make uncontrolled substance use a disorder.

Numerous drugs/substances can promote addiction. Thus far, scientists have identified the most common classes of addictive drugs/substances. According to the fifth revision of the Diagnostic and Statistical Manual of Mental Disorders (DSM-V) [5] and the Tenth Revision of the International Classification of Diseases and Related Health Problems (ICD-10) [83], the most common addictive drugs are: alcohol; caffeine; cannabis; hallucinogens; inhalants; opioids; sedatives, hypnotics and anxiolytics; cocaine; tobacco and other (or unknown) substances. In addition to these common drugs, DSM-V also list anxiolytics, amphetamine-type substances, and inhalants, while ICD-10 has volatile solvents on its list.

In the DSM-V, drug addiction is presented in the 'substance use disorders' section, which describes a cluster of cognitive, behavioral, and physiological symptoms indicating that the individual continues using the substance despite significant substance-related problems (see below). The detailed descriptions of these diagnostic criteria offer us a specific understanding of drug addiction. Generally, most drugs can fulfill 11 different diagnostic criteria terms, for the exception of caffeine.

Similarly, in ICD-10, the 'Dependence Syndrome' section also describes a cluster of physiological, behavioral, and cognitive phenomena in which the use of a substance or a class of substances, mainly fulfills 6 terms (see below). In contrast to the DSM-V however, the ICD-10 considers the desire (often strong, sometimes overpowering) to take psychoactive drugs as the central descriptive characteristic of a dependence syndrome.

2.2.2 Dependence Versus Addiction

The World Health Organization and the American Psychiatric Association once used the term "substance dependence" or "drug abuse", rather than "drug addiction", until the DSM-V was published [16, 57]. Drug dependence is a state of psychic or physical dependence, or both, on a drug, arising in a person following the administration of that drug on a periodic or continuous basis [26]. The transition from DSM-IV to DSM-V saw the preference for the word 'dependence' as a euphemism for addiction, reportedly as an attempt to help destigmatize addicted patients [69]. This however, resulted in confusion amongst clinicians, where 'dependence' in a DSM-sense was really 'addiction', yet dependence was known as the normal physiological adaptation to the repeated use of a drug or medication [58]. Thus, it important to highlight that pharmacological dependence is characterized by tolerance and/or withdrawal symptoms that arise from the continued exposure of the central nervous system to a drug. This is distinct from addiction, which is characterized by compulsive drug-seeking behavior.

2.2.3 How to Diagnose Substance Addiction?

Overall, drug addiction, or substance use disorder, may be diagnosed after thorough evaluation by a clinical psychologist, a psychiatrist, or licensed alcohol and drug counselor (http://www.mayoclinic.org). Current criteria for diagnosis are included in the Diagnostic and Statistical Manual of Mental Disorders (DSM-V) [5]. These include: ① Taking larger amounts or over a longer period than intended; ② A persistent desire or unsuccessful effort to cut down or control the use of the drug/substance; ③ A great deal of time is spent in activities necessary to obtain and use drug/substance or recover from its effects; ④ Craving, or a strong desire or urge to use the drug/substance; ⑤ Recurrent use resulting in a failure to fulfill major role obligations at work, school, or home; ⑥ Continued use despite having persistent or recurrent social or interpersonal problems caused or exacerbated by the effects of drugs; ⑦ Important social, occupational, or recreational activities are given up or reduced because of drug/substance use; ⑧ Recurrent use in situations in which it is physically hazardous; ⑨ Drug/substance use is continued despite knowledge of having a persistent or recurrent physical or psychological problem that is likely to have been caused or exacerbated by the drug; ⑩ Tolerance, as defined by either of the following: (a) a need for markedly increased amounts of drug/substance to achieve intoxication or desired effect, (b) a markedly diminished effect with continued use of the same amount of drug/substance; ⑪ Withdrawal, as manifested by either of the following: (a) the characteristic withdrawal syndrome for drug/substance, (b) drug/substance is taken to relieve or avoid withdrawal symptoms. These 11 criterion can be sorted into subgroupings of impaired control over substance (criterion 1 4), social impairment (criterion 5–7), risky use (criterion 8–9), and pharmacological criteria (criterion 10–11). However, it is necessary to note that different drug types fulfill different sets of criteria for withdrawal, and thus specific diagnoses should refer to drug-specific criteria sets of withdrawal.

The severity of substance use disorder can be from mild to severe, based on the number of symptom criteria endorsed. A mild substance use disorder can be referred by the presence of two to three symptoms, moderate by four to five symptoms, and severe by six or more symptoms within a 12-month period.

In ICD-10, a definite diagnosis of drug dependence should usually be made only if three or more of the following have been present together at some time during the previous year:

(a) a strong desire or sense of compulsion to take the substance;
(b) difficulties in controlling substance-taking behavior in terms of its onset, termination, or levels of use;

(c) a physiological withdrawal state when substance use has ceased or been reduced, as evidenced by: the characteristic withdrawal syndrome for the substance; or use of the same (or a closely related) substance with the intention of relieving or avoiding withdrawal symptoms;
(d) evidence of tolerance, such that increased doses of the psychoactive substance are required in order to achieve effects originally produced by lower doses (clear examples of this are found in alcohol- and opiate-dependent individuals who may take daily doses sufficient to incapacitate or kill non-tolerant users);
(e) progressive neglect of alternative pleasures or interests because of psychoactive substance use, increased amount of time necessary to obtain or take the substance or to recover from its effects;
(f) persisting with substance use despite clear evidence of overtly harmful consequences, such as harm to the liver through excessive drinking, depressive mood states consequent to periods of heavy substance use, or drug-related impairment of cognitive functioning; efforts should be made to determine that the user was actually, or could be expected to be, aware of the nature and extent of the harm [83].

2.3 Non-substance Addiction

2.3.1 Pathological Gambling

2.3.1.1 What Is Pathological Gambling?

Gambling, a widespread activity around the world, involves risking something of value in the hopes of obtaining something of greater value [28]. Most adults gamble without incurring problems. In the general population, the lifetime prevalence rate of pathological gambling is about 0.4–1.0% (DSM-V, [5]). Pathological gambling can be described as a disorder that consists of frequent, repeated episodes of gambling that dominate the patient's life to the detriment of social, occupational, material, and family values and commitments [28].

Excessive gambling was first officially recognized as a psychiatric disorder in the ninth edition of the International Classification of Diseases [82]. Three years later, pathological gambling (PG) was added to USA diagnostic coding of the Diagnostic and Statistical Manual of Mental Disorders, third edition (DSM-III, [3]). The original diagnostic criteria included in the DSM-III was based on several professors' clinical experience at that time, which put an emphasis on damage and disruption to the individual's family, personal or vocational pursuits, and money-related issues. In this edition, PG was classified as an impulse control disorder ('ICD'). In the next edition, the PG criteria were revised to reflect its similarity to substance dependence, such as the addition of "repeated unsuccessful attempts to control, cut back or stop gambling" (DSM-IV, [4]).

In the DSM-IV, PG was classified under the section of "Impulse Control Disorders Not Elsewhere Classified". As the growing scientific literature on PG reveals common elements with substance use disorders, PG was moved to the category "Substance-Related and Addictive Disorders" ('SAD') in the fifth revision of the DSM (DSM-V, [5]). Moreover, in the DSM-V, PG was renamed "gambling disorder". Today, PG is thus the only non-substance related disorder or behavior addiction in the SAD category.

2.3.1.2 Recreational Versus Pathological Gambling

It is known that when someone is too involved with gambling, he or she is prone to pathological gambling. But where should we draw the line? Fong et al., [30] summarized differences between a recreational gambler, and a pathological one. See two cases of gamblers: recreational versus pathological.

Gambler 1: Recreational Gambler

- Sixty-seven year-old retired physician who plays poker at the local casino 5 times per week and up to 5 h per session.
- Not increased gambling limits for the past 20 years.
- Never stayed at the casino for more than time planned.
- Allocates appropriate time for exercise and family.
- Financially comfortable with retirement account.
- Family is aware of gambling behavior.

Gambler 2: Pathological Gambler

- Twenty-year-old college student who gambles whenever he has money.
- Skips courses and assignments to gamble instead.
- Engages in bank fraud and steals from girlfriend to finance gambling.
- Has attempted to quit or reduce gambling 10 times in the last 2 years.
- Conceals gambling behavior from family and friends.
- Uses money from financial aid and scholarships to gamble.
- About to get kicked out of college for poor grades and financial status.

Compared with the first gambler, the second gambler cannot control his gambling, and experiences significant negative consequences from his gambling behavior.

2.3.1.3 How to Diagnose Pathological Gambling?

The definition of pathological gambling in the DSM-IV had been the most widely used diagnostic code for a long time. To be diagnosed as a pathological gambler according to the DSM-IV, an individual must meet at least 5 of the 10 diagnostic criteria, and all criteria were granted equal weight. Now we have the newest diagnostic code in the DSM-V, in which one major change of clinical description of gambling disorders includes the elimination of the "illegal acts" criterion. The rationale for this change is the low prevalence of illegal behavior among individuals with gambling disorder, and no studies have found that assessing criminal behavior helps distinguish between people with a gambling disorder and those without one [70].

To be diagnosed with gambling disorder according to the DSM-V, an individual must meet at least 4 of the 9 diagnostic criteria in a 12-month period. See below for the DSM-V's diagnostic criteria of gambling disorder:

1. Needs to gamble with increasing amounts of money in order to achieve the desired excitement.
2. Is restless or irritable when attempting to cut down or stop gambling.
3. Has made repeated unsuccessful efforts to control, cut back, or stop gambling.
4. Is often preoccupied with gambling (e.g., having persistent thoughts of reliving past gambling experiences, handicapping or planning the next venture, thinking of ways to get money with which to gamble).
5. Often gambles when feeling distressed (e.g., helpless, guilty, anxious, depressed).
6. After losing money gambling, often returns another day to get even ("chasing" one's losses).
7. Lies to conceal the extent of involvement with gambling.
8. Has jeopardized or lost a significant relationship, job, or educational or career opportunity because of gambling.
9. Relies on others to provide money to relieve desperate financial situations caused by gambling.

Apart from the DSM-V, screening tools such as the South Oaks Gambling Screen (SOGS), Problem Gambling Severity Index (PGSI), Lie/Bet Questionnaire, Gamblers Anonymous Twenty Questions (GA20) and the Massachusetts Gambling Screen (MAGS) can also help with diagnosis.

2.3.2 Food Addiction

2.3.2.1 What Is Food Addiction?

Food addiction (FA) was first proposed in the mid-1950s [61] as a loss of control over food intake with a persistent desire for food and unsuccessful attempts to curtail the amount of food eaten despite knowledge of adverse consequences [89]. Noting that obese and individuals who overeat, display characteristics reminiscent of addiction, an addiction model has been used to explain the abnormal eating patterns found in obese [77] and overweight individuals, as well as in patients with Eating Disorders (EDs) [44, 79].

The concept of food addiction has been controversial due to definitional and conceptual difficulties, as well as from a lack of rigorous scientific data [21]. Nevertheless, this concept still attracts much scientific and popular media interest [89], and the concrete attempts to operationalize the FA construct are quite recent.

H. Ziauddeen et al., reviewed food addiction as a phenotypic description, one that is based on an overlap between certain eating behaviors and substance dependence [89]. More recently, Imperatori et al., argued that FA seems to be a transnosografic construct that exists in all EDs [44]. They argued that FA refers to specific food related behaviors characterized by excessive and dysregulated consumption of high calorie food (i.e., foods with high sugar and/or fat). Most recently, FA is defined as a chronic and relapsing condition caused by the interaction of many complex variables that increase cravings for certain specific foods in order to achieve a state of high pleasure, energy or excitement, or to relieve negative emotional or physical states [44].

2.3.2.2 Food Addiction Verse Eating Disorders

It is crucial to distinguish the similarities and differences between FA and EDs, as well as between FA and other addiction models.

Several eating disorders have already been identified in DSM-V: (1) Bulimia Nervosa (BN); (2) Anorexia Nervosa (AN); (3) overweight and obesity; (4) Binge Eating Disorder (BED); (5) Night Eating Syndrome (NES); (6) Food Craving (FC) [5]. FA, as an abnormal consumption pattern, seems to have significant psychopathological overlaps with these EDs, especially with BED and BN. Reduced control over eating, continued use despite negative consequences, and elevated levels of impulsivity and psychopathology are several overlaps between FA and both BED and BN [32, 60].

However, there are also some crucial differences between FA and other EDs [8, 32, 60]. (1) Contrary to FA, BED is associated with elevated concerns with shape or weight. This crucial psychopathological core construct is not considered in patients with FA. (2) Contrary to FA, BED and BN diagnoses specify that binge eating episodes must occur during a discrete period of time. (3) FA has many typical symptoms of addiction, e.g., tolerance, withdrawal syndrome, devoting a certain amount of time to activities associated with eating and neglecting or abandoning other activities for food, which are not included in any ED. (4) The function of eating in BED is to reduce mental tension (caused by for example: cognitive distortions related to food, embarrassment caused by shape and weight, eating restraint), however, in FA, food is used to induce hedonistic satisfaction (pleasant psychophysiological feelings). (5) Contrary to BED, the body mass in FA is excessive or normal (e.g. when the dependence relates to one specific product).

2.3.2.3 How to Diagnose Food addiction?

It is important to note that the most widely employed definition of FA derives from the overlay with the DSM-IV-TR criteria [4] for drug addiction. These criteria include: (1) Substance taken in larger amount and for longer period than intended;

(2) persistent desire or repeated unsuccessful attempts to quit; (3) a large amount of time/activity necessary to obtain, to use or to recover; (4) important social, occupational, or recreational activities dismissed or reduced; (5) continuative use despite knowledge of adverse consequences; (6) tolerance; (7) withdrawal symptoms [4]. For a person to be considered addicted to any given specific food, at least three of the seven criteria must be met at any time within a given year [22].

Several questionnaires have also been developed to help diagnose FA. For example, in 2009, Merlo et al., developed the Eating Behaviors Questionnaire (EBQ) [55] to investigate, in a pediatric sample, the three crucial components of FA, the so-called "3 Cs" of addiction: Compulsive use, attempts to Cut down (quitting attempts), and Continued use despite adverse consequences. Despite its good psychometric properties, attempts to adapt this self-report to an adult population have not yet been pursued [44].

The Yale Food Addiction Scale (YFAS) [31], developed by Gearhardt et al., is the most commonly used tool to assess FA in clinical and non-clinical samples. It is a specific self-report questionnaire for FA evaluation and diagnosis, most recently presented as the 35-item YFAS 2.0 [34]. It investigates eating behaviors concerning hyper-palatable food consumed in the previous 12 months [44]. A symptom count can be calculated, which can range between 0 and 11 symptoms. Furthermore, a diagnostic score can be calculated ('food addiction' vs no 'food addiction'), and diagnosis can be further specified as mild, moderate, or severe, depending on the number of symptoms present [78]. Internal consistency has an $\alpha = 0.970$. A shorter version of the YFAS and a version for children have recently been developed [33, 44].

2.3.3 *Internet Addiction*

2.3.3.1 What Is Internet Addiction?

The phenomenon of Internet addiction has many related terms, including Internet Addiction (IA), Internet Addiction Disorder (IAD), Internet dependency or Internet dependence (ID), Pathological internet use or Problematic internet use (PIU), excessive internet use, and impulsive-compulsive internet usage disorder (IC-IUD).

Professor Kimberly Young, a leading American psychologist and international Internet addiction rehabilitation specialist, published a study on computer/internet addiction at the 1996 American Psychological Association conference, the world's first study on computer/internet addiction. In her report, she began to study internet addiction from a clinicopathological point of view, defining internet addiction as "an Impulse-Control Disorder that does not involve poisoning". Her study not only caused the attendees great interest and concern, but it also led to its study in-depth [86].

Armstrong et al., [6] described internet addiction in a more comprehensive way, saying: "Internet addiction is a very broad concept, addiction has a lot of behavior and impulse control problems, such as Internet addiction, Internet compulsive behavior, information gathering addiction, computer addiction, etc.". However, Davis et al., [24] advocated the use of pathological internet use (pathological internet use, PIU) to replace internet addiction, arguing that the term addiction refers to the psychological and physical dependence of an organism on a drug that is used or ingested as a chemical or narcotic.

In 2008, Chinese researcher Tao Ran proposed that internet addiction can be defined as: "An individual's overuse of the internet caused by a mental and behavioral disorder, where the re-use of the internet involves a strong desire to stop or reduce withdrawal from the internet. Similarly, it may be associated with mental and physical symptoms" [73]. Yang Hongmei [84] defines internet addiction as "a chronic or cyclical state of obsession resulting from the repeated use of the internet, creating an irresistible desire to re-use, while at the same time creating tensions and tolerances that increase spent time on it, involving psychological and physical dependence. As a result of this unreasonable over-use behavior, internet addiction can eventually lead to individual social and psychological damage, accompanied by somatic symptoms". These two definitions are both comprehensive descriptions of internet addiction and have been used widely.

Internet addiction can be divided into six types: (1) online game addiction, (2) cyber-relational addiction, (3) cyber-sexual addiction, (4) information overload, (5) cyber-impulse act, and (6) computer-technology addiction.

2.3.3.2 How to Diagnose Internet Addiction?

Young [86] identified eight questions for internet addiction according to 10 criteria for pathological gambling in the DSM-IV: ① Do you feel preoccupied with the internet (think about previous on-line activity or anticipate the next on-line session)? ② Do you feel the need to use the internet with increasing amounts of time in order to achieve satisfaction? ③ Have you repeatedly made unsuccessful efforts to control, cut back, or stop internet use? ④ Do you feel restless, moody, depressed, or irritable when attempting to cut down or stop internet use? ⑤ Do you stay on-line longer than originally intended? ⑥ Have you jeopardized or risked the loss of significant relationship, job, educational or career opportunity because of the internet? ⑦ Have you lied to family members, therapists, or others to conceal the extent of involvement with the internet? ⑧ Do you use the internet as a way of escaping from problems or for relieving a dysphoric mood (e.g., feelings of helplessness, guilt, anxiety, depression)? Patients were considered "addicted" when answering "yes" to five (or more) of the questions, and when their behavior could not be better accounted for by a Manic Episode. A cut-off score of "five" was consistent with the number of criteria used for Pathological Gambling, and was seen as an adequate number of criteria to differentiate normal from pathological addictive internet use.

It should also be noted that a patient's denial of addictive use is likely to be reinforced from the encouraged practice of utilizing the internet for academic or employment-related tasks [87]. Therefore, even if a patient meets all eight criteria, these symptoms can easily be masked as "I need this as part of my job," "It's just a machine," or "Everyone is using it" due to the internet's prominent role in our society.

Ivan Gordenberg put forward seven criteria for how to identify internet addiction, which coincide with Young's scale. He stresses that the following six are central to internet addiction: ① Salience: Internet use occupies the user's thinking and behavior; ② Tolerance: Internet users continue to increase time and effort in order to obtain satisfaction; ③ Withdrawal symptoms: Negative physiological response and negative emotions caused by a cessation from the internet; ④ Conflict: the use of the internet conflicts with daily activities or interpersonal communication; ⑤ Relapse: the internet addiction recurs even after remission and treatment; ⑥ Mood alteration: the internet is used to change a negative state of mind [35].

Shapira et al., argues that internet addiction is an impulse control hurdle, where the core of the problem lies in the individual's strong desire for the internet, thus weakening the individual's life in many aspects. His diagnostic criteria are: "not properly focused on the use of the internet, and have the following: ① an irresistible strong desire to use the internet, ② use of the internet for unexpected amounts of time, ③ use of the internet causes significant clinical pain or social occupational or other important functional impairment, ④ excessive use of the internet does not appear in a manic or hypomanic period, and cannot be explained by other diagnoses [67].

The China Youth Internet Association developed the following criteria for determining addiction in 2005. The criteria have one prerequisite and three conditions. The prerequisite is that the internet addiction must severely jeopardize a young person's social functioning and interpersonal communication. An individual would be classified as an internet addict as long as he or she meets any one of the following three conditions: (1) one would feel that it is easier to achieve self-actualization online than in real life, (2) one would experience dysphoria or depression whenever access to the internet is broken or ceases to function; (3) one would try to hide his or her true usage time from family members.

Professor Tao Ran, the framer of the "Internet addiction clinical diagnostic criteria", believes that the criteria to determine the degree of internet addiction must be combined with the following in order to form a comprehensive consideration: (1) standard course of disease (i.e., the average daily continuous use of internet time to reach or more than 6 h, and meet the symptomatic standard has reached or exceeded 3 months); (2) Social function (i.e., learning, work and communication skills) is damaged because of long-term Internet access; (3) symptomatic criteria. Specific symptom criteria include: long-term, repeated use of the internet, the purpose of using the internet not to learn and work or not conducive to their own learning and work, in line with the following symptoms: (1) having a strong desire or impulse to the use of internet; (2) whole body discomfort, irritability, inability to concentrate, disordered sleep, and other withdrawal reactions that appear when reducing or stopping internet use; the withdrawal reaction may also be eased via the use of other similar electronic media (such as television, handheld game, etc.); (3) at least meeting one of the following five: ① increasing use of internet time and input level to achieve satisfaction; ② difficulty controlling the beginning, end and duration of internet use even after repeated efforts to stop; ③ stubborn use of the internet regardless of its obvious harmful consequences; ④ reducing or abandoning other interests, entertainment or social activities because use of the internet; ⑤ use of the internet to escape problems or alleviate negative emotions [73].

Based on previous studies, Prof. Chen Shuhui compiled the "Chinese Internet Addiction Rating Scale (CIAS)", which includes the following five basic elements: "forced online behavior", "withdrawal behavior and withdrawal addiction", "internet addiction tolerance", "time management", and "interpersonal and health problems", forming a total of 26 items, on a four-level self-rating scale. The total score is the degree of addiction to the internet. The higher the score, the more severe the degree of internet addiction [17].

2.3.4 Mobile Phone Addiction

2.3.4.1 What Is Mobile Phone Addiction?

Mobile phone addiction (MPA) can also be called problematic mobile phone use [7, 71], excessive use of mobile phone [39], or mobile phone dependence [75]. All of these terms describe the uncontrolled use or overuse of a mobile phone.

Bianchi and Phillips [9] first proposed the Problematic Mobile Phone Use construct a decade ago. The authors found that mobile phone addicts show addictive behaviors, for example, obsession over mobile phones, substantial increase in the the time spent on mobile devices, failure to reduce or stop the use of mobile phone overuse. Since then, the number of related studies on this topic has grown substantially [10, 39, 42, 49, 50, 63]. In spite of this, it has received less attention than internet addiction [13].

Leung [50] regarded mobile phone addiction as an impulse control disorder, similar to pathological gambling. Furthermore, MPA can be considered as a form of technology addiction [85], which is operationally defined as non-chemical addictions with human-machine interaction [38]. Technological addiction is a branch of behavioral addictions [54]. Yen et al., [85] lists seven symptoms of MPA that may occur, such as tolerance, withdrawal, continued use regardless of adverse consequences, giving up or reducing important social activities, excessive time spent on mobile phone, and unsuccessful attempts to cut down mobile phone use. Furthermore, MPA may lead to social and psychological functional impairment [51].

Many MPA individuals report that they cannot help using their mobile phone even at inappropriate moments, while feeling uneasy when they have limited control over their phone or have to turn it off [48, 59]. In fact, researchers have shown that MPA is related to mental stress [20, 45], depression [15, 68], anxiety [23, 49], loneliness [72], self-control [46], and personality traits, like low self-esteem [42, 46], impulsivity [11], extroversion and neuroticism [7, 14]. Moreover, MPA may cause deficits in inhibitory control [18], decreased academic performance [49, 65] in college students, and even lead to impaired health risks, such as headaches [88], sleep disturbance and daytime fatigue [74]. The features mentioned above are similar to other addictive behaviors.

Despite no uniform definition of MPA in psychological circles, a more consistent view is that MPA, together with pathological gambling and internet gaming addiction, can be grouped into the spectrum of behavioral addiction [13, 62, 64].

Above all, we summarize that mobile phone addiction can be defined as: the uncontrolled use of a mobile phone, which causes a series of physiological, psychological and social problems, with symptoms of withdrawal, tolerance, mood modification, etc. It is a kind of behavioral addiction.

2.3.4.2 How to Diagnose Mobile Phone Addiction?

Questionnaires are employed to measure mobile phone addiction. More than half the scales used were developed on the basis of substance abuse literature [9, 80, 85]) or the criteria for internet addiction [50, 51].

Bianchi and Phillips [9] proposed the Mobile Phone Problem Use Scale (MPPUS), which was the first established questionnaire. The MPPUS contains 27 items, which covers the issues of tolerance, escape from other problems, withdrawal, and some negative life consequences (like social, work, and financial problems). All items are assigned 1–10 points. The MPPUS was revised into different versions and can be considered as a useful tool for mobile phone addiction assessment [15, 71, 77].

Afterwards, many investigators began to develop similar scales, for example, the Problematic Mobile Phone Use Questionnaire (PMPUQ) [11], the Problematic Cellular Phone Use Questionnaire (PCPU-Q) [85], as well as the Mobile Phone Addiction Index (MPAI) [50], etc. Nevertheless, the majority were not widely used, except for MPAI, which is established according to the diagnostic criteria for addiction on the DSM-IV. 17 items are included in the MPAI, and it's on a five-point Likert scale including four factors: inability to control craving, feeling anxious and lost, withdrawal and escape, and loss of productivity. The MPAI has been widely used [52, 81]. However, with the development of science and technology, traditional mobile phones have been replaced by smartphones, and the MPAI seems out of date for a smartphone society.

The MPAI being out of date led to the development of the Smartphone Addiction Scale (SAS) [47]. The SAS takes a smartphone's characteristics into consideration. The SAS consists of six factors, that is, withdrawal, tolerance, daily-life disturbance, positive anticipation, cyberspace-oriented relationship, and overuse. "Withdrawal" here, is represented as being impatient and intolerable without a smartphone, and becoming irritated when bothered while using a mobile phone. "Tolerance" is represented as always trying to control one's phone use but usually failing to do so. "Daily-life disturbance" can be defined as having a hard time concentrating in class or while working, pain on the wrists or at the back of the neck, and sleeping disturbance. "Positive anticipation" is described as feeling excited about smartphone use, even feeling empty without a phone. "Cyberspace-oriented relationship" mainly involves questions about one's relationships obtained via phone technology. "Overuse" refers to the uncontrollable use of a smartphone. This scale was proven to be relatively reliable and valid, and it has been extensively applied to different kinds of modified versions [19, 23, 25, 40, 53] around the world.

At present, the existing instruments do not use a cut-off point for mobile phone addiction, and most studies agree that higher scores indicate more serious addiction. However, the Short-Version of the Smartphone Addiction Scale for adolescents which contains ten items, provides a cut-off value to efficiently evaluate mobile phone addiction [47]. The cut-off point for boys is 31, and for girls 33, which means, a boy who scored higher than 31 may be addicted to a mobile phone.

In general, many MPA scales have emerged. Nevertheless, only a few validated scales are currently available for researchers [12]. Thus, researchers should translate and modify the available instruments with high reliability and validity, such as the SAS, and test them in different cultures around the world.

2.4 Conclusion

When defining substance addiction, or drug addiction, "loss of control" and "despite adverse consequences" are the key characteristics. Drug addiction is the outcome of continued drug use, and can be seen as a kind of brain disease caused by the repeated drug use. Though there is no drug taking in non-substance addiction, the symptoms and brain mechanisms are very similar to drug addiction. Thus, researchers have often defined and diagnosed them using the substance addiction model (see Table 2.2). However, differences among the various addiction disorders should not be neglected (see Table 2.1), and further studies are needed to explore the unique characteristics and neural mechanisms that underlie different kinds of addiction disorders.

Table 2.2 Definition and diagnosis of substance and non-substance addiction

	Definition	Concepts to be differentiated	Sub-types	Important diagnosis tools
Substance addiction	Loss of control over drug use, or the compulsive seeking and taking of drugs despite adverse consequences.	Dependence vs. addiction	Alcohol; caffeine; cannabis; opioids; hallucinogens; inhalants; sedatives, hypnotics and anxiolytics; cocaine; tobacco, anxiolytics, amphetamine-type substances, inhalants, volatile solvents	DSM-V (11 criteria); ICD-10 (6 criteria)
Pathological gambling	A disorder that consists of frequent, repeated episodes of gambling that dominate the patient's life to the detriment of social, occupational, material, and family values and commitments.	Recreational gambling vs. pathological gambling	–	DSM-V (9 criteria); Screening tools (e.g. SOGS, PGSI, GA20, MAGS)
Food addiction	A chronic and relapsing condition caused by the interaction of many complex variables that increase cravings for certain specific foods in order to achieve a state of high pleasure, energy or excitement, or to relieve negative emotional or physical states.	Food addiction vs. eating disorders	High calorie foods; foods with high sugar and/or fat	Overlay with the DSM-IV-TR criteria for drug addiction (7 criteria); YFAS; EBQ

Table 2.2 (continued)

	Definition	Concepts to be differentiated	Sub-types	Important diagnosis tools
Internet addiction	Overuse of the internet caused by a mental and behavioral disorder, characterized by a strong desire to use the internet, with unsuccessful attempts at stopping or reducing use, with withdrawal symptoms when the use of the internet is ceased. May also be associated with mental and physical symptoms.	Internet addiction vs. internet over-use for academic or working	1. online game addiction; 2. cyber-relational addiction; 3. cyber-sexual addiction; 4. information overload; 5. cyber-impulse act; 6. computer-technology addiction	According to 10 criteria for pathological gambling in the DSM-IV, [87] Comprehensive Internet addiction clinical diagnostic criteria [73]
Mobile phone addiction	The uncontrolled use of a mobile phone, causing a series of physiological, psychological and social problems, with symptoms of withdrawal, tolerance, mood modification. It is a kind of behavioral addiction.	–	–	MPPUS; PMPUQ; PCPU-Q; DSM-V; Smartphone Addiction Scale (SAS)

References

1. American Psychiatric Association (1952) Diagnostic and statistical manual of mental disorders, 1st edn. The American Psychiatric Association, Washington, DC. www.turkpsikiyatri.org/arsiv/dsm-1952.pdf
2. American Psychiatric Association (1968) Diagnostic and statistical manual of mental disorders, 2nd edn. The American Psychiatric Association, Washington, DC. www.behaviorismandmentalhealth.com/wp-content/uploads/2015/08/DSM-II.pdf
3. American Psychiatric Association (1980) Diagnostic and statistical manual of mental disorders, 3rd edn. The American Psychiatric Association, Washington, DC. displus.sk/DSM/subory/dsm3.pdf
4. American Psychiatric Association (1994) Diagnostic and statistical manual of mental disorders, 4th edn. The American Psychiatric Association, Washington, DC. https://justines2010blog.files.wordpress.com/2011/03/dsm-iv.pdf
5. American Psychiatric Association (2013) Diagnostic and statistical manual of mental disorders, 5th edn. The American Psychiatric Association, Washington, DC. www.psicovalero.files.wordpress.com/2014/11/dsm-v-ingles-manual-diagnc3b3stico-y-estadc3adstico-de-los-trastornos-mentales.pdf
6. Armstrong L (2001) How to beat addiction to Cyberspace. Vibrant Life 17(4):15
7. Augner C, Hacker GW (2012) Associations between problematic mobile phone use and psychological parameters in young adults. Int J Public Health 57(2):437–442
8. Bąk-Sosnowska M (2017) Differential criteria for binge eating disorder and food addiction in the context of causes and treatment of obesity. Psychiatr Polska 51(2):247–259
9. Bianchi A, Phillips JG (2005) Psychological predictors of problem mobile phone use. Cyberpsychol Behav 8(1):39–52
10. Billieux J, Van der Linden M, d'Acremont M, Ceschi G, Zermatten A (2007) Does impulsivity relate to perceived dependence on and actual use of the mobile phone? Appl Cogn Psychol 21(4):527–537
11. Billieux J, Van der Linden M, Rochat L (2008) The role of impulsivity in actual and problematic use of the mobile phone. Appl Cogn Psychol 22(9):1195–1211
12. Billieux J, Rochat L, Ceschi G, Carré A, Offerlin-Meyer I, Defeldre AC, Khazaal Y, Besche-Richard C, Van der Linden M (2012) Validation of a short French version of the UPPS-P Impulsive behavior scale. Compr Psychiatry 53(5):609–615
13. Billieux J, Maurage P, Lopez-Fernandez O, Kuss DJ, Griffiths MD (2016) Can disordered mobile phone use be considered a behavioral addiction? An update on current evidence and a comprehensive model for future research. Curr Addict Rep 2(2):156–162
14. Butt S, Phillips JG (2008) Personality and self reported mobile phone use. Comput Hum Behav 24(2):346–360
15. Çağan Ö, Ünsal A, Çelik N (2014) Evaluation of college students' the level of addiction to cellular phone and investigation on the relationsship between the addiction and the level of depression. Procedia-Soc Behav Sci 114:831–839
16. Camí J, Farré M (2003) Drug addiction. N Engl J Med 349:975–987
17. Chen SH, Weng LZ, SU YR et al (2003) Study on the establishment of Chinese Internet Addiction Scale and its psychometric characteristics. Chin J Psychol 45(4):279–294
18. Chen J, Liang Y, Mai C, Zhong X, Qu C (2016) General deficit in inhibitory control of excessive smartphone users: evidence from an event-related potential study. Front Psychol 7. http://doi.org/10.3389/fpsyg.2016.00511
19. Ching SM, Yee A, Ramachandran V, Lim SMS, Sulaiman WAW, Foo YL, kee Hoo F (2015) Validation of a Malay version of the smartphone addiction scale among medical students in Malaysia. PLoS One 10(10): e0139338. https://doi.org/10.1371/journal.pone.0139337
20. Chiu SI (2014) The relationship between life stress and smartphone addiction on taiwanese university student: a mediation model of learning self-Efficacy and social self-Efficacy. Comput Hum Behav 34:49–57

21. Corsica JA, Pelchat ML (2010) Food addiction: true or false? J Curr Opin Gastroenter 26(2):165–110
22. Corwin RL, Grigson PS (2009) Symposium overview—food addiction: fact or fiction? J Nutr 139(3):617–619
23. Darcin AE, Noyan C, Nurmedov S, Yilmaz O, Dilbaz N (2015) Smartphone addiction in relation with social anxiety and loneliness among University Students in Turkey. Eur Psychiatry 30:505. https://doi.org/10.1016/S0924-9338(15)30398-9
24. Davis RA (2001) A cognitive-behavioral model of pathological Internet use. Comput Hum Behav 17(2):187–195
25. Demirci K, Orhan H, Demirdas A, Akpınar A, Sert H (2014) Validity and reliability of the Turkish version of the smartphone addiction scale in a younger population. Bull Clin Psychopharmacol 24(3):226–235
26. Eddy NB, Halbach H, Isbell H, Seevers MH (1965) Drug dependence: its significance and characteristics. Bull World Health Organ 32(5):721–733
27. Eddy NB, Halbach H, Isbell H, Seevers MH (1966) Drug dependence: its significance and characteristics. Psychopharmacol Bull. 1966 Jul 3(3):1–13
28. Fauth-Bühler M, Mann K, Potenza MN (2017) Pathological gambling: a review of the neurobiological evidence relevant for its classification as an addictive disorder. Addict Biol 22:885–897
29. Flint AJ, Gearhardt AN, Corbin WR, Brownell KD, Field AE, Rimm EB (2014) Food-addiction scale measurement in 2 cohorts of middle-aged and older women. Am J Clin Nutr 99:578–587
30. Fong TW, Rory C, Parhami I (2012) Behavioral addictions where to draw the lines? Psychiatr Clin North Am 35:279–297
31. Gearhardt AN, Corbin WR, Brownell KD (2009) Preliminary validation of the Yale food addiction scale. Appetite 52:430–437
32. Gearhardt AN, White MA, Potenza MN (2011) Binge eating disorder and food addiction. Curr Drug Abuse Rev 4:201–208
33. Gearhardt AN, Roberto CA, Seamans MJ, Corbin WR, Brownell KD (2013) Preliminary validation of the Yale food addiction scale for children. Eat Behav 14(4):508–512
34. Gearhardt AN, Corbin WR, Brownell KD (2016) Development of the Yale Food Addiction Scale Version 2.0. Psychol Addict Behav 30:113–122
35. Goldberg I (1996) Internet addiction disorder. Retrieved November 24, 2005 from http://www.rider.edu/~suler/psycyber/supportgp.html
36. Goldstein RZ, Volkow ND (2002) Drug addiction and its underlying neurobiological basis: neuroimaging evidence for the involvement of the frontal cortex. Am J Psychiatry 159:1642–1653
37. Goldstein RZ, Volkow ND (2011) Dysfunction of the prefrontal cortex in addiction: neuroimaging findings and clinical implications. Nat Rev Neurosci 12:652–670
38. Griffiths M (1996) Gambling on the internet- A brief note. J Gambl Stud 12:471–474
39. Ha JH, Chin B, Park DH, Ryu SH, Yu J (2008) Characteristics of excessive cellular phone use in Korean adolescents. Cyberpsychol Behav 11(6):783–785
40. Haug S, Castro RP, Kwon M, Filler A, Kowatsch T, Schaub MP (2015) Smartphone use and smartphone addiction among young people in Switzerland. J Behav Addict 4(4):299–308
41. Heather N (1998) A conceptual framework for explaining drug addiction. J Psychopharmacol 12(1):3–7
42. Hong FY, Chiu SI, Huang DH (2012) A model of the relationship between psychological characteristics, mobile phone addiction and use of mobile phones by Taiwanese university female students. Comput Hum Behav 28(6):2152–2159
43. Hyman SE (2007) The neurobiology of addiction: implications for voluntary control of behavior. Am J Bioeth 7:8–12
44. Imperatori C, Fabbricatore M, Vumbaca V, Innamorati M, Contardi A, Farina B (2016) Food addiction: definition, measurement and prevalence in healthy subjects and in patients with eating disorders. Riv Psichiatr 51(2):60–66

45. Johansson A, Nordin S, Heiden M, Sandström M (2010) Symptoms, personality traits, and stress in people with mobile phone-related symptoms and electromagnetic hypersensitivity. J Psychosom Res 68(1):37–46
46. Khang H, Kim JK, Kim Y (2013) Self-traits and motivations as antecedents of digital media flow and addiction: the Internet, mobile phones, and video games. Comput Hum Behav 29(6):2416–2425
47. Kwon M, Lee J-Y, Won W-Y, Park J-W, Min J-A, Hahn C, Gu X, Choi J-H, Kim D-J (2013) Development and validation of a smartphone addiction scale (SAS). PLoS One 8(2): e56937. https://doi.org/10.1371/journal.pone.0056936
48. Lee, U., Lee, J., Ko, M., Lee, C., Kim, Y., Yang, S., ... & Song, J. (2014, April). Hooked on smartphones: an exploratory study on smartphone overuse among college students. In: Proceedings of the 32nd annual ACM conference on Human factors in computing systems (pp. 2327–2336). ACM. https://doi.org/10.1145/2556288.2557366
49. Lepp A, Barkley JE, Karpinski AC (2014) The relationship between cell phone use, academic performance, anxiety, and satisfaction with life in college students. Comput Hum Behav 31:343–351
50. Leung L (2008) Linking psychological attributes to addiction and improper use of the mobile phone among adolescents in Hong Kong. J Child Media 2(2):93–114
51. Lin YH, Chang LR, Lee YH, Tseng HW, Kuo TB, Chen SH (2014) Development and validation of the Smartphone Addiction Inventory (SPAI). PLoS One 9(6): e98312. https://doi.org/10.1371/journal.pone.0098312
52. Liu H, Wang HL (2011). The relationship among college students' mobile phone addiction, mobile phone use motivation and loneliness. J Psychol Sci 34(6):1453–1457. https://doi.org/10.16719/j.cnki.1671-6981.2011.06.017
53. Lopez-Fernandez O (2016) Short version of the smartphone addiction scale adapted to Spanish and French: towards a cross-cultural research in problematic mobile phone use. Addict Behav 64:275–280
54. Marks I (1990) Behavioural (non-chemical) addictions. Br J Addict 85(11):1389–1395
55. Merlo LJ, Klingman C, Malasanos TH, Silverstein JH (2009) Exploration of food addiction in pediatric patients: a preliminary investigation. J Addict Med 3:26–33
56. Nestler EJ (2001) Molecular basis of long-term plasticity underlying addiction. Nat Rev Neurosci 2:119–129
57. O'Brien C (2011) Addiction and dependence in DSM-V. Addiction 106:866–868
58. O'Brien CP, Volkow ND, Li T-K (2006) What's in a word? Addiction versus dependence in DSM-V. Am J Psychiatr 163:764–766
59. Oulasvirta A, Rattenbury T, Ma L, Raita E (2012) Habits make smartphone use more pervasive. Pers Ubiquit Comput 16(1):105–115
60. Pursey KM, Stanwell P, Gearhardt AN, Collins CE, Burrows TL (2014) The prevalence of food addiction as assessed by the Yale food addiction scale: a systematic review. Nutrients 6(10):4552–4590
61. Randolph TG (1956) The descriptive features of food addiction; addictive eating and drinking. Q J Stud Alcohol 17:198–225
62. Robbins T, Clark L (2015) Behavioral addictions. Curr Opin Neurobiol 30:66–73
63. Roberts JA, Pullig C, Manolis C (2015) I need my smartphone: a hierarchical model of personality and cell-phone addiction. Personal Individ Differ 79:13–20
64. Salehan M, Negahban A (2013) Social networking on smartphones: when mobile phones become addictive. Comput Hum Behav 29(6):2632–2640
65. Seo DG, Park Y, Kim MK, Park J (2016) Mobile phone dependency and its impacts on adolescents' social and academic behaviors. Comput Hum Behav 63:282–293
66. Shaffer HJ, Hall MN, Vander Bilt J (1999) Estimating the prevalence of disordered gambling behavior in the United States and Canada: a research synthesis. Am J Public Health 89(9):1369–1376. http://doi/abs/10.2105/AJPH.89.9.13
67. Shapira NA, Goldsmith TD, Keck PE, Khosla UM, McElroy SL (2000) Preliminary communication: psychiatric features of individuals with problematic internet use. J Affect Disord 57:267–273

68. Smetaniuk P (2014) A preliminary investigation into the prevalence and prediction of problematic cell phone use. J Behav Addict 3(1):41–54
69. Fainsinger RL, Thai V, Frank G, Fergusson J (2006) What's in a word? Addiction versus dependence in DSM-V. Am J Psychiatry 163(11):2014–2014. https://doi.org/10.1176/ajp.2006.163.11.2014a
70. Strong DR, Kahler CW (2007) Evaluation of the continuum of gambling problems using the DSM-IV. Addiction 102(5):713–722
71. Takao M, Takahashi S, Kitamura M (2009) Addictive personality and problematic mobile phone use. Cyberpsychol Behav 12(5):501–508
72. Tan Ç, Pamuk M, Dönder A (2013) Loneliness and mobile phone. Procedia Soc Behav Sci 103:606–611. https://doi.org/10.1016/j.sbspro.2013.10.378
73. Tao R, Wang JN, Huang XQ, Liu CY, Yao SM, Xiao LJ (2008) Nomenclature, definition and clinical diagnostic criterion of internet addiction. Med J Chin People's Armed Police Forces 19(09):773–777. https://doi.org/10.14010/j.cnki.wjyx.2008.09.028
74. Thomée S, Härenstam A, Hagberg M (2011) Mobile phone use and stress, sleep disturbances, and symptoms of depression among young adults-a prospective cohort study. BMC Public Health 11(1):66. https://doi.org/10.1186/1471-2458-11-66
75. Toda M, Ezoe S, Nishi A, Mukai T, Goto M, Morimoto K (2008) Mobile phone dependence of female students and perceived parental rearing attitudes. Soc Behav Personal 36(6):765–771
76. Van Deursen AJ, Bolle CL, Hegner SM, Kommers PA (2016) Modeling habitual and addictive smartphone behavior: the role of smartphone usage types, emotional intelligence, social stress, self-regulation, age, and gender. Comput Hum Behav 45:411–420
77. Volkow ND, Wang GJ, Fowler JS, Tomasi D (2012) Addiction circuitry in the human brain. Annu Rev Pharmacol Toxicol 52:321–336
78. Vries SK, Meule A (2016) Food addiction and bulimia nervosa: new data based on the Yale food addiction scale 2.0. Eur Eat Disord Rev 24(6):518–522
79. Volkow ND, Wang GJ, Tomasi D, Baler RD (2013) Obesity and addiction: neurobiological overlaps. J Obes Rev 1:2–19
80. Walsh SP, White KM, McD Young R (2010) Needing to connect: the effect of self and others on young people's involvement with their mobile phones. Aust J Psychol 62(4):194–204
81. Wang Y, Zou Z, Song H, Xu X, Wang H, Uquillas FDO, Huang X (2016) Altered gray matter volume and white matter integrity in college students with mobile phone dependence. Front Psychol 7. https://doi.org/10.3389/fpsyg.2016.00597
82. World Health Organization (1977) Manual of the international statistical classification of diseases, injuries, and causes of death: based on the recommendations of the ninth revision conference, 1975, and adopted by the Twenty-ninth World Health Assembly. http://www.who.int/iris/handle/10665/40492
83. World Health Organization (2004) Icd-10: international statistical classification of diseases and related health problems: tenth revision. Acta Chir Iugosl 56(3):65. http://apps.who.int/iris/bitstream/10665/42980/1/9241546530_eng.pdf
84. Yang HM (2008) A research on the influence of internet addiction on colleage students. D, Nanjing Normal University, Nanjing
85. Yen CF, Tang TC, Yen JY, Lin HC, Huang CF, Liu SC, Ko CH (2009) Symptoms of problematic cellular phone use, functional impairment and its association with depression among adolescents in Southern Taiwan. J Adolesc 32(4):863–873
86. Young KS (1996) Psychology of computer use: XL. Addictive use of the Internet: a case that breaks the stereotype. Psychol Rep 79(3):899–902
87. Young KS (1997, August) What makes on-line usage stimulating? Potential explanations for pathological Internet use. In: Symposia paper presented at the 105th annual meeting of the American Psychological Association
88. Zheng F, Gao P, He M, Li M, Wang C, Zeng Q, Zhou Z, Yu Z, Zhang L (2014) Association between mobile phone use and inattention in 7102 Chinese adolescents: a population-based cross-sectional study. BMC Public Health 14(1):1022. https://doi.org/10.1186/1471-2458-14-1022
89. Ziauddeen H, Farooqi IS, Fletcher PC (2012) Obesity and the brain: how convincing is the addiction model? Nat Rev Neurosci 4:279–287

Part II
Comparison Between Substance and Non-substance Addictions in Mechanism

Chapter 3
Similarities and Differences in Neurobiology

Manli Chen, Yan Sun, Lin Lu, and Jie Shi

Abstract Substance addiction is a chronic, relapsing brain disease characterized by compulsive drug seeking and use despite harmful consequences. Non-substance addiction is defined recently that people may compulsively engage in an activity despite any negative consequences to their lives. Despite differences with respect to their addictive object, substance addiction and non-substance addiction may share similarities with respect to biological, epidemiological, clinical, genetic and other features. Here we review the similarities and differences in neurobiology between these two addictions with a focus on dopamine, serotonin, opioid, glutamate and norepinephrine systems. Studies suggest the involvement of all these systems in both substance addiction and non-substance addiction while differences may exist with respect to their contributions.

Keywords Substance addiction • Non-substance addiction • Neurobiology

M. Chen
Department of Pharmacology, School of Basic Medical Sciences,
Peking University Health Science Center, Beijing 100191, China

National Institute on Drug Dependence, Peking University,
No. 38, Xueyuan Road, Haidian District, Beijing 100191, China

Y. Sun • J. Shi (✉)
National Institute on Drug Dependence, Peking University,
No. 38, Xueyuan Road, Haidian District, Beijing 100191, China
e-mail: shijie@bjmu.edu.cn

L. Lu
Institute of Mental Health/Peking University Sixth Hospital and National Clinical Research
Center for Mental Disorders & Key Laboratory of Mental Health, Peking University,
Beijing 100191, China

X. Zhang et al. (eds.), *Substance and Non-substance Addiction*, Advances in
Experimental Medicine and Biology 1010, DOI 10.1007/978-981-10-5562-1_3

3.1 Introduction

Substance addiction, is a chronic, relapsing brain disease characterized by compulsive substance seeking and use, loss of control when limiting intake and emergence of a negative emotional state (e.g. dysphoria, anxiety, irritability) reflecting a motivational withdrawal syndrome during discontinuation or dosage reduction of the drug [45]. Gambling disorder (previously termed 'pathological gambling') was reclassified from the Impulse-Control Disorders category into a new category labeled 'Substance-Related and Addictive Disorders' and became the only non-substance addiction recognized in the fifth edition of the Diagnostic and Statistical Manual (DSM-5). This decision ratifies the concept of non-substance addiction that people may compulsively engage in an activity that does not involve exogenous substance administration despite any negative consequences to their physical, mental, social and/or financial well-being.

Non-substance addiction shares important elements with substance addictions which include impaired control over engagement, continuous engagement despite harmful consequences and urges or cravings [76]. Similarities between substance addiction and non-substance addiction extend to biological, epidemiological, clinical, genetic and other domains [6, 21, 76]. In the DSM-5, the inclusionary criteria for non-substance addiction share similarities with those for substance addictions. For example, criteria targeting tolerance, withdrawal, repeated unsuccessful efforts to control, cut back, or stop and impairment in important areas of life functioning are contained in the criteria for both non-substance and substance addictions [78]. Moreover, non-substance and substance addictions frequently co-occur [52] and there are similarities in the progression of both disorders such as high rates of the conditions in adolescents and young adults and a "telescoping" phenomenon observed in females [77] which means that women are more likely than men to develop rapidly in the progression of addictions. In addition, substance addiction and non-substance addiction manifest in similar psychological and behavioral patterns which include craving, impaired control over the behavior, tolerance, withdrawal and high rates of relapse [68]. A variety of evidence has suggested important commonalities in the neurobiology of substance addiction and non-substance addiction which includes several neurotransmitter systems. Here we review the similarities and differences between substance addiction and non-substance addiction with a focus on dopamine (DA), serotonin (5HT), opioid, glutamate and norepinephrine (NE) systems.

3.2 Similarities and Differences

3.2.1 Dopamine

Dopamine has become characterized as the 'pleasure' neurotransmitter in the human brain that produces reward [38, 105, 108] and there is a prevailing view that the dopamine system has a central role in addiction.

Similarities exist with respect to the involvement of dopamine in substance addiction and non-substance addiction. As with gambling [12], substance addiction has been shown to be associated with dopamine release [81], supported by a recent positron emission tomography (PET) study that oral alcohol induces dopamine release in nonalcoholic individuals [94]. As dopamine is implicated in rewarding and reinforcing effects of substance [20], studies have also suggested a role for dopamine in non-substance-related reward and reinforcement. In a loss-chasing game from a recent study, D2-like receptor agonist pramipexole significantly increased the value of losses chased and decreased the value of losses surrendered which means increasing the perceived value of rewards and minimization of punishment [8].

Differences between substance addiction and non-substance addiction have been found regarding two markers of abnormal dopamine function in substance addiction: the lower availability of striatal dopamine receptors and the diminished release of striatal dopamine in response to a pharmacological challenge. Early studies showed that chronic cocaine abuse was associated with decreased striatal dopamine D2 and D3 receptor availability [97]. This was attributed to the effects of cocaine, which was expected to induce dopamine release, downregulate postsynaptic dopamine receptors and lead to reduced receptor availability. This result has been replicated in other studies of cocaine [56–58, 63, 98, 102] and methamphetamine [49, 101, 106] users. Decreased dopamine receptor availability has also been found in individuals with alcohol dependence [31, 33, 55, 99, 104] and higher striatal dopamine receptor availability has been reported to protect against alcoholism in high-risk individuals [103]. Moreover, in many substance addictions, individuals show a decreased release of striatal dopamine after a pharmacological challenge. A marked reduction in dopamine release has been demonstrated in participants with cocaine dependence after a methylphenidate or amphetamine administration [56, 58, 100], which has also been reported in opiate dependence after a methylphenidate challenge [59] and in alcohol dependence after an amphetamine challenge [55, 104].

Unlike with substance addiction, there appeared to be no marked differences in baseline D2/D3 receptor availability in individuals with gambling problems compared with healthy subjects [4, 11]. In one of these gambling disorder studies, striatal D2/D3 receptor availability was inversely correlated with mood-related impulsivity [11], while in another, D2/D3 receptor availability was positively correlated with impulsiveness in the substantia nigra, a brain region rich of dopamine D3 receptors [4]. Moreover, in contrast with the blunted stimulant-related dopamine release that was seen in substance addiction, dopamine release was found to increase in the dorsal striatum after amphetamine administration in gamblers compared with healthy controls [5]. This increase in dorsal striatal dopamine was predicted by D3 receptor levels and it was hypothesized that dopaminergic sensitization involving D3-related mechanisms might contribute to the pathophysiology of this non-substance addiction [5].

Non-substance addiction may differ from some substance addictions with respect to dopaminergic response to particular manipulations. Unlike the finding that amphetamine failed to increase motivation to drink in problem drinkers, amphetamine administration has been found to prime gambling urges in individuals with gambling problems [109], which suggested that prodopaminergic agents or states may promote gambling and underlie the pathophysiology observed in gambling disorder. However, unlike the consistent findings regarding the relationship between D2 receptor and cocaine-seeking behavior that the administration of D2-like receptor agonists reinstate cocaine-seeking behavior [14] and D2-like receptor antagonists attenuate cocaine priming-induced drug-seeking behavior [41], there have been seemingly opposing results with dopamine antagonists regarding gambling disorder. The D2-like receptor antagonist haloperidol decreased the motivation in people with gambling problems to gamble more aggressively in a slot machine task [93], while in another study, haloperidol significantly enhanced self-reported rewarding effects of gambling and primed motivation to gamble in gambling addicts [109]. These results may help us to better understand the negative clinical trial findings for drugs with D2-like receptor antagonism [18, 60], and may also lead us to a conclusion that dopamine plays a different role in substance addiction and non-substance addiction.

3.2.2 Serotonin

Neurochemical studies suggest serotonergic similarities in non-substance addiction and substance addiction. As for non-substance addiction, the cerebrospinal fluid levels of the serotonin metabolite 5-hydroxyindoleacetic acid (5-HIAA) was found significantly lower in gamblers compared with healthy controls [65]. In another study, the density of the platelet serotonin transporter (SERT), a protein that regulates synaptic serotonin concentration through reuptake mechanisms, was decreased in participants with gambling problems, suggesting the involvement of serotonin in gambling disorder [54]. As in gambling disorder, low levels of 5-HIAA were also found in alcohol dependence [17, 80], which indicates an important role for serotonin in both substance addiction and non-substance addiction.

Similarities also exist with response to serotonin agonists. Compared with control, individuals with gambling problems had a significantly increased "high" sensation after administration of meta-chlorophenylpiperazine (m-CPP), a mixed serotonin agonist with significant affinities for 5HT1 and 5HT2 receptors [71]. This response was similar to the finding with alcoholic subjects after receiving the drug [48]. Moreover, a blunted growth hormone response has been observed in participants with gambling problems compared with healthy controls after administration of sumatriptan, a selective 5HT1B receptor agonist, which was similar to the finding with abstinent alcoholics [96]. PET studies with the selective 5HT1B receptor antagonist radioligand [(11)C]P943 have also indicated a role for 5HT1B receptor in gambling disorder and alcohol dependence [36, 79]. Altogether, these findings suggest that increased 5HT1B receptor availability may contribute across substance addiction and non-substance addiction.

In clinical trials, the effect of serotonergic agents on gambling disorder is controversial, which includes findings of selective serotonin reuptake inhibitor (SSRI) (e.g., paroxetine and fluvoxamine) (positive: [35, 43]; negative: [22, 84]) and results with olanzapine, a drug acts as 5HT2 receptor antagonist [18, 60]. While the positive findings support a role for serotonin in gambling disorder, the mixed findings help to explain the individual differences that contribute to the variable efficacy of SSRIs in the treatment of gambling disorder. Similar to these, both SSRIs [92] and olanzapine [1, 30] have shown limited efficacy when treating substance addictions.

Differences in the contribution of serotonin in non-substance addiction and substance addiction may also exist. Tryptophan depletion, which reduced central serotonin levels and altered serotonin neurotransmission, significantly reduced the number of decisions made to chase losses in simulated gambling [8], while among those with a positive family history of alcoholism, tryptophan depletion impaired performance on the behavioral inhibition task [13, 50]. These studies suggest that tryptophan depletion, and by extension, serotonin function may influence decision-making and impulsivity differently.

3.2.3 Opioid

The opioid system consists of several types of receptors (μ, δ and κ) and peptides (β-endorphin, enkephalins and dynorphins). Mu- and δ-opioid receptor ligands may produce rewarding effects, while κ-opioid receptor ligands may have aversive effects [32]. Preclinical evidence indicates that opioid receptors are distributed widely in the mesolimbic system and are implicated in the hedonic aspects of reward processing [3, 73]. Opioid function may influence dopamine release in the mesolimbic pathway extending from the ventral tegmental area to the nucleus accumbens or ventral striatum [90].

Gambling has been associated with elevated blood levels of the endogenous opioid β-endorphin [86] and modulation of the opioid system through opioid receptor antagonists [26] and partial agonists [23, 28] has shown significant promise in the treatment of gambling disorder. An fMRI study of the μ-opioid antagonist naloxone found attenuated reward-related responses in the ventral striatum and enhanced loss-related activity in the medial prefrontal cortex on a wheel of fortune task in healthy volunteers [75]. In a multicenter trial of the opioid antagonist nalmefene in the treatment of gambling disorder, participants who received nalmefene had a statistically significant reduction in severity of gambling disorder and low-dose nalmefene was found to be associated with few adverse events [23]. Subjects who reported strong gambling urges at treatment onset responded better to naltrexone than to placebo [42], which was proven in another trial that significant reductions in problem gambling severity, self-reported urges and gambling behavior were found among patients reporting gambling urges primarily [26]. Subsequent studies revealed that a family history of alcoholism can predict positive treatment response to an opiate antagonist in gambling disorder [25]. Thus, these results indicate the positive effects of opioid receptor antagonists on reducing gambling severity which seems particularly robust among those with strong gambling urges or a family history of alcoholism.

Clinical trial results with opioid antagonists for substance addiction, particularly alcohol and opiate dependence, share similarities with those for gambling disorder. μ- opioid receptor antagonists such as naltrexone and nalmefene have been examined as potential therapies for opioid addiction for many years, which have been shown somewhat effective in treating alcoholism [64]. Naltrexone, in conjunction with psychosocial therapy, appears to be an effective and safe strategy in alcoholism treatment [82], mechanisms for which include reducing the urges to drink [61, 70]. Similar to treating gambling disorder, evidence suggests naltrexone particularly effective in reducing alcohol drinking [46] among those with a family history of alcoholism. Further, opioid receptors have been shown to mediate or modulate other rewarding or motivational behaviors that many 'natural' rewarding stimuli such as palatable food-seeking, social behavior and maternal reward have a opioid receptor-mediated component [53]. Opioid receptors also appear to play a role in mediating psychostimulant-induced behaviors. Local intracerebral injections of the selective opioid receptors antagonist, CTAP, have been shown to inhibit cocaine-induced reinstatement to self-administration [87, 91] and similar effects were seen in both cue-induced and drug-induced reinstatement to alcohol seeking [74].

3.2.4 Glutamate

A persuasive body of preclinical evidence has indicated a critical role for glutamate transmission and glutamate receptors in drug reward, reinforcement and relapse. Glutamate appears to be implicated in long-lasting neuroadaptations in the cortico-striatal brain circuitry and the imbalance in glutamate homeostasis engenders changes in neuroplasticity that impair communication between the prefrontal cortex and the nucleus accumbens and thereby result in reward-seeking behaviors [39, 40]. Glutamate is also involved in associative learning between stimuli and promotes the immediate approach response through its link to the dopamine reward system [67]. Data from cerebrospinal fluid studies also suggest a dysfunctional glutamate system in disordered gamblers [66].

Medications that alter glutamate neurotransmission may affect both substance addiction and non-substance addiction. Open-label N-acetyl cysteine (NAC), an amino acid, seems to restore extracellular glutamate concentration in the nucleus accumbens and affect neurotransmission in regions including the ventral striatum. Studies have revealed that administration of NAC may reduce heroin-induced drug seeking in rats [110] as well as block cocaine-induced reinstatement produced by a low or high dose of cocaine [2]. Moreover, NAC administration was found to decreases cigarette smoking [44] and marijuana use in humans [29]. Paralleling these findings with substance addiction, NAC has also been shown to decrease gambling severity significantly [24], suggesting that pharmacological manipulation of the glutamate system might target core symptoms of reward-seeking addictive behaviors in both substance addiction and non-substance addiction.

In addition, open-label memantine, an N-methyl-D-aspartate receptor antagonist, has shown promise in reducing gambling severity and cognitive inflexibility in gambling disorder [27]. Among participants with gambling problems, money spent gambling and hours spent gambling per week both decreased significantly and there was a significant improvement in intradimensional/extradimensional set-shifting (IDED) performance at the end of treatment. It was hypothesized that this finding may due to the reduction of glutamate and glutamine concentrations in the in prefrontal cortex after administration of memantine, which was associated with a reduction in brain activation [95]. Similarly, memantine may also have a role in treating substance addiction. Human studies have shown that memantine administration may attenuate alcohol cue-induced craving and produce ethanol-like subjective effects without adverse cognitive or behavioral effects [47]. Moreover, evidence supported that an imbalance in glutamate neurotransmission may induce impulsive and compulsive behavior in both gambling disorder [27] and substance addictions [39], showing that pharmacological manipulation of the glutamate system may be an effective treatment for both substance addiction and non- substance addiction [27, 39].

3.2.5 *Norepinephrine/Noradrenaline*

Norepinephrine is a catecholamine structurally related to epinephrine which is released in response to stress and affects the response of the sympathetic nervous system. Norepinephrine can be synthesized from dopamine and can have systemic (central and peripheral) effects [62, 69]. The noradrenergic system, which uses norepinephrine as the main chemical messenger, serves multiple brain functions including arousal, attention, mood, learning, memory and stress response [89].

In preclinical models of substance addiction, norepinephrine is critically involved in mediating stimulant effects including sensitization [15] and reinstatement of drug seeking. For example, both clonidine and lofexidine, medications that inhibit the adrenergic activity by stimulating α_2 receptors, attenuate stress-induced reinstatement of cocaine seeking in rats [16, 34]. Beta-adrenergic receptors may also participate in stress-induced reinstatement since both β1- and β2-adrenergic receptor antagonists block stress-induced reinstatement in rats [51]. In clinical studies, adrenergic blockers have shown promise as treatments for cocaine abuse and dependence, especially in patients experiencing severe withdrawal symptoms. For example, disulfiram, which blocks norepinephrine synthesis, has been proven to increase the number of cocaine-negative urines in several randomized clinical trials [9, 10, 19]. In addition, lofexidine, an α_2-adrenergic agonist, has also been shown to reduce stress-induced and drug-induced craving in opioid-dependent patients [88].

Similar to substance addiction, non-substance addiction also involves a norepinephrine-related arousal which can mimic a "high" feeling. Studies have found that individuals with gambling disorder have significantly higher cerebrospinal fluid levels of 3-methoxy-4-hydroxy-phenylglycol which is the main metabolite of the noradrenergic system [83]. In addition, individuals with gambling disorder maintained significantly higher noradrenergic levels throughout an entire gambling session while healthy controls exhibited elevated levels only at the onset of the gambling session [72]. Noradrenergic function has been linked to sensation seeking behavior in gambling disorder, which shares some similarities with substance addiction. Overactivity of the noradrenergic system in gambling disorder may reinforce and/or maintain gambling behavior through influences on arousal and individuals with gambling disorder, particularly men, often report excitement as an important reason for gambling [7]. Given that noradrenergic system may mediate rewarding effects of addictive behaviors [107] and noradrenergic drugs may influence treatment outcomes in substance addictions [37, 85, 88], further study is needed to investigate the relationships between noradrenergic system and non-substance addiction.

3.3 Conclusions

Evidence supports the involvement of dopamine, serotonin, opioids, glutamate and norepinephrine in both substance addiction and non-substance addiction while differences may exist with respect to the contributions of these systems. Dopamine is implicated in rewarding and reinforcing effects in both substance addiction and non-substance addiction but they may differ from each other with respect to dopaminergic response to particular manipulations. Low level of the serotonin metabolite has been found in both substance addiction and non-substance addiction, which show similar response towards serotonin agonists. Clinical trial results with opioid antagonists for substance addiction, particularly alcohol and opiate dependence, share similarities with those for gambling disorder. Preliminary evidence has proven the efficacy of medications that alter glutamate neurotransmitter for both substance addiction and non-substance addiction. Similar to non-substance addiction, a norepinephrine-related arousal was involved in substance addiction and noradrenergic function has been linked to substance seeking behavior. All these findings support the conclusion that the neurobiology of non-substance addiction share similarities with those of substance addiction while differences may also exit between them. Further studies are still needed to understand the similarities and differences more precisely, which can facilitate the understanding of neurobiology underlying these addictions as well as the development of treatment across substance addiction and non-substance addiction.

References

1. Amato L, Minozzi S, Pani PP, Davoli M (2007) Antipsychotic medications for cocaine dependence. Cochrane Database Syst Rev 3:CD006306
2. Baker DA, McFarland K, Lake RW, Shen H, Toda S, Kalivas PW (2003) N-acetyl cysteine-induced blockade of cocaine-induced reinstatement. Ann N Y Acad Sci 1003:349–351
3. Barbano MF, Cador M (2007) Opioids for hedonic experience and dopamine to get ready for it. Psychopharmacology 191(3):497–506
4. Boileau I, Payer D, Chugani B, Lobo D, Behzadi A, Rusjan PM, Houle S, Wilson AA, Warsh J, Kish SJ, Zack M (2013) The D2/3 dopamine receptor in pathological gambling: a positron emission tomography study with [11C]-(+)-propyl-hexahydro-naphtho-oxazin and [11C] raclopride. Addiction 108(5):953–963
5. Boileau I, Payer D, Chugani B, Lobo DS, Houle S, Wilson AA, Warsh J, Kish SJ, Zack M (2014) In vivo evidence for greater amphetamine-induced dopamine release in pathological gambling: a positron emission tomography study with [(11)C]-(+)-PHNO. Mol Psychiatry 19(12):1305–1313
6. Brewer JA, Potenza MN (2008) The neurobiology and genetics of impulse control disorders: relationships to drug addictions. Biochem Pharmacol 75(1):63–75
7. Bullock SA, Potenza MN (2012) Pathological gambling: neuropsychopharmacology and treatment. Curr Psychopharmacol 1(1):67–85
8. Campbell-Meiklejohn D, Wakeley J, Herbert V, Cook J, Scollo P, Ray MK, Selvaraj S, Passingham RE, Cowen P, Rogers RD (2011) Serotonin and dopamine play complementary roles in gambling to recover losses. Neuropsychopharmacology 36(2):402–410
9. Carroll KM, Nich C, Ball SA, McCance E, Rounsavile BJ (1998) Treatment of cocaine and alcohol dependence with psychotherapy and disulfiram. Addiction 93(5):713–727
10. Carroll KM, Fenton LR, Ball SA, Nich C, Frankforter TL, Shi J, Rounsaville BJ (2004) Efficacy of disulfiram and cognitive behavior therapy in cocaine-dependent outpatients: a randomized placebo-controlled trial. Arch Gen Psychiatry 61(3):264–272
11. Clark L, Stokes PR, Wu K, Michalczuk R, Benecke A, Watson BJ, Egerton A, Piccini P, Nutt DJ, Bowden-Jones H, Lingford-Hughes AR (2012) Striatal dopamine D(2)/D(3) receptor binding in pathological gambling is correlated with mood-related impulsivity. NeuroImage 63(1):40–46
12. Clark L, Averbeck B, Payer D, Sescousse G, Winstanley CA, Xue G (2013) Pathological choice: the neuroscience of gambling and gambling addiction. J Neurosci 33(45):17617–17623
13. Crean J, Richards JB, de Wit H (2002) Effect of tryptophan depletion on impulsive behavior in men with or without a family history of alcoholism. Behav Brain Res 136(2):349–357
14. De Vries TJ, Schoffelmeer AN, Binnekade R, Raaso H, Vanderschuren LJ (2002) Relapse to cocaine- and heroin-seeking behavior mediated by dopamine D2 receptors is time-dependent and associated with behavioral sensitization. Neuropsychopharmacology 26(1):18–26
15. Drouin C, Darracq L, Trovero F, Blanc G, Glowinski J, Cotecchia S, Tassin JP (2002) Alpha1b-adrenergic receptors control locomotor and rewarding effects of psychostimulants and opiates. J Neurosci 22(7):2873–2884
16. Erb S, Hitchcott PK, Rajabi H, Mueller D, Shaham Y, Stewart J (2000) Alpha-2 adrenergic receptor agonists block stress-induced reinstatement of cocaine seeking. Neuropsychopharmacology 23(2):138–150
17. Fils-Aime ML, Eckardt MJ, George DT, Brown GL, Mefford I, Linnoila M (1996) Early-onset alcoholics have lower cerebrospinal fluid 5-hydroxyindoleacetic acid levels than late-onset alcoholics. Arch Gen Psychiatry 53(3):211–216
18. Fong T, Kalechstein A, Bernhard B, Rosenthal R, Rugle L (2008) A double-blind, placebo-controlled trial of olanzapine for the treatment of video poker pathological gamblers. Pharmacol Biochem Behav 89(3):298–303

19. George TP, Chawarski MC, Pakes J, Carroll KM, Kosten TR, Schottenfeld RS (2000) Disulfiram versus placebo for cocaine dependence in buprenorphine-maintained subjects: a preliminary trial. Biol Psychiatry 47(12):1080–1086
20. Goldstein RZ, Volkow ND (2002) Drug addiction and its underlying neurobiological basis: neuroimaging evidence for the involvement of the frontal cortex. Am J Psychiatry 159(10):1642–1652
21. Goudriaan AE, Oosterlaan J, de Beurs E, Van den Brink W (2004) Pathological gambling: a comprehensive review of biobehavioral findings. Neurosci Biobehav Rev 28(2):123–141
22. Grant JE, Kim SW, Potenza MN, Blanco C, Ibanez A, Stevens L, Hektner JM, Zaninelli R (2003) Paroxetine treatment of pathological gambling: a multi-centre randomized controlled trial. Int Clin Psychopharmacol 18(4):243–249
23. Grant JE, Potenza MN, Hollander E, Cunningham-Williams R, Nurminen T, Smits G, Kallio A (2006) Multicenter investigation of the opioid antagonist nalmefene in the treatment of pathological gambling. Am J Psychiatry 163(2):303–312
24. Grant JE, Kim SW, Odlaug BL (2007) N-acetyl cysteine, a glutamate-modulating agent, in the treatment of pathological gambling: a pilot study. Biol Psychiatry 62(6):652–657
25. Grant JE, Kim SW, Hollander E, Potenza MN (2008) Predicting response to opiate antagonists and placebo in the treatment of pathological gambling. Psychopharmacology (Berl) 200(4):521–527
26. Grant JE, Kim SW, Hartman BK (2008a) A double-blind, placebo-controlled study of the opiate antagonist naltrexone in the treatment of pathological gambling urges. J Clin Psychiatry 69(5):783–789
27. Grant JE, Chamberlain SR, Odlaug BL, Potenza MN, Kim SW (2010) Memantine shows promise in reducing gambling severity and cognitive inflexibility in pathological gambling: a pilot study. Psychopharmacology 212(4):603–612
28. Grant JE, Odlaug BL, Potenza MN, Hollander E, Kim SW (2010) Nalmefene in the treatment of pathological gambling: multicentre, double-blind, placebo-controlled study. Br J Psychiatry 197(4):330–331
29. Gray KM, Watson NL, Carpenter MJ, Larowe SD (2010) N-acetylcysteine (NAC) in young marijuana users: an open-label pilot study. Am J Addict 19(2):187–189
30. Guardia J, Segura L, Gonzalvo B, Iglesias L, Roncero C, Cardus M, Casas M (2004) A double-blind, placebo-controlled study of olanzapine in the treatment of alcohol-dependence disorder. Alcohol Clin Exp Res 28(5):736–745
31. Heinz A, Siessmeier T, Wrase J, Hermann D, Klein S, Grusser SM, Flor H, Braus DF, Buchholz HG, Grunder G, Schreckenberger M, Smolka MN, Rosch F, Mann K, Bartenstein P (2004) Correlation between dopamine D(2) receptors in the ventral striatum and central processing of alcohol cues and craving. Am J Psychiatry 161(10):1783–1789
32. Herz A (1997) Endogenous opioid systems and alcohol addiction. Psychopharmacology (Berl) 129(2):99–111
33. Hietala J, West C, Syvalahti E, Nagren K, Lehikoinen P, Sonninen P, Ruotsalainen U (1994) Striatal D2 dopamine receptor binding characteristics in vivo in patients with alcohol dependence. Psychopharmacology (Berl) 116(3):285–290
34. Highfield D, Yap J, Grimm JW, Shalev U, Shaham Y (2001) Repeated lofexidine treatment attenuates stress-induced, but not drug cues-induced reinstatement of a heroin-cocaine mixture (speedball) seeking in rats. Neuropsychopharmacology 25(3):320–331
35. Hollander E, DeCaria CM, Finkell JN, Begaz T, Wong CM, Cartwright C (2000) A randomized double-blind fluvoxamine/placebo crossover trial in pathologic gambling. Biol Psychiatry 47(9):813–817
36. Hu J, Henry S, Gallezot JD, Ropchan J, Neumaier JF, Potenza MN, Sinha R, Krystal JH, Huang Y, Ding YS, Carson RE, Neumeister A (2010) Serotonin 1B receptor imaging in alcohol dependence. Biol Psychiatry 67(9):800–803
37. Jobes ML, Ghitza UE, Epstein DH, Phillips KA, Heishman SJ, Preston KL (2011) Clonidine blocks stress-induced craving in cocaine users. Psychopharmacology (Berl) 218(1):83–88
38. Kalivas PW (2001) Drug addiction: to the cortex and beyond! Am J Psychiatry 158(3):349–350

39. Kalivas PW (2009) The glutamate homeostasis hypothesis of addiction. Nat Rev Neurosci 10(8):561–572
40. Kalivas PW, Volkow ND (2011) New medications for drug addiction hiding in glutamatergic neuroplasticity. Mol Psychiatry 16(10):974–986
41. Khroyan TV, Barrett-Larimore RL, Rowlett JK, Spealman RD (2000) Dopamine D1- and D2-like receptor mechanisms in relapse to cocaine-seeking behavior: effects of selective antagonists and agonists. J Pharmacol Exp Ther 294(2):680–687
42. Kim SW, Grant JE, Adson DE, Shin YC (2001) Double-blind naltrexone and placebo comparison study in the treatment of pathological gambling. Biol Psychiatry 49(11):914–921
43. Kim SW, Grant JE, Adson DE, Shin YC, Zaninelli R (2002) A double-blind placebo-controlled study of the efficacy and safety of paroxetine in the treatment of pathological gambling. J Clin Psychiatry 63(6):501–507
44. Knackstedt LA, LaRowe S, Mardikian P, Malcolm R, Upadhyaya H, Hedden S, Markou A, Kalivas PW (2009) The role of cystine-glutamate exchange in nicotine dependence in rats and humans. Biol Psychiatry 65(10):841–845
45. Koob GF, Le Moal M (1997) Drug abuse: hedonic homeostatic dysregulation. Science 278(5335):52–58
46. Krishnan-Sarin S, Krystal JH, Shi J, Pittman B, O'Malley SS (2007) Family history of alcoholism influences naltrexone-induced reduction in alcohol drinking. Biol Psychiatry 62(6):694–697
47. Krupitsky EM, Neznanova O, Masalov D, Burakov AM, Didenko T, Romanova T, Tsoy M, Bespalov A, Slavina TY, Grinenko AA, Petrakis IL, Pittman B, Gueorguieva R, Zvartau EE, Krystal JH (2007) Effect of memantine on cue-induced alcohol craving in recovering alcohol-dependent patients. Am J Psychiatry 164(3):519–523
48. Krystal JH, Webb E, Cooney N, Kranzler HR, Charney DS (1994) Specificity of ethanol-like effects elicited by serotonergic and noradrenergic mechanisms. Arch Gen Psychiatry 51(11):898–911
49. Lee B, London ED, Poldrack RA, Farahi J, Nacca A, Monterosso JR, Mumford JA, Bokarius AV, Dahlbom M, Mukherjee J, Bilder RM, Brody AL, Mandelkern MA (2009) Striatal dopamine d2/d3 receptor availability is reduced in methamphetamine dependence and is linked to impulsivity. J Neurosci 29(47):14734–14740
50. LeMarquand DG, Benkelfat C, Pihl RO, Palmour RM, Young SN (1999) Behavioral disinhibition induced by tryptophan depletion in nonalcoholic young men with multigenerational family histories of paternal alcoholism. Am J Psychiatry 156(11):1771–1779
51. Leri F, Flores J, Rodaros D, Stewart J (2002) Blockade of stress-induced but not cocaine-induced reinstatement by infusion of noradrenergic antagonists into the bed nucleus of the stria terminalis or the central nucleus of the amygdala. J Neurosci 22(13):5713–5718
52. Lobo DS, Kennedy JL (2009) Genetic aspects of pathological gambling: a complex disorder with shared genetic vulnerabilities. Addiction 104(9):1454–1465
53. Lutz PE, Kieffer BL (2013) The multiple facets of opioid receptor function: implications for addiction. Curr Opin Neurobiol 23(4):473–479
54. Marazziti D, Golia F, Picchetti M, Pioli E, Mannari P, Lenzi F, Conversano C, Carmassi C, Catena Dell'Osso M, Consoli G, Baroni S, Giannaccini G, Zanda G, Dell'Osso L (2008) Decreased density of the platelet serotonin transporter in pathological gamblers. Neuropsychobiology 57(1–2):38–43
55. Martinez D, Gil R, Slifstein M, Hwang DR, Huang Y, Perez A, Kegeles L, Talbot P, Evans S, Krystal J, Laruelle M, Abi-Dargham A (2005) Alcohol dependence is associated with blunted dopamine transmission in the ventral striatum. Biol Psychiatry 58(10):779–786
56. Martinez D, Narendran R, Foltin RW, Slifstein M, Hwang DR, Broft A, Huang Y, Cooper TB, Fischman MW, Kleber HD, Laruelle M (2007) Amphetamine-induced dopamine release: markedly blunted in cocaine dependence and predictive of the choice to self-administer cocaine. Am J Psychiatry 164(4):622–629

57. Martinez D, Greene K, Broft A, Kumar D, Liu F, Narendran R, Slifstein M, Van Heertum R, Kleber HD (2009) Lower level of endogenous dopamine in patients with cocaine dependence: findings from PET imaging of D(2)/D(3) receptors following acute dopamine depletion. Am J Psychiatry 166(10):1170–1177
58. Martinez D, Carpenter KM, Liu F, Slifstein M, Broft A, Friedman AC, Kumar D, Van Heertum R, Kleber HD, Nunes E (2011) Imaging dopamine transmission in cocaine dependence: link between neurochemistry and response to treatment. Am J Psychiatry 168(6):634–641
59. Martinez D, Saccone PA, Liu F, Slifstein M, Orlowska D, Grassetti A, Cook S, Broft A, Van Heertum R, Comer SD (2012) Deficits in dopamine D(2) receptors and presynaptic dopamine in heroin dependence: commonalities and differences with other types of addiction. Biol Psychiatry 71(3):192–198
60. McElroy SL, Nelson EB, Welge JA, Kaehler L, Keck PE Jr (2008) Olanzapine in the treatment of pathological gambling: a negative randomized placebo-controlled trial. J Clin Psychiatry 69(3):433–440
61. Monti PM, Rohsenow DJ, Hutchison KE, Swift RM, Mueller TI, Colby SM, Brown RA, Gulliver SB, Gordon A, Abrams DB (1999) Naltrexone's effect on cue-elicited craving among alcoholics in treatment. Alcohol Clin Exp Res 23(8):1386–1394
62. Moore RY, Bloom FE (1979) Central catecholamine neuron systems: anatomy and physiology of the norepinephrine and epinephrine systems. Annu Rev Neurosci 2:113–168
63. Narendran R, Martinez D, Mason NS, Lopresti BJ, Himes ML, Chen CM, May MA, Price JC, Mathis CA, Frankle WG (2011) Imaging of dopamine D2/3 agonist binding in cocaine dependence: a [11C]NPA positron emission tomography study. Synapse 65(12):1344–1349
64. Niciu MJ, Arias AJ (2013) Targeted opioid receptor antagonists in the treatment of alcohol use disorders. CNS Drugs 27(10):777–787
65. Nordin C, Eklundh T (1999) Altered CSF 5-HIAA disposition in pathologic male gamblers. CNS Spectr 4(12):25–33
66. Nordin C, Gupta RC, Sjodin I (2007) Cerebrospinal fluid amino acids in pathological gamblers and healthy controls. Neuropsychobiology 56(2–3):152–158
67. Nussbaum D, Honarmand K, Govoni R, Kalahani-Bargis M, Bass S, Ni X, Laforge K, Burden A, Romero K, Basarke S, Courbasson C, Deamond W (2011) An eight component decision-making model for problem gambling: a systems approach to stimulate integrative research. J Gambl Stud 27(4):523–563
68. Olsen CM (2011) Natural rewards, neuroplasticity, and non-drug addictions. Neuropharmacology 61(7):1109–1122
69. Padgett DA, Glaser R (2003) How stress influences the immune response. Trends Immunol 24(8):444–448
70. Palfai T, Davidson D, Swift R (1999) Influence of naltrexone on cue-elicited craving among hazardous drinkers: the moderational role of positive outcome expectancies. Exp Clin Psychopharmacol 7(3):266–273
71. Pallanti S, Bernardi S, Quercioli L, DeCaria C, Hollander E (2006) Serotonin dysfunction in pathological gamblers: increased prolactin response to oral m-CPP versus placebo. CNS Spectr 11(12):956–964
72. Pallanti S, Bernardi S, Allen A, Chaplin W, Watner D, DeCaria CM, Hollander E (2010) Noradrenergic function in pathological gambling: blunted growth hormone response to clonidine. J Psychopharmacol 24(6):847–853
73. Pecina S, Smith KS, Berridge KC (2006) Hedonic hot spots in the brain. Neuroscientist 12(6):500–511
74. Perry CJ, McNally GP (2013) A role for the ventral pallidum in context-induced and primed reinstatement of alcohol seeking. Eur J Neurosci 38(5):2762–2773
75. Petrovic P, Pleger B, Seymour B, Kloppel S, De Martino B, Critchley H, Dolan RJ (2008) Blocking central opiate function modulates hedonic impact and anterior cingulate response to rewards and losses. J Neurosci 28(42):10509–10516

76. Potenza MN (2008) Review. The neurobiology of pathological gambling and drug addiction: an overview and new findings. Philos Trans R Soc Lond Ser B Biol Sci 363(1507):3181–3189
77. Potenza MN, Steinberg MA, McLaughlin SD, Wu R, Rounsaville BJ, O'Malley SS (2001) Gender-related differences in the characteristics of problem gamblers using a gambling helpline. Am J Psychiatry 158(9):1500–1505
78. Potenza MN, Hong KI, Lacadie CM, Fulbright RK, Tuit KL, Sinha R (2012) Neural correlates of stress-induced and cue-induced drug craving: influences of sex and cocaine dependence. Am J Psychiatry 169(4):406–414
79. Potenza MN, Walderhaug E, Henry S, Gallezot JD, Planeta-Wilson B, Ropchan J, Neumeister A (2013) Serotonin 1B receptor imaging in pathological gambling. World J Biol Psychiatry 14(2):139–145
80. Ratsma JE, Van Der Stelt O, Gunning WB (2002) Neurochemical markers of alcoholism vulnerability in humans. Alcohol Alcohol 37(6):522–533
81. Ritz MC, Lamb RJ, Goldberg SR, Kuhar MJ (1987) Cocaine receptors on dopamine transporters are related to self-administration of cocaine. Science 237(4819):1219–1223
82. Rosner S, Hackl-Herrwerth A, Leucht S, Vecchi S, Srisurapanont M, Soyka M (2010) Opioid antagonists for alcohol dependence. Cochrane Database Syst Rev 12:Cd001867
83. Roy A, DeJong J, Ferraro T, Adinoff B, Gold P, Rubinow D, Linnoila M (1989) CSF GABA and neuropeptides in pathological gamblers and normal controls. Psychiatry Res 30(2):137–144
84. Saiz-Ruiz J, Blanco C, Ibanez A, Masramon X, Gomez MM, Madrigal M, Diez T (2005) Sertraline treatment of pathological gambling: a pilot study. J Clin Psychiatry 66(1):28–33
85. Shaham Y, Erb S, Stewart J (2000) Stress-induced relapse to heroin and cocaine seeking in rats: a review. Brain Res Brain Res Rev 33(1):13–33
86. Shinohara K, Yanagisawa A, Kagota Y, Gomi A, Nemoto K, Moriya E, Furusawa E, Furuya K, Terasawa K (1999) Physiological changes in Pachinko players; beta-endorphin, catecholamines, immune system substances and heart rate. Appl Hum Sci 18(2):37–42
87. Simmons D, Self DW (2009) Role of mu- and delta-opioid receptors in the nucleus accumbens in cocaine-seeking behavior. Neuropsychopharmacology 34(8):1946–1957
88. Sinha R, Kimmerling A, Doebrick C, Kosten TR (2007) Effects of lofexidine on stress-induced and cue-induced opioid craving and opioid abstinence rates: preliminary findings. Psychopharmacology (Berl) 190(4):569–574
89. Sofuoglu M, Sewell RA (2009) Norepinephrine and stimulant addiction. Addict Biol 14(2):119–129
90. Spanagel R, Herz A, Shippenberg TS (1992) Opposing tonically active endogenous opioid systems modulate the mesolimbic dopaminergic pathway. Proc Natl Acad Sci U S A 89(6):2046–2050
91. Tang XC, McFarland K, Cagle S, Kalivas PW (2005) Cocaine-induced reinstatement requires endogenous stimulation of mu-opioid receptors in the ventral pallidum. J Neurosci 25(18):4512–4520
92. Torrens M, Fonseca F, Mateu G, Farre M (2005) Efficacy of antidepressants in substance use disorders with and without comorbid depression. A systematic review and meta-analysis. Drug Alcohol Depend 78(1):1–22
93. Tremblay AM, Desmond RC, Poulos CX, Zack M (2011) Haloperidol modifies instrumental aspects of slot machine gambling in pathological gamblers and healthy controls. Addict Biol 16(3):467–484
94. Urban NB, Kegeles LS, Slifstein M, Xu X, Martinez D, Sakr E, Castillo F, Moadel T, O'Malley SS, Krystal JH, Abi-Dargham A (2010) Sex differences in striatal dopamine release in young adults after oral alcohol challenge: a positron emission tomography imaging study with [(1)(1)C]raclopride. Biol Psychiatry 68(8):689–696
95. van Wageningen H, Jorgensen HA, Specht K, Hugdahl K (2010) A 1H-MR spectroscopy study of changes in glutamate and glutamine (Glx) concentrations in frontal spectra after administration of memantine. Cereb Cortex 20(4):798–803

96. Vescovi PP, Coiro V (1997) Persistence of defective serotonergic and GABAergic controls of growth hormone secretion in long-term abstinent alcoholics. Alcohol Alcohol 32(1):85–90
97. Volkow ND, Fowler JS, Wolf AP, Schlyer D, Shiue CY, Alpert R, Dewey SL, Logan J, Bendriem B, Christman D et al (1990) Effects of chronic cocaine abuse on postsynaptic dopamine receptors. Am J Psychiatry 147(6):719–724
98. Volkow ND, Fowler JS, Wang GJ, Hitzemann R, Logan J, Schlyer DJ, Dewey SL, Wolf AP (1993) Decreased dopamine D2 receptor availability is associated with reduced frontal metabolism in cocaine abusers. Synapse 14(2):169–177
99. Volkow ND, Wang GJ, Fowler JS, Logan J, Hitzemann R, Ding YS, Pappas N, Shea C, Piscani K (1996) Decreases in dopamine receptors but not in dopamine transporters in alcoholics. Alcohol Clin Exp Res 20(9):1594–1598
100. Volkow ND, Wang GJ, Fowler JS, Logan J, Gatley SJ, Hitzemann R, Chen AD, Dewey SL, Pappas N (1997) Decreased striatal dopaminergic responsiveness in detoxified cocaine-dependent subjects. Nature 386(6627):830–833
101. Volkow ND, Chang L, Wang GJ, Fowler JS, Ding YS, Sedler M, Logan J, Franceschi D, Gatley J, Hitzemann R, Gifford A, Wong C, Pappas N (2001) Low level of brain dopamine D2 receptors in methamphetamine abusers: association with metabolism in the orbitofrontal cortex. Am J Psychiatry 158(12):2015–2021
102. Volkow ND, Wang GJ, Ma Y, Fowler JS, Wong C, Ding YS, Hitzemann R, Swanson JM, Kalivas P (2005) Activation of orbital and medial prefrontal cortex by methylphenidate in cocaine-addicted subjects but not in controls: relevance to addiction. J Neurosci 25(15):3932–3939
103. Volkow ND, Wang GJ, Begleiter H, Porjesz B, Fowler JS, Telang F, Wong C, Ma Y, Logan J, Goldstein R, Alexoff D, Thanos PK (2006) High levels of dopamine D2 receptors in unaffected members of alcoholic families: possible protective factors. Arch Gen Psychiatry 63(9):999–1008
104. Volkow ND, Wang GJ, Telang F, Fowler JS, Logan J, Jayne M, Ma Y, Pradhan K, Wong C (2007) Profound decreases in dopamine release in striatum in detoxified alcoholics: possible orbitofrontal involvement. J Neurosci 27(46):12700–12706
105. Volkow ND, Wang GJ, Fowler JS, Tomasi D, Telang F (2011) Addiction: beyond dopamine reward circuitry. Proc Natl Acad Sci U S A 108(37):15037–15042
106. Wang GJ, Smith L, Volkow ND, Telang F, Logan J, Tomasi D, Wong CT, Hoffman W, Jayne M, Alia-Klein N, Thanos P, Fowler JS (2012) Decreased dopamine activity predicts relapse in methamphetamine abusers. Mol Psychiatry 17(9):918–925
107. Weinshenker D, Schroeder JP (2007) There and back again: a tale of norepinephrine and drug addiction. Neuropsychopharmacology 32(7):1433–1451
108. Welberg L (2011) Addiction: from mechanisms to treatment. Nat Rev Neurosci 12(11):621
109. Zack M, Poulos CX (2004) Amphetamine primes motivation to gamble and gambling-related semantic networks in problem gamblers. Neuropsychopharmacology 29(1):195–207
110. Zhou W, Kalivas PW (2008) N-acetylcysteine reduces extinction responding and induces enduring reductions in cue- and heroin-induced drug-seeking. Biol Psychiatry 63(3):338–340

Chapter 4
Similarities and Differences in Genetics

Yang Zhang, Yan Sun, Jie Liang, Lin Lu, and Jie Shi

Abstract Similar symptomatology manifestations and high co-morbidity in substance and non-substance addictions suggest that there may be a common pathogenesis between them. Associated with impulse control and emotional processing, the monoamine neurotransmitter system genes are suggested to be related to both substance and non-substance addictions, such as dopamine (DA) system, 5-hydroxytryptamine/serotonin (5-HT) system, the endogenous opioid system and so on. Here we reviewed the similarities and differences in genetics between classic substance addiction and common types of non-substance addiction, e.g. pathological gambling, Internet addiction, binge-eating disorder etc. It is necessary to systematically compare genetic mechanisms of non-substance addiction and substance addiction, which could reveal similarities and differences of substance addiction and non-addictive substances essentially, enhance our understanding of addiction theory and improve clinical practice with research results.

Keywords Substance addiction • Non-substance addiction • Genetics

---

Y. Zhang • J. Liang
Department of Pharmacology, School of Basic Medical Sciences,
Peking University Health Science Center, Beijing 100191, China

National Institute on Drug Dependence, Peking University,
No. 38, Xueyuan Road, Haidian District, Beijing 100191, China

Y. Sun • J. Shi (✉)
National Institute on Drug Dependence, Peking University,
No. 38, Xueyuan Road, Haidian District, Beijing 100191, China
e-mail: shijie@bjmu.edu.cn

L. Lu
Institute of Mental Health/Peking University Sixth Hospital and National Clinical Research
Center for Mental Disorders & Key Laboratory of Mental Health, Peking University,
Beijing 100191, China

X. Zhang et al. (eds.), *Substance and Non-substance Addiction*, Advances in
Experimental Medicine and Biology 1010, DOI 10.1007/978-981-10-5562-1_4

4.1 Introduction

Since 1980s, the studies by psychologists and psychiatrists indicated that similar to drug, outside non-drug stimulation shared important elements with substance addictions. These include indulgency in certain behaviors, continued engagement despite negative consequences, impaired control over the craving so that the behavior of repeated, many times unsuccessful withdrawal experience and continuing to increase the intensity of use etc. [47, 52, 62]. Therefore, the researchers began to reconsider the meaning of addiction, and proposed other terms to refer to the physical and psychological changes similar to substance addiction, triggered by environmental cues and mediated by behaviors, such as "behavioral addictions", "non-substance-related addiction", "non-drug addiction" and so on.

Non-substance addiction manifestations vary and are scattered classified in the different classifications in the "Diagnostic and Statistical Manual of mental disorders" 5th Edition (DSM-V) established by the American Psychiatric Association, "International Classification of Diseases" 10th Edition (ICD-10) developed by World Health Organization (WHO) and other guides, e.g., pathological gambling (PG), Internet addiction, binge-eating disorder (BED). Non-substance addiction also results into huge adverse effects to individuals, families and society, such as dropouts or unemployment to individual physical and mental health of individuals, family conflicts and interpersonal problems, and even increasing the risk of crime. Therefore, non-substance addiction is not only related to personal mental health problems, but also the whole society.

There are some similarities between the progression of the non-substance addiction and substance addiction. As widely abused as nicotine addiction and alcohol dependence (AD), the incidences of PG, Internet addiction disorder and BED are high in adolescents and young adults [27]. Besides, a study based on 7,869 male twins indicated that genetic and environmental factors contributed to co-occurrence in substance addiction and non-substance addiction [75]. What's more, scientists hypothesize that substance addiction and non-substance addictions could "transfer to" or "replace" each other. The authors gave an assumption on genetics predictive capabilities. In this assumption, the researchers believe it could be relative on neurochemical similarities [6]. Comorbidity among substance addiction, non-substance addiction and other psychiatric conditions appears to involve shared genetic factors [6, 27]. Genetic findings provide further evidence for similarities and differences in pathophysiological mechanisms between non-substance and substance addictions [6, 49].

There is important theoretical and clinical value for investigating the similarities and differences in genetic mechanism between substance addiction and non-substance addiction. Firstly, for the purpose to further understand the mechanisms of addiction, the role of non-substance addiction may just exclude the direct influences and damage of addictive substances. Secondly, owing to that rewards from non-substance addiction is not classified as drug reward, the study about whether natural reward or pharmaceutical reward initiated similar physiological processes may greatly promote the understanding of addiction. Thirdly, clarifying the genetic similarities and differences could promote more distinct criteria for the diagnosis of both non-substance addiction and substance addiction, which might help integrate consideration and treatment of individuals [34, 62, 63]. Thanks to the development of theory and technology in genetics and neuroscience, and lower cost of the experiment, more and more addiction related genes would be discovered by candidate gene studies, such as genome-wide association study, whole genome sequencing, and even epigenetic research.

4.2 Similarities in Genetics

Existing researches indicated that genes especially in multiple neurotransmitter systems related with substance addiction were also correlated with non-substance addiction, which suggested parallels between non-substance addiction and substance addiction. Inextricably linked with impulse control and emotional processing, the monoamine neurotransmitter systems are paid attention in genetic studies related to substance addiction and non-substance addiction, especially dopamine (DA) system. Others contain 5-hydroxytryptamine/serotonin (5-HT) system, the endogenous opioid system, norepinephrine systems and other related genes. Here, we describe the common genetic risk factors between substance addiction and non-substance addiction in each related neurotransmitter system.

4.2.1 DA System

4.2.1.1 Dopamine Receptor (DR) Genes

DA is an important neurotransmitter in the brain, and is closely related to spontaneous activity, euphoria, etc., which plays an important role in occurrence, maintenance and relapse in addiction. It implements strengthening and rewarding effects of drugs or behaviors through a variety of mechanisms to the signal conditions association, strengthening the memory of the stimulus event and motivational behavior [4]. The dopamine receptors include 5 types and are classified as D1-like (DRD1 and DRD5) or D2-like (DRD2, DRD3 and DRD4) receptors based on sequence homology and pharmacology.

DRD1 mediates reward and motivation sensitization of substance addiction, which plays a leading role in the initial treatment of neural adaptation. Involved in the reward process of alcohol, cocaine addiction and other substances addiction, DRD1 is critical for impulse control disorders [14, 26, 67]. Currently, some studies also have found *DRD1* is associated with non-substance addiction, e.g. the results of 140 PG patients suggest the association of PG with *DRD1* -800 T/C allele T in Brazil [15].

DRD2 is also associated with reward and pleasure, which may be more involved in conditional reinforcement and compulsive drug-seeking behaviors, and has effects on the neural adaptation of the latter part of medication. Subjects were genotyped for a *TaqIA* Single Nucleotide Polymorphisms (SNP, rs1800497) of the adjacent gene Ankyrin repeat and kinase domain containing 1 (*ANKK1*), located 10 kb downstream from exon VIII, which include two alleles A1 and A2. The *TaqIA* A1 allele (A1/A1 homozygote and A1/A2 heterozygous) was associated with reduced D2 receptor density and with altered substrate-binding specificity [40], which may predict the risk of substance dependence disorders [42, 58], such as AD [30, 53]. *DRD2* is also associated with non-substance addiction [17]. Davis et al., reported that compared with the control group, 56 BED patients who carried *TaqI* A1 allele were in a higher ratio and less sensitive to reward [16]. A study for the online game addiction also showed that, Internet video game play addicts carried a higher proportion of *TaqI* A1 allele, and the *TaqI* A1 allele carriers showed higher reward-dependence scores [31]. Moreover, researchers also found the association between *TaqI* A1 allele and PG in Canada, USA and Spain [12, 22, 51].

DRD3 mainly exists in the limbic system, which inhibits intracellular cAMP function [61]. DRD3 deficient mice exhibit increased sensitivity and impulses to reward, which easily lead to drug taking [45]. Studies have shown that *DRD3* rs3773678 and rs7638876 were associated with nicotine addiction [36, 71], and PG studies have similar findings [29].

DRD4, the 48 bp variable number of tandem repeats (VNTR) located at exon III, has an impact on its distribution and function. There is evidence that the 7-repeat allele alters the pharmacological profile of the receptor and is associated with alcohol dependence [30]. What's more, European and American studies have shown that this VNTR was associated with PG [10, 20, 29]. The significantly correlate result was also reported in a mixing sample study, which contains PG, drug addiction, attention deficit hyperactivity disorder (ADHD) and Tourette's syndrome in the United States [11]. Similar results have also been found in a US research on BED [66] and a PG study has indicated that this phenomenon occurred in women in Spain [60].

4.2.1.2 Dopamine Transporter (DAT) Gene

One of the implicated genetic polymorphism in *DAT* (solute carrier family 6, member 3, *SLC6A3*) for addiction is the 40 bp VNTR located downstream from exon 15 in the 3′ untranslated region. Guo et al. reported that this variation was associated with AD [30]. Spanish and American scientists also found that the *DAT* VNTR polymorphism was related with PG [10, 22]. However, in a genetic study in Italian, Mignini et al. found there was no significant difference in the polymorphism distribution between 280 alcohol addicts and controls [53].

4.2.2 5-HT System

4.2.2.1 5-HT Transporter (5-HTT) Gene (*5-HTT/SLC18A2*)

5-HTT is distributed in the presynaptic membrane, which controls the 5-HT level in synaptic cleft as well as the duration of 5-HT and its receptor interaction by reuptake. The *SLC18A2* linked polymorphic region (*5-HTTLPR*) polymorphism locate in promoter region of *SLC18A2*, which plays an important role on *SLC18A2* transcription. There are two alleles including short allele (S) and long allele (L) of the site. Compared with the L genotype, the person who carried S genotype had a lower transcription efficiency, resulting in less re-uptake of 5-HT [32].

There has always been a central issue that, the relationship between *5-HTTLPR* and addictions. Numerous studies have showed that those who carried S allele were more associated with alcohol or other drug addiction [8, 23, 30, 48, 50]. Recently, some scholars have pointed out that *5-HTTLPR* and non-substance addiction also had a certain relationship [24, 29]. Compared with the *5-HTTLPR* genotype between 68 PG patients and 68 normal rational, Pérez de Castro et al. found that male patients were significatly related to S allele, but not found in female subjects [59]. Lee et al. reported that the proportion of Internet addiction in S/S homozygotes was significantly higher than that of controls [46]. These results suggested that the S allele was associated with both substance addiction and non-substance addiction.

4.2.2.2 5-HT Receptor Gene

The serotonergic receptors are divided into seven types (5-HT1–7). The type 2 receptors are categorized into three sub-types (A, B, and C). The 5-HT2A receptor gene is located at chromosome 13. A T102C (rs6313) polymorphism located on *5-HT2A*, which does not determine a change of amino acid in the receptor molecule, resulted in a diminished synthesis of 5-HT2A receptors. Mechanisms underlying these expression differences remain unclear, although it has been speculated that the T102C polymorphism affects the stability of the respective mRNA. Another hypothesis concerns potential methylation differences in the promoter region of HTR2A gene. Individuals with the C/C genotype in the *HTR2A* T102C polymorphism probably have significantly lower expression of the gene. The rs6313 C/C genotype have been found that it is the risk factor for addictive disorders, including PG [73], nicotine dependence and AD [19, 37, 39, 57, 72].

4.2.3 Other Related Genes

4.2.3.1 μ Opioid Receptor (OPRM1) Gene

OPRM1 activation increases the release of DA in nucleus accumbens (Hadad and Knackstedt). As a primary target for opioid drugs and peptides, OPRM1 plays a key role in pain perception and addiction. Genetic variants of *OPRM1* have been implicated in predisposition to drug addiction, in particular the SNP A118G. In transfected cells, *OPRM1*-Asp^{40} was reported to have threefold higher affinity for β-endorphin than *OPRM1*-Asn^{40} [7], suggesting a gain of function, but subsequent studies have failed to corroborate these results [3, 5]. In vitro transfection studies, however, indicated that the G118 allele might be associated with lower OPRM1 protein expression than the A118 allele [5]. Therefore, the functional significance of the A118G variant of *OPRM1* remains unresolved.

Nevertheless, serving as a primary target for opioid drugs and peptides, OPRM1 mediates the effects of morphine and heroin [2, 44, 56]. Some researches have shown that *OPRM1* (A118G) polymorphism affects the ability of opioid peptides binding to ligands [77], suggesting that the G allele may be strongly associated with susceptibility to opiate addiction. For instance, heroin abuse susceptibility may be associated with the increased OPRM1-drug affinity of drug users [56]. By impinging on dopaminergic pathways, *OPRM1* also plays a role in addiction to other drugs of abuse, such as cocaine, nicotine, and alcohol [33, 43]. In addition, a BED study in Canadian also indicated that this "gain-of-function" G allele of A118G was related with highly visible and easily accessible surfeit of sweet and fatty foods [17].

4.2.3.2 Catecholamine-O-Methyltransferase (COMT) Gene

COMT has a key role in the degradation pathway of DA. There are 2 allelic forms of the *COMT*, expression of which results in a valine variant ($COMT^H$) and methionine variant ($COMT^L$). These two variants have different functional activity; $COMT^H$ has been demonstrated to have 3 to 4 times higher enzymatic activity than $COMT^L$. Many scientific researchers noticed that *COMT* rs4680 (val158met) polymorphism was related with heroin, nicotine, alcohol and other substance addiction in China, the United States and South Korea [1, 9, 21, 41, 68, 74, 76]. The similar discoveries were also reported in the non-substance addiction research, such as the Korea research has shown that this polymorphic loci was related to excessive internet video game play (EIPG) [31]. Moreover, PG study had similar results in the United States [28].

4.2.3.3 Monoamine Oxidase A (MAOA) Gene

MAOA is an important monoamine neurotransmitter metabolic enzyme, which could degrade 5-HT, DA, epinephrine, norepinephrine and other monoamine neurotransmitters. *MAOA* locates on the X chromosome. In *MAOA* promoter region, there is a VNTR of the 30 bp repeating unit and the number of repetitions ranges from 2 to 5 times. On the same site, different variable number tandem repeats would form different nucleotide sequences, which could affect the transcriptional activity of *MAOA*. Previous studies have found that transcription efficient of 3 times repeat allele was lower than that of 3.5 or 4 times repeat allele [64].

Due to transcription inefficiency caused by the 3 times repeat allele, MAOA synthesis reduction results in increasing concentration of monoamine neurotransmitters, which could lead to consequences of enhancing the individual response to stress and vulnerable to drug addiction [70]. In the past few years many researchers have reported that this polymorphism was significantly associated with AD in United States, Germany and Brazil [13, 30, 65]. Some studies on the relationship between non-substance addiction and this polymorphism have also shown the similar results. For instance, a Spanish study has demonstrated that the radio (44.9 %) of 3 times repeat allele in PG patients was higher than that of controls (32.6 %). Furthermore, this difference is more significant in male: the rate of 3 times repeat allele was 55.3 % in male gamblers, while that of the male control group was only 37.2 % [38, 59].

4.3 Differences in Genetics Between Substance Addiction and Non-substance Addiction

As mentioned above, most results are consistent in genetics research involving both substance addiction and non-substance addiction, which suggests that they may have a common genetic or neurobiological mechanism. However, a few studies revealed the differences between them, which were mainly related to *5-HTT* and brain-derived neurotrophic factor (BDNF) gene.

4.3.1 5-HTT

Based on the above findings on *5-HTT*, it seems that S allele was related with substance and non-substance addictions. However, Wilson D et al. did not find this difference in 140 PG patients [73]. Studies on BED and shopping addiction did not show the consistent results either. For example, a study indicated no association between *5-HTTLPR* and symptoms of shopping addiction in 21 patients conducted by Devor et al. [18]. Some studies even got opposite results, such as a study showed that L allele was related with BED in 77 female patients, by which patients with L allele and L/L genotype were significantly higher than that of the control group [54]. Gorwood P et al. found that S allele was a risk factor of anorexia nervosa [25]. In summary, the association between *5-HTTLPR* and substance addiction is consistent rather than non-substance addiction. Therefore, more research is needed to ascertain the relationship between *5-HTTLPR* and non-substance addiction.

4.3.2 BDNF

BDNF is a member of the neurotrophic factors family, and the most abundant brain neurotrophic factor, which leads it as a key factor in survival and differentiation of dopaminergic neurons. Evidence from animal and clinical studies have shown that *BDNF* activity was related with the pathogenesis of substance addiction, it may be owing to the effects of *BDNF* Val66Met polymorphism on BDNF secretion. Some studies have shown that *BDNF* 66Val homozygous was more in substance addicts [69]. Besides, one Chinese heroin addiction study found that Val66Met carriers had earlier onset of heroin abuse than that of Val66Val and Met66Met carriers [35]. However, a study of 210 Caucasian women has shown that the 66Met homozygous carriers had more frequency and severity of binge eating [55].

4.4 Summary and Prospect

Research on the genetics of substance addiction and non-substance addiction has accelerated in recent years benefit from the development of molecular biology techniques. Molecular genetics of classic substance addiction provided some trials for non-substance addiction-related genes. Exactly, a plurality of genes related with substance addiction especially monoamine neurotransmitter gene is correlated with non-substance addiction, including *DRD1-DRD4*, *SLC6A3*, *5-HT2A*, *OPRM1*, *COMT* and *MAOA*, most of which were consistent with each other, suggesting that there may be a common biological mechanism between them. However, there were still some inconsistent results (for instance, *SLC18A2* and *BDNF*) even in various types of non-substance addiction, which may be due to different mechanism between substance addiction and non-substance addiction, or the different research methods, such as diagnostic criteria for different disorders or ethnic differences. The following aspects could be focused on in future research:

First, the concept and definition of non-substance addiction need to be further clarified. Since the animal models of non-substance addiction are difficult to build, the clinical research become very important. It is unclear or dispersed that diagnostic criteria of many approved types of non-substance addiction [47, 52, 62], which impede further studies and the comparison between the different studies now. Recently, the genetic mechanism studies of non-substance addiction were mainly focused on PG, BED and Internet addiction. However, these types of non-substance addiction have a significant relationship with gender and age, which makes the influence on representation and promotion of the findings. As heritability estimation may be effected by age and gender, future studies should be carried out on other types of non-substance addiction in different gender and age groups, such as adolescents and young adults.

In addition, for the best of our knowledge, there are few reports on candidate genes and gene interactions, in particular on non-addictive substance addiction in the last few years. Just like substance addiction, non-substance addiction is also affected by multiple genes [6, 27, 49]. Moreover, the current evidence suggests that non-substance addiction may be also affected by both disease-causing genes similar with substance addiction and environmental risk factors, e.g. poor family or other social environment. Therefore, the researchers should pay more attention on investigating the influence of environmental factors and mechanisms in the future. Furthermore, the effect of the epigenetics may also be a way to interact with the environment, so we believe that it will be an interesting topic on investigating the relationship with specific genes and non-substance addiction by interactions of gene-environment or gene-gene.

References

1. Ashare RL, Valdez JN, Ruparel K, Albelda B, Hopson RD, Keefe JR, Loughead J, Lerman C (2013) Association of abstinence-induced alterations in working memory function and COMT genotype in smokers. Psychopharmacology 230(4):653–662
2. Basbaum AI, Fields HL (1984) Endogenous pain control systems: brainstem spinal pathways and endorphin circuitry. Ann Rev Neurosci 7:309–338
3. Befort K, Filliol D, Decaillot FM, Gaveriaux-Ruff C, Hoehe MR, Kieffer BL (2001) A single nucleotide polymorphic mutation in the human mu-opioid receptor severely impairs receptor signaling. J Biol Chem 276(5):3130–3137
4. Berridge KC, Robinson TE (1998) What is the role of dopamine in reward: hedonic impact, reward learning, or incentive salience? Brain Res Brain Res Rev 28(3):309–369
5. Beyer A, Koch T, Schroder H, Schulz S, Hollt V (2004) Effect of the A118G polymorphism on binding affinity, potency and agonist-mediated endocytosis, desensitization, and resensitization of the human mu-opioid receptor. J Neurochem 89(3):553–560
6. Blum K, Bailey J, Gonzalez AM, Oscar-Berman M, Liu Y, Giordano J, Braverman E, Gold M (2011) Neuro-genetics of Reward Deficiency Syndrome (RDS) as the root cause of "Addiction Transfer": a new phenomenon common after bariatric surgery. J Genet Syndr Gene Ther 2012(1)
7. Bond C, LaForge KS, Tian M, Melia D, Zhang S, Borg L, Gong J, Schluger J, Strong JA, Leal SM, Tischfield JA, Kreek MJ, Yu L (1998) Single-nucleotide polymorphism in the human mu opioid receptor gene alters beta-endorphin binding and activity: possible implications for opiate addiction. Proc Natl Acad Sci U S A 95(16):9608–9613
8. Brody GH, Beach SR, Philibert RA, Chen YF, Lei MK, Murry VM, Brown AC (2009) Parenting moderates a genetic vulnerability factor in longitudinal increases in youths' substance use. J Consult Clin Psychol 77(1):1–11

9. Choi HD, Shin WG (2015) Association between catechol-O-methyltransferase (COMT) Val/Met genotype and smoking cessation treatment with nicotine: a meta-analysis. Pharmacogenomics 16(16):1879–1885
10. Comings DE, Gade-Andavolu R, Gonzalez N, Wu S, Muhleman D, Chen C, Koh P, Farwell K, Blake H, Dietz G, MacMurray JP, Lesieur HR, Rugle LJ, Rosenthal RJ (2001) The additive effect of neurotransmitter genes in pathological gambling. Clin Genet 60(2):107–116
11. Comings DE, Gonzalez N, Wu S, Gade R, Muhleman D, Saucier G, Johnson P, Verde R, Rosenthal RJ, Lesieur HR, Rugle LJ, Miller WB, MacMurray JP (1999) Studies of the 48 bp repeat polymorphism of the DRD4 gene in impulsive, compulsive, addictive behaviors: Tourette syndrome, ADHD, pathological gambling, and substance abuse. Am J Med Genet 88(4):358–368
12. Comings DE, Rosenthal RJ, Lesieur HR, Rugle LJ, Muhleman D, Chiu C, Dietz G, Gade R (1996) A study of the dopamine D2 receptor gene in pathological gambling. Pharmacogenetics 6(3):223–234
13. Contini V, Marques FZ, Garcia CE, Hutz MH, Bau CH (2006) MAOA-uVNTR polymorphism in a Brazilian sample: further support for the association with impulsive behaviors and alcohol dependence. Am J M Genet B, Neuropsychiatr Genet 141B(3):305–308
14. D'Souza MS, Ikegami A, Olsen CM, Duvauchelle CL (2003) Chronic D1 agonist and ethanol coadministration facilitate ethanol-mediated behaviors. Pharmacol Biochem Behav 76(2):335–342
15. da Silva Lobo DS, Vallada HP, Knight J, Martins SS, Tavares H, Gentil V, Kennedy JL (2007) Dopamine genes and pathological gambling in discordant sib-pairs. J Gambl Stud/ Co-Sponsored Nat Counc Probl Gambl Inst Study Gambl Commer Gaming 23(4):421–433
16. Davis C, Levitan RD, Kaplan AS, Carter J, Reid C, Curtis C, Patte K, Hwang R, Kennedy JL (2008) Reward sensitivity and the D2 dopamine receptor gene: a case-control study of binge eating disorder. Prog Neuro-Psychopharmacol Biol Psychiatry 32(3):620–628
17. Davis CA, Levitan RD, Reid C, Carter JC, Kaplan AS, Patte KA, King N, Curtis C, Kennedy JL (2009) Dopamine for "wanting" and opioids for "liking": a comparison of obese adults with and without binge eating. Obesity (Silver Spring, Md) 17(6):1220–1225
18. Devor EJ, Magee HJ, Dill-Devor RM, Gabel J, Black DW (1999) Serotonin transporter gene (5-HTT) polymorphisms and compulsive buying. Am J Med Genet 88(2):123–125
19. do Prado-Lima PA, Chatkin JM, Taufer M, Oliveira G, Silveira E, Neto CA, Haggstram F, Bodanese LC, da Cruz IB (2004) Polymorphism of 5HT2A serotonin receptor gene is implicated in smoking addiction. Am J Med Genet B Neuropsychiatr Genet 128b(1):90–93
20. Eisenegger C, Knoch D, Ebstein RP, Gianotti LR, Sandor PS, Fehr E (2010) Dopamine receptor D4 polymorphism predicts the effect of L-DOPA on gambling behavior. Biol Psychiatry 67(8):702–706
21. Enoch MA (2006) Genetic and environmental influences on the development of alcoholism: resilience vs. risk. Ann N Y Acad Sci 1094:193–201
22. Fagundo AB, Fernandez-Aranda F, de la Torre R, Verdejo-Garcia A, Granero R, Penelo E, Gene M, Barrot C, Sanchez C, Alvarez-Moya E, Ochoa C, Aymami MN, Gomez-Pena M, Menchon JM, Jimenez-Murcia S (2014) Dopamine DRD2/ANKK1 Taq1A and DAT1 VNTR polymorphisms are associated with a cognitive flexibility profile in pathological gamblers. J Psychopharmacol (Oxford, England) 28(12):1170–1177
23. Feinn R, Nellissery M, Kranzler HR (2005) Meta-analysis of the association of a functional serotonin transporter promoter polymorphism with alcohol dependence. Am J Med Genet B Neuropsychiatr Genet 133b(1):79–84
24. Goodman A (2008) Neurobiology of addiction. An integrative review. Biochem Pharmacol 75(1):266–322
25. Gorwood P (2004) Eating disorders, serotonin transporter polymorphisms and potential treatment response. Am J Pharmacogenomics 4(1):9–17

26. Graham DL, Hoppenot R, Hendryx A, Self DW (2007) Differential ability of D1 and D2 dopamine receptor agonists to induce and modulate expression and reinstatement of cocaine place preference in rats. Psychopharmacology 191(3):719–730
27. Grant JE, Brewer JA, Potenza MN (2006) The neurobiology of substance and behavioral addictions. CNS Spectr 11(12):924–930
28. Grant JE, Odlaug BL, Chamberlain SR, Hampshire A, Schreiber LR, Kim SW (2013) A proof of concept study of tolcapone for pathological gambling: relationships with COMT genotype and brain activation. Eur Neuropsychopharmacol 23(11):1587–1596
29. Gray JC, MacKillop J (2014) Genetic basis of delay discounting in frequent gamblers: examination of a priori candidates and exploration of a panel of dopamine-related loci. Brain Behav 4(6):812–821
30. Guo G, Wilhelmsen K, Hamilton N (2007) Gene-lifecourse interaction for alcohol consumption in adolescence and young adulthood: five monoamine genes. Am J Med Genet B Neuropsychiatr Genet 144b(4):417–423
31. Han DH, Lee YS, Yang KC, Kim EY, Lyoo IK, Renshaw PF (2007) Dopamine genes and reward dependence in adolescents with excessive internet video game play. J Addict Med 1(3):133–138
32. Heils A, Teufel A, Petri S, Stober G, Riederer P, Bengel D, Lesch KP (1996) Allelic variation of human serotonin transporter gene expression. J Neurochem 66(6):2621–2624
33. Herz A (1997) Endogenous opioid systems and alcohol addiction. Psychopharmacology 129(2):99–111
34. Holden C (2001) 'Behavioral' addictions: do they exist? Science (New York, NY) 294(5544):980–982
35. Hou H, Qing Z, Jia S, Zhang X, Hu S, Hu J (2010) Influence of brain-derived neurotrophic factor (val66met) genetic polymorphism on the ages of onset for heroin abuse in males. Brain Res 1353:245–248
36. Huang W, Payne TJ, Ma JZ, Li MD (2008) A functional polymorphism, rs6280, in DRD3 is significantly associated with nicotine dependence in European-American smokers. Am J Med Genet B Neuropsychiatr Genet 147b(7):1109–1115
37. Hwu HG, Chen CH (2000) Association of 5HT2A receptor gene polymorphism and alcohol abuse with behavior problems. Am J Med Genet 96(6):797–800
38. Ibanez A, Perez de Castro I, Fernandez-Piqueras J, Blanco C, Saiz-Ruiz J (2000) Pathological gambling and DNA polymorphic markers at MAO-A and MAO-B genes. Mol Psychiatry 5(1):105–109
39. Jakubczyk A, Wrzosek M, Lukaszkiewicz J, Sadowska-Mazuryk J, Matsumoto H, Sliwerska E, Glass J, Burmeister M, Brower KJ, Wojnar M (2012) The CC genotype in HTR2A T102C polymorphism is associated with behavioral impulsivity in alcohol-dependent patients. J Psychiatr Res 46(1):44–49
40. Jonsson EG, Nothen MM, Grunhage F, Farde L, Nakashima Y, Propping P, Sedvall GC (1999) Polymorphisms in the dopamine D2 receptor gene and their relationships to striatal dopamine receptor density of healthy volunteers. Mol Psychiatry 4(3):290–296
41. Kibitov AO, Voskoboeva E, Brodianskii VM, Chuprova NA, Smirnova EV (2010) Association study of the Val158Met polymorphism of the catechol-O-methyltransferase gene and alcoholism and heroin dependence: the role of a family history. Zh nevrologii Psikhiatr Im SS Korsakova/Ministerstvo zdravookhraneniia Med Promyshlennosti Ross Federatsii Vserossiiskoe obshchestvo nevrologov [i] Vserossiiskoe obshchestvo Psikhiatr 110(4):84–88
42. Kirsch P, Reuter M, Mier D, Lonsdorf T, Stark R, Gallhofer B, Vaitl D, Hennig J (2006) Imaging gene-substance interactions: the effect of the DRD2 TaqIA polymorphism and the dopamine agonist bromocriptine on the brain activation during the anticipation of reward. Neurosci Lett 405(3):196–201
43. Kreek MJ (1996) Cocaine, dopamine and the endogenous opioid system. J Addict Dis 15(4):73–96

44. Kreek MJ (1996) Opioid receptors: some perspectives from early studies of their role in normal physiology, stress responsivity, and in specific addictive diseases. Neurochem Res 21(11):1469–1488
45. Le Foll B, Goldberg SR, Sokoloff P (2005) The dopamine D3 receptor and drug dependence: effects on reward or beyond? Neuropharmacology 49(4):525–541
46. Lee YS, Han DH, Yang KC, Daniels MA, Na C, Kee BS, Renshaw PF (2008) Depression like characteristics of 5HTTLPR polymorphism and temperament in excessive internet users. J Affect Disord 109(1–2):165–169
47. Lejoyeux M, Weinstein A (2010) Compulsive buying. Am J Drug Alcohol Abuse 36(5):248–253
48. Levran O, Peles E, Randesi M, Correa da Rosa J, Ott J, Rotrosen J, Adelson M, Kreek MJ (2015) Susceptibility loci for heroin and cocaine addiction in the serotonergic and adrenergic pathways in populations of different ancestry. Pharmacogenomics 16(12):1329–1342
49. Lim S, Ha J, Choi SW, Kang SG, Shin YC (2012) Association study on pathological gambling and polymorphisms of dopamine D1, D2, D3, and D4 receptor genes in a Korean population. J Gambl Stud/Co-Sponsored Nat Counc Probl Gambl Inst Study Gambl Commer Gaming 28(3):481–491
50. Limosin F, Loze JY, Boni C, Hamon M, Ades J, Rouillon F, Gorwood P (2005) Male-specific association between the 5-HTTLPR S allele and suicide attempts in alcohol-dependent subjects. J Psychiatr Res 39(2):179–182
51. Lobo DS, Souza RP, Tong RP, Casey DM, Hodgins DC, Smith GJ, Williams RJ, Schopflocher DP, Wood RT, el Guebaly N, Kennedy JL (2010) Association of functional variants in the dopamine D2-like receptors with risk for gambling behaviour in healthy Caucasian subjects. Biol Psychol 85(1):33–37
52. Marks I (1990) Behavioural (non-chemical) addictions. Br J Addict 85(11):1389–1394
53. Mignini F, Napolioni V, Codazzo C, Carpi FM, Vitali M, Romeo M, Ceccanti M (2012) DRD2/ANKK1 TaqIA and SLC6A3 VNTR polymorphisms in alcohol dependence: association and gene-gene interaction study in a population of Central Italy. Neurosci Lett 522(2):103–107
54. Monteleone P, Tortorella A, Castaldo E, Maj M (2006) Association of a functional serotonin transporter gene polymorphism with binge eating disorder. Am J Med Genet B Neuropsychiatr Genet: Off Publ Int Soc Psychiatr Genet 141B(1):7–9
55. Monteleone P, Zanardini R, Tortorella A, Gennarelli M, Castaldo E, Canestrelli B, Maj M (2006) The 196G/A (val66met) polymorphism of the BDNF gene is significantly associated with binge eating behavior in women with bulimia nervosa or binge eating disorder. Neurosci Lett 406(1–2):133–137
56. Nagaya D, Ramanathan S, Ravichandran M, Navaratnam V (2012) A118G mu opioid receptor polymorphism among drug addicts in Malaysia. J Integr Neurosci 11(1):117–122
57. Nakamura T, Matsushita S, Nishiguchi N, Kimura M, Yoshino A, Higuchi S (1999) Association of a polymorphism of the 5HT2A receptor gene promoter region with alcohol dependence. Mol Psychiatry 4(1):85–88
58. Noble EP (2003) D2 dopamine receptor gene in psychiatric and neurologic disorders and its phenotypes. Am J Med Genet B Neuropsychiatr Genet 116B(1):103–125
59. Perez de Castro I, Ibanez A, Saiz-Ruiz J, Fernandez-Piqueras J (2002) Concurrent positive association between pathological gambling and functional DNA polymorphisms at the MAO-A and the 5-HT transporter genes. Mol Psychiatry 7(9):927–928
60. Perez de Castro I, Ibanez A, Torres P, Saiz-Ruiz J, Fernandez-Piqueras J (1997) Genetic association study between pathological gambling and a functional DNA polymorphism at the D4 receptor gene. Pharmacogenetics 7(5):345–348
61. Pierce RC, Kumaresan V (2006) The mesolimbic dopamine system: the final common pathway for the reinforcing effect of drugs of abuse? Neurosci Biobehav Rev 30(2):215–238
62. Potenza MN (2006) Should addictive disorders include non-substance-related conditions? Addiction (Abingdon, England) 101(Suppl 1):142–151
63. Potenza MN (2008) Review. The neurobiology of pathological gambling and drug addiction: an overview and new findings. Philos Trans R Soc Lond Ser B Biol Sci 363(1507):3181–3189

64. Sabol SZ, Hu S, Hamer D (1998) A functional polymorphism in the monoamine oxidase A gene promoter. Hum Genet 103(3):273–279
65. Schmidt LG, Sander T, Kuhn S, Smolka M, Rommelspacher H, Samochowiec J, Lesch KP (2000) Different allele distribution of a regulatory MAOA gene promoter polymorphism in antisocial and anxious-depressive alcoholics. J Neural Transm 107(6):681–689
66. Sobik L, Hutchison K, Craighead L (2005) Cue-elicited craving for food: a fresh approach to the study of binge eating. Appetite 44(3):253–261
67. Sorg BA, Li N, Wu W, Bailie TM (2004) Activation of dopamine D1 receptors in the medial prefrontal cortex produces bidirectional effects on cocaine-induced locomotor activity in rats: effects of repeated stress. Neuroscience 127(1):187–196
68. Su H, Li Z, Du J, Jiang H, Chen Z, Sun H, Zhao M (2015) Predictors of heroin relapse: personality traits, impulsivity, COMT gene Val158met polymorphism in a 5-year prospective study in Shanghai, China. Am J Med Genet B Neuropsychiatr Genet 168(8):712–719
69. Tsai SJ (2007) Increased central brain-derived neurotrophic factor activity could be a risk factor for substance abuse: implications for treatment. Med Hypotheses 68(2):410–414
70. Vanyukov MM, Maher BS, Devlin B, Kirillova GP, Kirisci L, Yu LM, Ferrell RE (2007) The MAOA promoter polymorphism, disruptive behavior disorders, and early onset substance use disorder: gene-environment interaction. Psychiatr Genet 17(6):323–332
71. Wei J, Chu C, Wang Y, Yang Y, Wang Q, Li T, Zhang L, Ma X (2012) Association study of 45 candidate genes in nicotine dependence in Han Chinese. Addict Behav 37(5):622–626
72. White MJ, Young RM, Morris CP, Lawford BR (2011) Cigarette smoking in young adults: the influence of the HTR2A T102C polymorphism and punishment sensitivity. Drug Alcohol Depend 114(2–3):140–146
73. Wilson D, da Silva Lobo DS, Tavares H, Gentil V, Vallada H (2013) Family-based association analysis of serotonin genes in pathological gambling disorder: evidence of vulnerability risk in the 5HT-2A receptor gene. J Mol Neurosci : MN 49(3):550–553
74. Wojnar M, Brower KJ, Strobbe S, Ilgen M, Matsumoto H, Nowosad I, Sliwerska E, Burmeister M (2009) Association between Val66Met brain-derived neurotrophic factor (BDNF) gene polymorphism and post-treatment relapse in alcohol dependence. Alcohol Clin Exp Res 33(4):693–702
75. Xian H, Giddens JL, Scherrer JF, Eisen SA, Potenza MN (2014) Environmental factors selectively impact co-occurrence of problem/pathological gambling with specific drug-use disorders in male twins. Addiction (Abingdon, England) 109(4):635–644
76. Zhang X, Lee MR, Salmeron BJ, Stein DJ, Hong LE, Geng X, Ross TJ, Li N, Hodgkinson C, Shen PH, Yang Y, Goldman D, Stein EA (2013) Prefrontal white matter impairment in substance users depends upon the catechol-o-methyl transferase (COMT) val158met polymorphism. NeuroImage 69:62–69
77. Zhang Y, Wang D, Johnson AD, Papp AC, Sadee W (2005) Allelic expression imbalance of human mu opioid receptor (OPRM1) caused by variant A118G. J Biol Chem 280(38):32618–32624

Chapter 5
Similarities and Differences in Neuroimaging

Yan-Kun Sun, Yan Sun, Xiao Lin, Lin Lu, and Jie Shi

Abstract Addiction is a chronically relapsing disease characterized by drug intoxication, craving, bingeing, and withdrawal with loss of control. An increasing number of studies have indicated that non-substance addiction, like internet addiction and pathological gambling, share clinical, phenomenological, and biological features with substance addiction. With the development of imaging technology in the past three decades, neuroimaging studies have provided information on the neurobiological effects, and revealed neurochemical and functional changes in the brains of both drug-addicted and non-substance addicted subjects. Imaging techniques play a more critical role in understanding the neuronal processes of addiction and will lead the direction in future research for medication development of addiction treatment, especially for non-substance addiction, which shares an increasing percentage of addiction disorder. This article will review the similarities and differences between substance and non-substance addiction based on neuroimaging studies that may provide clues for future study on these two main kinds of addiction, especially the growing non-substance addiction.

Keywords Substance addiction • Non-substance addiction • Neuroimaging

Y.-K. Sun
Department of Pharmacology, School of Basic Medical Sciences, Peking University Health Science Center, Beijing 100191, China

National Institute on Drug Dependence, Peking University,
No. 38, Xueyuan Road, Haidian District, Beijing 100191, China

Y. Sun • J. Shi (✉)
National Institute on Drug Dependence, Peking University,
No. 38, Xueyuan Road, Haidian District, Beijing 100191, China
e-mail: shijie@bjmu.edu.cn

X. Lin
Peking-Tsinghua Center for Life Sciences and PKU-IDG/McGovern Institute for Brain Research, Peking University, Beijing, China

L. Lu
Institute of Mental Health/Peking University Sixth Hospital and National Clinical Research Center for Mental Disorders & Key Laboratory of Mental Health, Peking University, Beijing 100191, China

X. Zhang et al. (eds.), *Substance and Non-substance Addiction*, Advances in Experimental Medicine and Biology 1010, DOI 10.1007/978-981-10-5562-1_5

5.1 Introduction

Addiction is considered as a brain disease [48]. Modern imaging techniques enable researchers to observe actions and consequences of addiction as they occur and persist in the brains of addicted individuals [25]. With the development of imaging technology in the past three decades, imaging studies have provided information on the neurobiological effects of drugs, revealed neurochemical and functional changes in the brains of both drug-addicted and non-substance addicted subjects, and yielded important insights into individuals' subjective experiences and behaviors, which provide new evidences for the mechanisms underlying addiction and give clinicians opportunities to assess addiction, assign patients to appropriate care interventions, and to monitor response to therapy [96].

There are several brain imaging techniques used by researchers to explore the neurophysiology mechanisms for both drug addiction and non-substance addiction. Computed tomography (CT), which is developed in 1970s, is clinically used to identify brain diseases, such as addiction. The most commonly adopted imaging technique is structural magnetic resonance imaging (MRI) that can reflect the information on the location, shapes and sizes of the brains' various regions and subregions and demonstrate the presence of abnormal tissue and changes in tissue composition [25]. Indeed, the functional brain imaging technologies are widely accepted to reflect the functional location and reflection of specific brain areas. The functional magnetic resonance imaging (fMRI) aim to visualize the changes in oxygenation and blood flow associated with brain activates, and to detect the response of regional brain patterns to the drug with vulnerability to drug abuse, addictive symptoms and behaviors and long-term cognitive capacity. In addition to structural MRT and fMRI, magnetic resonance technology (MRS) is used to illustrate the location and concentrations of target chemicals in the brain tissues, such as N-acetyl-aspartic acid, choline, creatine etc. [97]. Normally, it focuses on brain areas that preclinical models or the neuropathology of drug abusers being affected. Like MRS, positronemission tomography (PET) and single photon emission computed tomography (SPECT) reveal the concentrations of molecules in the brain that provides researchers opportunity to figure out the drugs' effects on main components of cell-to-cell communication [94]. PET assesses rates of glucose metabolism and measures the changes in blood oxygen levels for determining cellular activity, which are used to quantify biochemical and pharmacological processes, investigate the influence of substance on different regions and functions of the brain, and figure out the changes of functions in return to the behaviors of addiction, craving and relapse, while SPECT is commonly used to measure receptor ligand interaction, physiological function, biochemical and pharmacological processes [25]. Besides, the magneto encephalography (MEG) is a functional neuroimaging technique for mapping brain activity by recording magnetic fields produced by electrical currents occurring naturally in the brain, which can be combined with fMRI to create functional maps of human cortex during more complex cognitive tasks.

The brain imaging techniques have been well used to identify the mechanism of substance addiction and changes of the regions in brain. For substance addiction, especially opioids, cocaine, ketamine, alcohol and nicotine, a great number of imaging studies have illustrated the structural and functional changes on the white matter in reward and craving circuits areas, such as bilateral amygdala and nucleus accumbens [52, 53, 92]; and the gray matter in decision-making and learning circuits regions, such as prefrontal cortex, cingulate gyrus, insula and temporal lobe, for instance, among opioids addiction [39, 51, 65]. In prescription opioid-dependent subjects, significant decreases in functional connectivity were observed for brain regions that included the anterior insula, nucleus accumbens and amygdala subdivisions, beyond that, longer duration of prescription opioid exposure was associated with greater changes in functional connectivity. These findings suggested that prescription opioid dependence was associated with structural and functional changes in brain regions implicated in the regulation of emotion and impulse control, as well as in reward and motivational functions [92]. Besides, in cocaine dependents, increased regional cerebral blood flow were revealed in the superior temporal gyrus, dorsal anterior and posterior cingulate cortex, nucleus accumbens area, and the central sulcus, [39] and, specifically, alterations in dopamine, serotonin, opioid, and GABA systems in cocaine, alcohol, nicotine, and heroin dependence have been examined [14]. Some impairment caused by substance addiction such as dopamine receptor may induce some psychiatric symptoms [82].

Additionally, brain imaging techniques are also used to explore the structural and functional changes on brain caused by non-substance addiction such as internet addiction disorder (IAD) and pathological gambling (PG), which are specified by The Diagnostic and Statistical Manual of Mental Disorders Fifth Edition (DSM-V). According to the neuroimaging studies, like substance addicts, subjects with behavioral addiction experience the similar structural and functional damages on brain regions related to reward, decision-making and emotion processes [22, 76, 101, 107]. While some studies also indicate the similar cerebral activation patterns on subjects with non-substance addiction when they get enhanced cue reactivity [76]. Although there still some variations among the results of these studies and unidentified mechanisms for the non-substance addiction, most of the neuroimaging studies reveal similar neurobiological mechanism between substance and on non-substance addiction.

Imaging techniques play a more critical role in understanding the neuronal processes of addiction and will lead the direction in future research for medication development of addiction treatment, especially for non-substance addiction, which shares an increasing percentage of addiction disorder. The aim of this article is to review the similarities and differences between substance and non-substance addiction based on neuroimaging studies, which may provide clues for future study on these two kinds of addiction, especially the growing non-substance addiction.

5.2 Structural Changes

A great number of neuroimaging researches illustrating that, under the exposure of substance, the structure or volume of specific brain regions can be changed greatly. While, even studies focusing on the influences of non-substance addiction on structural changes of brain are relatively few, some demonstrate similar changes comparing with substance addiction. There are two commonly used methods to measure the structural changes of brain: voxel-based morphometry (VBM) and diffusion tensor imaging (DTI).

Voxel-based morphometry (VBM) is a neuroimaging analysis technique that can be used to detect regional cerebral volume and tissue concentration differences in structural magnetic resonance images [5]. It is a fully automated alternative to the techniques that require volumetric samples to characterize the differences between groups, without the need for a priori selection of the regions of interest (ROIs). Compared with traditional morphometric approaches which rely on measuring brain volumes manually, it is a time-saving technique and not specific to particular brain regions. VBM is widely used in the research of both substance and non-substance addiction.

Diffusion tensor imaging (DTI) assesses integrity of white matter microstructures by indexing the degree to which water diffusion deviates from isotropic diffusion in the white matter, with greater deviations from isotropic diffusion indicating more uniform directionality of water diffusion along the axon, implying greater white matter integrity [9]. A fractional anisotropy (FA) value is calculating from the normalized standard deviation of axial eigenvalue and radial eigenvalue [1, 104]. FA reduction indicated disrupted white matter integrity in these areas.

5.2.1 Gray Matter Changes

According to VBM studies, it demonstrated that both substance and non-substance addiction can lead to structural changes in brain that substance and non-substance addiction group experience reduced gray matter density (GMD) and volume in consistent brain regions including cingulate cortex, insula and frontal gyrus (including dorsal lateral prefrontal cortex, DLPFC) [23, 54, 60, 90, 100, 107, 109]. A great number of studies indicate the crucial role of PFC in addiction [29], while the PFC and anterior cingulate cortex (ACC) have been proved to be involved in cognitive control [44, 61]. The uncontrolled behavior in addiction subjects is associated with reduced gray matter volume (GMV) in the PFC. The insula is regarded as a region related to decision-making processes that involve uncertain risk and reward [66].

Besides the consistent brain areas, reduced GMD are also found in different areas between substance and non-substance addiction group. In non-substance addiction subjects, some particular brain areas with gray matter deficits are identified by VBM studies, which are not consistent with those in substance abusers. The internet gaming addiction (IGA) participants showed significant lower GMD in the right precuneus and left lingual gyrus comparing with healthy controls [54, 109]. In addition, Wang et al. found that IGA showed decreased GMV in brain areas including precuneus, supplementary motor area (SMA), superior parietal cortex, left insula, and bilateral cerebellum [100]. Using the same technique, decreased GMV in the SMA, the orbitofrontal cortex (OFC) and the cerebellum in adolescents with IAD have been discovered by Yuan et al. Moreover, they indicated that the decreased GMV of SMA were significantly correlated with the duration of internet addiction in the adolescents with IAD [107].

Indeed, different findings are reported by Sun et al. that IGA had higher GMV in the right inferior and middle temporal gyri, and right parahippocampal gyrus, and lower GMV in the left precentral gyrus [86]. Furthermore, some scholars demonstrated the exist of sex differences on gray matter in IGA that female problematic internet use was associated with increased GMV in the brain areas of reward system and decreased GMV of OFC [2].

There are several specific impaired brain areas revealed by VBM only in substance dependents. A recent neuroimaging meta-analysis on cocaine and methamphetamine dependence measured by VBM indicated that both substances was associated with reduced gray matter in subcortical reductions in the thalamus, along with cortical reductions in bilateral insula, ACC and bilateral frontal lobes [23]. Additionally, GMD was found in bilateral superior temporal cortex, left fusiform cortex, and right uncus in opiate-dependent subjects comparing with healthy subjects [60], while severe gray matter deficits were demonstrated in the cingulate, limbic, and paralimbic cortices in methamphetamine-dependent group [90].

5.2.2 White Matter Changes

Non-substance abusers experience impaired white matter integrity in particular brain areas. With tract-based spatial statistics (TBSS) analysis, a research indicated that online game addiction (OGA) group had significantly reduced FA in the right genu of corpus callosum, bilateral frontal lobe white matter, and right external capsule [102]. Using same technique, Lin et al. found that the IAD had significantly lower FA than controls throughout the brain, including the orbito-frontal, corpus callosum, cingulum, inferior fronto-occipital fasciculus, and corona radiation, internal and external capsules, while exhibiting no areas of higher FA [53]. Reduced FA value was also identified in white matter in the right parahippocampal gyrus of internet addicts [107].

Some researchers conducted DTI in heroin abusers and found that heroin addiction reduced the FA in multiple pathways including the corpus callosum, thalamic radiation, inferior longitudinal fasciculus, bilateral frontal sub-gyral regions, right precentral and left cingulate gyrus [11, 55]. Similar results have been found in other substance abusers, including cocaine addiction and alcoholics. Moeller et al. conducted DTI in cocaine dependents and indicated significantly reduced FA in the genu and rostral body of the anterior corpus callosum in cocaine-dependent subjects compared to controls. Besides, it demonstrated that reduced integrity of anterior corpus callosum white matter in cocaine users was related to impulse control impairment and discriminability reduction between target and catch stimuli, which were consistent with prior theories regarding frontal cortical involved in inhibitory control impairment in cocaine dependents [64]. The results are consistent with what found in alcohol dependents, providing in evidences that alcoholism disrupts white-matter microstructure and suggested that the interruption of both intra- and inter-voxel coherence contributes to deficits in attention and working memory associated with chronic alcoholism [68–71].

In contrast, enhanced white matter FA value also examined by DTI in some brain areas. For instance, IAD subjects were found with higher FA value in the left posterior limb of the internal capsule (PLIC). Indeed, the enhanced white matter FA of the PLIC were significantly correlated with the duration of internet addiction in the adolescents with IAD [107]. Moreover, Dong et al. found higher FA, indicating greater white matter integrity, in the thalamus and left posterior cingulate cortex between the IGA and the healthy subjects, and the higher FA in the thalamus was associated with greater severity of internet addiction [19]. Comparing with healthy controls, Jeong et al. also found increased FA in a wider range of brain regions in internet game disorder (IGD), including the forceps minor, right anterior thalamic radiation, right corticospinal tract, right inferior longitudinal fasciculus, right cingulum to hippocampus and right inferior fronto-occipital fasciculus (IFOF) [38].

However, the white matter changes from VBM in substance abusers are inconsistent. Some of the researches indicate no significant changes in white matter density of substance users comparing with the controls [26, 60, 62]. There is significant white-matter hypertrophy has been detected in methamphetamine abusers than the healthy controls [90], while greater lesion severity of deep and insular white matter hyperintensities (WMH) had been detected in cocaine-dependent group than the opiate-dependent group and the healthy subjects. Similar finding has been reported by Bae et al. with additional result insisting that male abusers had greater severity of WMH than female abusers [7]. Another imaging study on chronic substance users who abused heroin, cocaine and cannabis found that substance abusers had significantly less frontal white-matter volume percentage than the controls [81]. Similar to the findings on white matter in substance abusers, the changes of white matter in non-substance addiction group are ambiguous. Some studies showed that IGA participants experienced significant lower white matter density in the inferior frontal gyrus, insula, amygdala, and anterior cingulate than healthy controls [53], however, others insist no significant differences between non-substance addiction group and the control group [109].

5.3 Functional Changes

5.3.1 PET and SPECT

5.3.1.1 Glucose Metabolism

It is found that most drugs of abuse decreased regional cerebral metabolic rate(s) for glucose (rCMRglc) in human subjects [12, 84]. For instance, Chang et al. adopted PET in testing brain activities in methamphetamine addicts and found that altered brain glucose metabolism in the limbic and orbitofrontal regions was correlated with severity of psychiatric symptoms Similar dysregulation of glucose metabolism has been reported in non-substance addiction. Park et al. investigated the regional cerebral glucose metabolism at resting state between internet game over-users and normal ones. They found that the IGA subjects had significantly increased resting glucose metabolism in the right middle orbitofrontal gyrus, the left caudate nucleus, and the right insula, compared to the normal users, whereas significantly decreased glucose metabolism in the postcentral and precentral gyrus, the superior parietal lobule, and the occipital gyrus [67]. As these areas are implicated in impulsivity and inhibitory control, reward processing, and somatic representation of previous experiences, IGA showed greater impulsiveness than the normal subjects with a positive correlation between the severity of Internet game overuse and impulsiveness. With PET on the resting state, Tian and his colleagues indicated that there were increased glucose metabolism in some brain areas, such as the right supplementary motor area, middle cingulum and thalamus; decreased glucose metabolism in the right orbitofrontal gyrus and bilateral temporal poles, compared to the normal controls [91].

5.3.1.2 Dopamine

The dysregulation of dopamine system was revealed in substance addiction. The reduced dopamine transporter (DAT) density and reduced dopamine D2 receptors had been found in the striatum of methamphetamine subjects [12]. Besides, subjects who enjoyed the effects of intravenous methylphenidate had significantly lower D2 receptor levels than subjects who disliked its effects, while the higher the D2 levels were found, the more intense were methylphenidate's unpleasant effects. These results indicated that D2 receptor levels predict response to psychostimulants in humans and that low D2 receptors may lead to psychostimulant abuse by favoring pleasant response [93]. As well, the higher DAT levels in cocaine dependent subjects compared to controls at the anterior putamen, posterior putamen and caudate [15]. Consistent dysregulations of dopamine system are revealed in non-substance addiction. Tian et al. reported a low level of D2 receptors in the striatum was significantly associated with decreased glucose metabolism in the orbitofrontal cortex in IGD subjects, which indicates that D2/5-HT2A receptor-mediated dysregulation of the orbitofrontal cortex could underlie a mechanism for loss of control and compulsive behavior in IGD subjects [91].The similar results have been found that individuals with Internet addiction showed reduced levels of dopamine D2 receptor availability in subdivisions of the striatum including the bilateral dorsal caudate and right putamen [40]. Besides, using Tc-99m-TRODAT-1 SPECT to determine the changes of the striatal DAT levels in individuals with IAD, it indicated that DAT expression level of striatum was significantly decreased and the uptake ratio of corpus striatum/the whole brain were greatly reduced in the individuals with IAD compared to controls [36].

5.3.2 Resting-State fMRI

In the present resting state fMRI study, regional homogeneity (ReHo) method was employed to analyze the BOLD signal of the brain. Findings from resting-state fMRI studies are inconsistent. With fMRI test, it was reported that, compared with normal controls, the IAD group showed increased ReHo brain regions, which distributed over the cerebellum, brainstem, right cingulate gyrus, bilateral parahippocampus, right frontal lobe (rectal gyrus, inferior frontal gyrus and middle frontal gyrus), left superior frontal gyrus, left precuneus, right postcentral gyrus, right middle occipital gyrus, right inferior temporal gyrus, left superior temporal gyrus and middle temporal gyrus [56]. The results indicated that the functional change of brain in IAD subjects, while the connections between the enhancement of synchronization among cerebellum, brainstem, limbic lobe, frontal lobe and apical lobe might be relative to reward pathways. Moreover, comparing to the healthy controls, IGA subjects show enhanced ReHo in brainstem, inferior parietal lobule, left posterior cerebellum, and left middle frontal gyrus. All of these regions are thought related with sensory-motor coordination. Indeed, it reported decreased ReHo in temporal, occipital and parietal brain regions, which were thought to be responsible for visual and auditory functions [20]. Lower ReHo were also reported in brain areas of internet adolescents (IA), such as right parahippocampa gyrus, right posterior cingulated, left insula, right postcentral gyrus, left superior parietal lobule [75].

Additionally, with resting-state fMRI in studying IGD comparing with alcohol use disorder (AUD) and healthy controls, significantly increased ReHo were measured in the posterior cingulate cortex (PCC) of the IGD and AUD groups, and decreased ReHo in the right superior temporal gyrus (STG) of those with IGD, compared with the AUD and HC groups. As well, decreased ReHo was indicated in the ACC of patients with AUD. Scores on Internet addiction severity were positively correlated with ReHo in the medial frontal cortex, precuneus/PCC, and left inferior temporal cortex (ITC) among those with IGD. It revealed that increased ReHo in the PCC may be a common neurobiological feature of IGD and AUD and that reduced ReHo in the STG may be a candidate neurobiological marker for IGD, differentiating individuals with this disorder from those with AUD and healthy controls [41].

5.3.3 *Task-State fMRI*

Substance addiction is characterized by compulsive, even uncontrollable, behaviors that occur at the expense of other activities and intensify with repeated access [6], meanwhile, it is well recognized that the development and maintenance of drug addiction is closely related to the substance-related cue might provoke enhanced value of substance in the reward, motivation and memory circuits [78, 95, 98]. Besides, several evidence indicate that there exists a deficient reward system in drug addicts and the behavior of drug intake is an attempt to compensate for this deficit [78]. Similarly, in non-substance addiction subjects, consistent brain activations and neurobiological mechanisms were indicated, such as cue reactivity, control system, reward process and etc.

5.3.3.1 Cue-Induced Brain Activation

The exposure of substance-related cues in substance addicts can induce craving activities to drug with activation in several brain regions. Non-substance addiction experienced similar reactions to cues. Ko and her colleagues tested the brain correlates of craving for online gaming and found that bilateral DLPFC, precuneus, left parahippocampus, posterior cingulate and right anterior cingulate were activated in response to gaming cues in the IGA and their activation was stronger than the control group. These activated brain areas represent the brain circuit corresponding to the mechanism of substance use disorder. Thus, it would suggest that the mechanism of IGA is similar to substance use disorder [43]. It is consistent with the findings that, in internet addiction disorder group, they show increased brain activities in the inferior frontal cortex, insula, anterior cingulate cortex and decreased activation in the caudate and posterior cingulate cortex after continuous wins than healthy controls.

Using gaming pictures as cues to evaluate the brain activities in online gaming addicts in contrast to the control group, it found that several brain areas were activated, such as right orbitofrontal cortex, right nucleus accumbens, bilateral anterior cingulate and medial frontal cortex, right dorsolateral prefrontal cortex, and right caudate nucleus [42]. Some of these brain regions contributed to the craving in online gaming addiction have been reported to be related to the craving in substance addiction. Indeed, in pathological computer game players (PCGPs), higher connectivity strength was reported between right inferior frontal gyrus and cue reactivity-related regions (left orbitofrontal cortex and ventral striatum) in PCGPs. These findings suggest that top-down inhibitory processes might suppress the cue reactivity-related neural activity [57].

According to a meta-analysis on the substance and non-substance addiction [28], for the processing of the problematic stimuli in internet/gaming addicts and pathological gamblers, increased activations are found in regions such as inferior frontal gyrus, posterior cingulate cortex, temporal horn, amygdala and cerebellum, while no significant decreased activation are illustrated [16, 17, 34, 42, 43, 57, 63, 72, 74, 86].

Besides, when processing of substance stimuli, in substance addicts, enhanced blood-oxygen-level-dependent (BOLD) fMRI activities were demonstrated in areas such as the thalamus, posterior cingulate cortex, superior and inferior frontal gyrus, putamen, caudate, anterior insula, accumbens/putamen, anterior cingulate cortex, super frontal gyrus, amygdala/parahippocampal gyrus, lateral orbitofrontal cortex, middle temporal gyrus and brainstem, whereas decreased activation in intracalcarine cortex and superior frontal gyrus [3, 4, 10, 27, 30–33, 35, 37, 47, 49, 50, 58, 59, 73, 79, 83, 87–89, 99, 105, 106, 108, 110].

These results indicated that the neural substrate of cue-induced craving in non-substance addiction was consistent to that in substance addiction. Therefore, it suggested that the craving in OGA and that in substance dependence might share the same neurobiological mechanism [24, 45, 77, 103].

5.3.3.2 Reward Task

Reward system has been revealed to be related to occurrence and maintenance of substance addiction. To examine the activation of reward system in substance and non-substance addiction, reward paradigms adopted including general rewarding stimuli (e.g. monetary reward pleasant soft touch) and problematic stimuli (e.g. like drug related pictures). Dong et al. accessed the brain activations during guessing task in internet addiction compared with healthy controls and found increased activation in orbitofrontal cortex in gain trials and decreased anterior cingulate activation in loss trials than normal controls. It suggested that Internet addicts experienced enhanced reward sensitivity and decreased loss sensitivity than normal ones [18]. Consistently, higher superior frontal gyrus activations were reported after continuous wins for IAD than for healthy controls, while no significant differences disturbed by their losses. Indeed, IAD subjects showed decreased posterior cingulate activation after continuous losses. These results indicated that IAD participants showed preference to win while neglecting their losses [21].

Participants with addictions (both substance addiction and behavioral addiction) exhibited similar BOLD fMRI hyperactivity in the reward system when processing either general rewarding stimuli or the problematic stimuli [18, 37, 46]. When processing general rewarding stimuli, compared with healthy participants, participants with addictions exhibits increased BOLD activity in brain areas including executive related brain regions like frontal gyrus, posterior cingulate cortex, emotion and memory related regions like amygdala, parahippocampal gyrus and temporal cortex meanwhile, substance addicts showed decreased activation in brain areas related to cognitive control such as anterior cingulate cortex, superior frontal gyrus, caudate, putamen and anterior insula, whereas the behavioral addicts only showed decreased activation in hippocampus, inferior parietal gyrus and postcentral gyrus [8, 13, 18, 34, 80]. Even though the similarities exist in the reward system between non-substance and substance addiction, there are some slight differences demonstrated by several studies.

5.4 Conclusion

Non-substance addiction has been investigated by many behavioral and neuroimaging studies. Recent years, more and more researches devoted to exploring the neuroimaging mechanism in the IAD, pathological gamblers, and internet porn, and the substance addictions, such as cocaine addicts, opioid dependence, and heroin addicts. The neural basis of non-substance addiction has been more extensively studied and is better established compared to other forms of "addiction" (e.g., drug addictions). Both substance and non-substance addiction subjects share similar structural and functional changes in brain areas according to neuroimaging studies. For instance, reduced gray volume in DLPFC, ACC, insula etc. Besides, these two groups experience impaired white matter integrity in particular brain areas, however, the identified regions are ambiguous particularly in non-substance addicts with fewer studies. PET and SPECT studies identified the dysregulation of glucose metabolic rates in brain areas such as frontal gyrus, insula etc., while dysregulation of dopamine in frontal gyrus, dorsal caudate, putamen etc. Task-state fMRI studies have indicated changes in reward- and control-related brain networks among substance and non-substance users, including areas like DLPFC, posterior cingulate cortex, temporal horn, amygdala and cerebellum etc.

Growing evidence indicates that behavioral addictions resemble substance addictions in neurobiological mechanisms with roles for brain glutamatergic, opioidergic, serotonergic and dopamine mesolimbic systems. It needs to be noticed that there may exist some differences between substance and non-substance addiction in both structural and functional changes in neurobiological mechanism, as non-substance addition are behavioral and neural changes without drug-taking, which can be regarded more likely as psychological addiction. However, with relatively fewer neuroimaging studies in non-substance addictions or direct comparison on the two groups, the differences between these two kinds of addictions are not well-identified where current results are not quite inconsistent.

In the future research, more studies need to be conducted in order to deeply figure out the similarities and differences in neurobiological mechanism between substance and non-substance, and provide more evidences to establish the diagnostic criteria and treatment guidelines, particularly for non-substance addiction.

References

1. Alexander AL, Lee JE, Lazar M, Field AS (2007) Diffusion tensor imaging of the brain. Neurotherapeutics 4(3):316–329. doi:10.1016/j.nurt.2007.05.011
2. Altbacker A, Plozer E, Darnai G, Perlaki G, Horvath R, Orsi G, Nagy SA, Bogner P, Schwarcz A, Kovacs N, Komoly S, Clemens Z, Janszky J (2015) Problematic internet use is associated with structural alterations in the brain reward system in females. Brain Imaging Behav. doi:10.1007/s11682-015-9454-9

3. Ames SL, Grenard JL, Stacy AW, Xiao L, He QH, Wong SW, Xue G, Wiers RW, Bechara A (2013) Functional imaging of implicit marijuana associations during performance on an implicit association test (IAT). Behav Brain Res 256:494–502
4. Ames SL, Grenard JL, He QH, Stacy AW, Wong SW, Xiao L, Xue G, Bechara A (2014) Functional imaging of an alcohol-implicit association test (IAT). Addict Biol 19(3):467–481
5. Ashburner J, Friston KJ (2000) Voxel-based morphometry – the methods. NeuroImage 11(6 Pt 1):805–821. doi:10.1006/nimg.2000.0582
6. Avena NM, Rada P, Hoebel BG (2008) Evidence for sugar addiction: behavioral and neurochemical effects of intermittent, excessive sugar intake. Neurosci Biobehav Rev 32(1):20–39. doi:10.1016/j.neubiorev.2007.04.019
7. Bae SC, Lyoo IK, Sung YH, Yoo J, Chung A, Yoon SJ, Kim DJ, Hwang J, Kim SJ, Renshaw PF (2006) Increased white matter hyperintensities in male methamphetamine abusers. Drug Alcohol Depend 81(1):83–88. doi:10.1016/j.drugalcdep.2005.05.016
8. Balodis IM, Kober H, Worhunsky PD, Stevens MC, Pearlson GD, Potenza MN (2012) Diminished frontostriatal activity during processing of monetary rewards and losses in pathological gambling. Biol Psychiatry 71(8):749–757. doi:10.1016/j.biopsych.2012.01.006
9. Basser PJ, Mattiello J, LeBihan D (1994) MR diffusion tensor spectroscopy and imaging. Biophys J 66(1):259–267. doi:10.1016/S0006-3495(94)80775-1
10. Beck A, Schlagenhauf F, Wustenberg T, Hein J, Kienast T, Kahnt T, Schmack K, Hagele C, Knutson B, Heinz A, Wrase J (2009) Ventral striatal activation during reward anticipation correlates with impulsivity in alcoholics. Biol Psychiatry 66(8):734–742
11. Bora E, Yucel M, Fornito A, Pantelis C, Harrison BJ, Cocchi L, Pell G, Lubman DI (2012) White matter microstructure in opiate addiction. Addict Biol 17(1):141–148. doi:10.1111/j.1369-1600.2010.00266.x
12. Chang L, Alicata D, Ernst T, Volkow N (2007) Structural and metabolic brain changes in the striatum associated with methamphetamine abuse. Addiction 102:16–32
13. Choi JS, Shin YC, Jung WH, Jang JH, Kang DH, Choi CH, Choi SW, Lee JY, Hwang JY, Kwon JS (2012) Altered brain activity during reward anticipation in pathological gambling and obsessive-compulsive disorder. PLoS One 7(9):e45938. doi:10.1371/journal.pone.0045938
14. Cosgrove KP (2010) Imaging receptor changes in human drug abusers. Curr Top Behav Neurosci 3:199–217. doi:10.1007/7854_2009_24
15. Crits-Christoph P, Newberg A, Wintering N, Ploessl K, Gibbons MBC, Ring-Kurtz S, Gallop R, Present J (2008) Dopamine transporter levels in cocaine dependent subjects. Drug Alcohol Depend 98(1–2):70–76
16. Crockford DN, Goodyear B, Edwards J, Quickfall J, El-Guebaly N (2005) Cue-induced brain activity in pathological gamblers. Biol Psychiatry 58(10):787–795. doi:10.1016/j.biopsych.2005.04.037
17. de Ruiter MB, Veltman DJ, Goudriaan AE, Oosterlaan J, Sjoerds Z, van den Brink W (2009) Response perseveration and ventral prefrontal sensitivity to reward and punishment in male problem gamblers and smokers. Neuropsychopharmacology 34(4):1027–1038. doi:10.1038/npp.2008.175
18. Dong G, Huang J, Du X (2011) Enhanced reward sensitivity and decreased loss sensitivity in Internet addicts: an fMRI study during a guessing task. J Psychiatr Res 45(11):1525–1529. doi:10.1016/j.jpsychires.2011.06.017
19. Dong G, DeVito E, Huang J, Du X (2012a) Diffusion tensor imaging reveals thalamus and posterior cingulate cortex abnormalities in internet gaming addicts. J Psychiatr Res 46(9):1212–1216. doi:10.1016/j.jpsychires.2012.05.015
20. Dong G, Huang J, Du X (2012b) Alterations in regional homogeneity of resting-state brain activity in internet gaming addicts. Behav Brain Funct 8:41. doi:10.1186/1744-9081-8-41
21. Dong G, Hu Y, Lin X (2013a) Reward/punishment sensitivities among internet addicts: Implications for their addictive behaviors. Prog Neuro-Psychopharmacol Biol Psychiatry 46:139–145. doi:10.1016/j.pnpbp.2013.07.007

22. Dong G, Hu Y, Lin X, Lu Q (2013b) What makes Internet addicts continue playing online even when faced by severe negative consequences? Possible explanations from an fMRI study. Biol Psychol 94(2):282–289. doi:10.1016/j.biopsycho.2013.07.009
23. Ersche KD, Williams GB, Robbins TW, Bullmore ET (2013) Meta-analysis of structural brain abnormalities associated with stimulant drug dependence and neuroimaging of addiction vulnerability and resilience. Curr Opin Neurobiol 23(4):615–624. doi:10.1016/j.conb.2013.02.017
24. Field M, Cox WM (2008) Attentional bias in addictive behaviors: a review of its development, causes, and consequences. Drug Alcohol Depend 97(1–2):1–20
25. Fowler JS, Volkow ND, Kassed CA, Chang L (2007) Imaging the addicted human brain. Sci Pract Perspect Publ Natl Inst Drug Abuse Natl Inst Health 3(2):4–16
26. Franklin TR, Acton PD, Maldjian JA, Gray JD, Croft JR, Dackis CA, O'Brien CP, Childress AR (2002) Decreased gray matter concentration in the insular, orbitofrontal, cingulate, and temporal cortices of cocaine patients. Biol Psychiatry 51(2):134–142
27. Fryer SL, Jorgensen KW, Yetter EJ, Daurignac EC, Watson TD, Shanbhag H, Krystal JH, Mathalon DH (2013) Differential brain response to alcohol cue distractors across stages of alcohol dependence. Biol Psychol 92(2):282–291
28. Garcia-Garcia I, Horstmann A, Jurado MA, Garolera M, Chaudhry SJ, Margulies DS, Villringer A, Neumann J (2014) Reward processing in obesity, substance addiction and non-substance addiction. Obes Rev 15(11):853–869. doi:10.1111/obr.12221
29. Goldstein RZ, Volkow ND (2011) Dysfunction of the prefrontal cortex in addiction: neuroimaging findings and clinical implications. Nat Rev Neurosci 12(11):652–669. doi:10.1038/nrn3119
30. Goldstein RZ, Alia-Klein N, Tomasi D, Carrillo JH, Maloney T, Woicik PA, Wang RL, Telang F, Volkow ND (2009a) Anterior cingulate cortex hypoactivations to an emotionally salient task in cocaine addiction. Proc Natl Acad Sci USA 106(23):9453–9458
31. Goldstein RZ, Tomasi D, Alia-Klein N, Carrillo JH, Maloney T, Woicik PA, Wang RL, Telang F, Volkow ND (2009b) Dopaminergic response to drug words in cocaine addiction. J Neurosci 29(18):6001–6006
32. Goldstein RZ, Woicik PA, Maloney T, Tomasi D, Alia-Klein N, Shan JT, Honorio J, Samaras D, Wang RL, Telang F, Wang GJ, Volkow ND (2010) Oral methylphenidate normalizes cingulate activity in cocaine addiction during a salient cognitive task. Proc Natl Acad Sci USA 107(38):16667–16672
33. Goudriaan AE, Veltman DJ, van den Brink W, Dom G, Schmaal L (2013) Neurophysiological effects of modafinil on cue-exposure in cocaine dependence: a randomized placebo-controlled cross-over study using pharmacological fMRI. Addict Behav 38(2):1509–1517
34. Han DH, Hwang JW, Renshaw PF (2010) Bupropion sustained release treatment decreases craving for video games and cue-induced brain activity in patients with Internet video game addiction. Exp Clin Psychopharmacol 18(4):297–304. doi:10.1037/a0020023
35. Heinz A, Wrase J, Kahnt T, Beck A, Bromand Z, Grusser SM, Kienast T, Smolka MN, Flor H, Mann K (2007) Brain activation elicited by affectively positive stimuli is associated with a lower risk of relapse in detoxified alcoholic subjects. Alcohol Clin Exp Res 31(7):1138–1147. doi:10.1111/j.1530-0277.2007.00406.x
36. Hou HF, Jia SW, Hu S, Fan R, Sun W, Sun TT, Zhang H (2012) Reduced striatal dopamine transporters in people with internet addiction disorder. J Biomed Biotechnol 2012:854584
37. Ihssen N, Cox WM, Wiggett A, Fadardi JS, Linden DEJ (2011) Differentiating heavy from light drinkers by neural responses to visual alcohol cues and other motivational stimuli. Cereb Cortex 21(6):1408–1415
38. Jeong BS, Han DH, Kim SM, Lee SW, Renshaw PF (2015) White matter connectivity and internet gaming disorder. Addict Biol. doi:10.1111/adb.12246
39. Kilts CD, Gross RE, Ely TD, Drexler KP (2004) The neural correlates of cue-induced craving in cocaine-dependent women. Am J Psychiatry 161(2):233–241. doi:10.1176/appi.ajp.161.2.233
40. Kim SH, Baik SH, Park CS, Kim SJ, Choi SW, Kim SE (2011) Reduced striatal dopamine D2 receptors in people with Internet addiction. Neuroreport 22(8):407–411

41. Kim H, Kim YK, Gwak AR, Lim JA, Lee JY, Jung HY, Sohn BK, Choi SW, Kim DJ, Choi JS (2015) Resting-state regional homogeneity as a biological marker for patients with Internet gaming disorder: a comparison with patients with alcohol use disorder and healthy controls. Prog Neuro-Psychopharmacol Biol Psychiatry 60:104–111. doi:10.1016/j.pnpbp.2015.02.004
42. Ko CH, Liu GC, Hsiao S, Yen JY, Yang MJ, Lin WC, Yen CF, Chen CS (2009) Brain activities associated with gaming urge of online gaming addiction. J Psychiatr Res 43(7):739–747. doi:10.1016/j.jpsychires.2008.09.012
43. Ko CH, Liu GC, Yen JY, Chen CY, Yen CF, Chen CS (2013) Brain correlates of craving for online gaming under cue exposure in subjects with internet gaming addiction and in remitted subjects. Addict Biol 18(3):559–569. doi:10.1111/j.1369-1600.2011.00405.x
44. Krawczyk DC (2002) Contributions of the prefrontal cortex to the neural basis of human decision making. Neurosci Biobehav Rev 26(6):631–664
45. Kuhn S, Gallinat J (2011) Common biology of craving across legal and illegal drugs – a quantitative meta-analysis of cue-reactivity brain response. Eur J Neurosci 33(7):1318–1326
46. Kuss DJ, Griffiths MD (2012) Internet and gaming addiction: a systematic literature review of neuroimaging studies. Brain Sci 2(3):347–374. doi:10.3390/brainsci2030347
47. Lee E, Ku J, Jung YC, Lee H, An SK, Kim KR, Yoon KJ, Namkoong K (2013) Neural evidence for emotional involvement in pathological alcohol craving. Alcohol Alcohol 48(3):288–294
48. Leshner AI (1997) Addiction is a brain disease, and it matters. Science (New York, NY) 278(5335):45–47
49. Li Q, Wang YR, Zhang Y, Li W, Yang WC, Zhu J, Wu N, Chang HF, Zheng Y, Qin W, Zhao LY, Yuan K, Liu JX, Wang W, Tian J (2012) Craving correlates with mesolimbic responses to heroin-related cues in short-term abstinence from heroin: an event-related fMRI study. Brain Res 1469:63–72
50. Li Q, Yang WC, Wang YR, Huang YF, Li W, Zhu J, Zhang Y, Zhao LY, Qin W, Yuan K, von Deneen KM, Wang W, Tian J (2013) Abnormal function of the posterior cingulate cortex in heroin addicted users during resting-state and drug-cue stimulation task. Chin Med J 126(4):734–739
51. Li M, Tian J, Zhang R, Qiu Y, Wen X, Ma X, Wang J, Xu Y, Jiang G, Huang R (2014) Abnormal cortical thickness in heroin-dependent individuals. NeuroImage 88:295–307. doi:10.1016/j.neuroimage.2013.10.021
52. Lim KO, Choi SJ, Pomara N, Wolkin A, Rotrosen JP (2002) Reduced frontal white matter integrity in cocaine dependence: a controlled diffusion tensor imaging study. Biol Psychiatry 51(11):890–895
53. Lin F, Zhou Y, Du Y, Qin L, Zhao Z, Xu J, Lei H (2012) Abnormal white matter integrity in adolescents with internet addiction disorder: a tract-based spatial statistics study. PLoS One 7(1):e30253. doi:10.1371/journal.pone.0030253
54. Lin X, Dong G, Wang Q, Du X (2015) Abnormal gray matter and white matter volume in 'Internet gaming addicts'. Addict Behav 40:137–143. doi:10.1016/j.addbeh.2014.09.010
55. Liu H, Li L, Hao Y, Cao D, Xu L, Rohrbaugh R, Xue Z, Hao W, Shan B, Liu Z (2008) Disrupted white matter integrity in heroin dependence: a controlled study utilizing diffusion tensor imaging. Am J Drug Alcohol Abuse 34(5):562–575. doi:10.1080/00952990802295238
56. Liu J, Gao XP, Osunde I, Li X, Zhou SK, Zheng HR, Li LJ (2010) Increased regional homogeneity in internet addiction disorder: a resting state functional magnetic resonance imaging study. Chin Med J Peking 123(14):1904–1908. doi:10.3760/cma.j.issn.0366-6999.2010.14.014
57. Lorenz RC, Kruger JK, Neumann B, Schott BH, Kaufmann C, Heinz A, Wustenberg T (2013) Cue reactivity and its inhibition in pathological computer game players. Addict Biol 18(1):134–146. doi:10.1111/j.1369-1600.2012.00491.x
58. Luijten M, Veltman DJ, van den Brink W, Hester R, Field M, Smits M, Franken IHA (2011) Neurobiological substrate of smoking-related attentional bias. NeuroImage 54(3):2374–2381
59. Luijten M, Veltman DJ, Hester R, Smits M, Pepplinkhuizen L, Franken IH (2012) Brain activation associated with attentional bias in smokers is modulated by a dopamine antagonist. Neuropsychopharmacology 37(13):2772–2779. doi:10.1038/npp.2012.143

60. Lyoo IK, Pollack MH, Silveri MM, Ahn KH, Diaz CI, Hwang J, Kim SJ, Yurgelun-Todd DA, Kaufman MJ, Renshaw PF (2006) Prefrontal and temporal gray matter density decreases in opiate dependence. Psychopharmacology 184(2):139–144. doi:10.1007/s00213-005-0198-x
61. MacDonald AW 3rd, Cohen JD, Stenger VA, Carter CS (2000) Dissociating the role of the dorsolateral prefrontal and anterior cingulate cortex in cognitive control. Science (New York, NY) 288(5472):1835–1838
62. Matochik JA, London ED, Eldreth DA, Cadet JL, Bolla KI (2003) Frontal cortical tissue composition in abstinent cocaine abusers: a magnetic resonance imaging study. NeuroImage 19(3):1095–1102
63. Miedl SF, Fehr T, Meyer G, Herrmann M (2010) Neurobiological correlates of problem gambling in a quasi-realistic blackjack scenario as revealed by fMRI. Psychiatry Res 181(3):165–173. doi:10.1016/j.pscychresns.2009.11.008
64. Moeller FG, Hasan KM, Steinberg JL, Kramer LA, Dougherty DM, Santos RM, Valdes I, Swann AC, Barratt ES, Narayana PA (2005) Reduced anterior corpus callosum white matter integrity is related to increased impulsivity and reduced discriminability in cocaine-dependent subjects: diffusion tensor imaging. Neuropsychopharmacology 30(3):610–617
65. Nakama H, Chang L, Fein G, Shimotsu R, Jiang CS, Ernst T (2011) Methamphetamine users show greater than normal age-related cortical gray matter loss. Addiction 106(8):1474–1483. doi:10.1111/j.1360-0443.2011.03433.x
66. Naqvi NH, Bechara A (2009) The hidden island of addiction: the insula. Trends Neurosci 32(1):56–67. doi:10.1016/j.tins.2008.09.009
67 Park HS, Kim SH, Bang SA, Yoon EJ, Cho SS, Kim SE (2010) Altered regional cerebral glucose metabolism in internet game overusers: a 18F-fluorodeoxyglucose positron emission tomography study. CNS Spectr 15(3):159–166
68. Pfefferbaum A, Sullivan EV (2002) Microstructural but not macrostructural disruption of white matter in women with chronic alcoholism. NeuroImage 15(3):708–718. doi:10.1006/nimg.2001.1018
69. Pfefferbaum A, Sullivan EV, Hedehus M, Adalsteinsson E, Lim KO, Moseley M (2000a) In vivo detection and functional correlates of white matter microstructural disruption in chronic alcoholism. Alcohol Clin Exp Res 24(8):1214–1221
70. Pfefferbaum A, Sullivan EV, Hedehus M, Lim KO, Adalsteinsson E, Moseley M (2000b) Age-related decline in brain white matter anisotropy measured with spatially corrected echo-planar diffusion tensor imaging. Magn Reson Med 44(2):259–268
71. Pfefferbaum A, Rosenbloom M, Serventi KL, Sullivan EV (2002) Corpus callosum, pons, and cortical white matter in alcoholic women. Alcohol Clin Exp Res 26(3):400–406
72. Potenza MN, Steinberg MA, Skudlarski P, Fulbright RK, Lacadie CM, Wilber MK, Rounsaville BJ, Gore JC, Wexler BE (2003) Gambling urges in pathological gambling: a functional magnetic resonance imaging study. Arch Gen Psychiatry 60(8):828–836. doi:10.1001/archpsyc.60.8.828
73. Potenza MN, Hong KI, Lacadie CM, Fulbright RK, Tuit KL, Sinha R (2012) Neural correlates of stress-induced and cue-induced drug craving: influences of sex and cocaine dependence. Am J Psychiatry 169(4):406–414. doi:10.1176/appi.ajp.2011.11020289
74. Power Y, Goodyear B, Crockford D (2012) Neural correlates of pathological gamblers preference for immediate rewards during the iowa gambling task: an fMRI study. J Gambl Stud Co-sponsored Natl Counc Probl Gambl Inst Study Gambl Commer Gaming 28(4):623–636. doi:10.1007/s10899-011-9278-5
75. QIN L, ZHOU Y, Zhimin Z, Lu Q, Ge X, Li L, Du Y, Xu J (2011) Preliminary study of brain activity in internet addiction adolescents: revealed by resting state functional MRI. J Clin Radiol 30(1):7–10
76. Quester S, Romanczuk-Seiferth N (2015) Brain imaging in gambling disorder. Curr Addict Rep 2(3):220–229. doi:10.1007/s40429-015-0063-x
77. Robbins SJ, Ehrman RN (2004) The role of attentional bias in substance abuse. Behav Cogn Neurosci Rev 3(4):243–260. doi:10.1177/1534582305275423

78. Robbins TW, Everitt BJ (1999) Drug addiction: bad habits add up. Nature 398(6728):567–570. doi:10.1038/19208
79. Roberts GM, Garavan H (2013) Neural mechanisms underlying ecstasy-related attentional bias. Psychiatry Res 213(2):122–132. doi:10.1016/j.pscychresns.2013.03.011
80. Romanczuk-Seiferth N, Koehler S, Dreesen C, Wustenberg T, Heinz A (2015) Pathological gambling and alcohol dependence: neural disturbances in reward and loss avoidance processing. Addict Biol 20(3):557–569. doi:10.1111/adb.12144
81. Schlaepfer TE, Lancaster E, Heidbreder R, Strain EC, Kosel M, Fisch HU, Pearlson GD (2006) Decreased frontal white-matter volume in chronic substance abuse. Int J Neuropsychopharmacol 9(2):147–153. doi:10.1017/S1461145705005705
82. Sekine Y, Iyo M, Ouchi Y, Matsunaga T, Tsukada H, Okada H, Yoshikawa E, Futatsubashi M, Takei N, Mori N (2001) Methamphetamine-related psychiatric symptoms and reduced brain dopamine transporters studied with PET. Am J Psychiatry 158(8):1206–1214. doi:10.1176/appi.ajp.158.8.1206
83. Seo DJ, Lacadie CM, Tuit K, Hong KI, Constable RT, Sinha R (2013) Disrupted ventromedial prefrontal function, alcohol craving, and subsequent relapse risk. JAMA Psychiat 70(7):727–739
84. Stapleton JM, Gilson SF, Wong DF, Villemagne VL, Dannals RF, Grayson RF, Henningfield JE, London ED (2003) Intravenous nicotine reduces cerebral glucose metabolism: a preliminary study. Neuropsychopharmacology 28(4):765–772. doi:10.1038/sj.npp.1300106
85. Sun Y, Ying H, Seetohul RM, Xuemei W, Ya Z, Qian L, Guoqing X, Ye S (2012) Brain fMRI study of crave induced by cue pictures in online game addicts (male adolescents). Behav Brain Res 233(2):563–576. doi:10.1016/j.bbr.2012.05.005
86. Sun Y, Sun J, Zhou Y, Ding W, Chen X, Zhuang Z, Xu J, Du Y (2014) Assessment of in vivo microstructure alterations in gray matter using DKI in Internet gaming addiction. Behav Brain Funct 10:37. doi:10.1186/1744-9081-10-37
87. Tabatabaei-Jafari H, Ekhtiari H, Ganjgahi H, Hassani-Abharian P, Oghabian MA, Moradi A, Sadighi N, Zarei M (2014) Patterns of brain activation during craving in heroin dependents successfully treated by methadone maintenance and abstinence-based treatments. J Addict Med 8(2):123–129. doi:10.1097/ADM.0000000000000022
88. Tapert SF, Cheung EH, Brown GG, Frank LR, Paulus MP, Schweinsburg AD, Meloy MJ, Brown SA (2003) Neural response to alcohol stimuli in adolescents with alcohol use disorder. Arch Gen Psychiatry 60(7):727–735. doi:10.1001/archpsyc.60.7.727
89. Tapert SF, Brown GG, Baratta MV, Brown SA (2004) fMRI BOLD response to alcohol stimuli in alcohol dependent young women. Addict Behav 29(1):33–50
90. Thompson PM, Hayashi KM, Simon SL, Geaga JA, Hong MS, Sui YH, Lee JY, Toga AW, Ling W, London ED (2004) Structural abnormalities in the brains of human subjects who use methamphetamine. J Neurosci 24(26):6028–6036
91. Tian M, Chen Q, Zhang Y, Du F, Hou H, Chao F, Zhang H (2014) PET imaging reveals brain functional changes in internet gaming disorder. Eur J Nucl Med Mol Imaging 41(7):1388–1397. doi:10.1007/s00259-014-2708-8
92. Upadhyay J, Maleki N, Potter J, Elman I, Rudrauf D, Knudsen J, Wallin D, Pendse G, McDonald L, Griffin M, Anderson J, Nutile L, Renshaw P, Weiss R, Becerra L, Borsook D (2010) Alterations in brain structure and functional connectivity in prescription opioid-dependent patients. Brain 133(Pt 7):2098–2114. doi:10.1093/brain/awq138
93. Volkow ND, Wang GJ, Fowler JS, Logan J, Gatley SJ, Gifford A, Hitzemann R, Ding YS, Pappas N (1999) Prediction of reinforcing responses to psychostimulants in humans by brain dopamine D2 receptor levels. Am J Psychiatry 156(9):1440–1443. doi:10.1176/ajp.156.9.1440
94. Volkow ND, Chang L, Wang GJ, Fowler JS, Ding YS, Sedler M, Logan J, Franceschi D, Gatley J, Hitzemann R, Gifford A, Wong C, Pappas N (2001) Low level of brain dopamine D2 receptors in methamphetamine abusers: association with metabolism in the orbitofrontal cortex. Am J Psychiatry 158(12):2015–2021. doi:10.1176/appi.ajp.158.12.2015

95. Volkow ND, Fowler JS, Wang GJ, Goldstein RZ (2002) Role of dopamine, the frontal cortex and memory circuits in drug addiction: insight from imaging studies. Neurobiol Learn Mem 78(3):610–624
96. Volkow ND, Fowler JS, Wang GJ (2003a) The addicted human brain: insights from imaging studies. J Clin Invest 111(10):1444–1451. doi:10.1172/JCI18533
97. Volkow ND, Fowler JS, Wang GJ (2003b) Positron emission tomography and single-photon emission computed tomography in substance abuse research. Semin Nucl Med 33(2):114–128. doi:10.1053/snuc.2003.127300
98. Volkow ND, Wang GJ, Fowler JS, Tomasi D, Telang F, Baler R (2010) Addiction: decreased reward sensitivity and increased expectation sensitivity conspire to overwhelm the brain's control circuit. BioEssays 32(9):748–755
99. Vollstadt-Klein S, Wichert S, Rabinstein J, Buhler M, Klein O, Ende G, Hermann D, Mann K (2010) Initial, habitual and compulsive alcohol use is characterized by a shift of cue processing from ventral to dorsal striatum. Addiction 105(10):1741–1749
100. Wang H, Jin C, Yuan K, Shakir TM, Mao C, Niu X, Niu C, Guo L, Zhang M (2015a) The alteration of gray matter volume and cognitive control in adolescents with internet gaming disorder. Front Behav Neurosci 9:64. doi:10.3389/fnbeh.2015.00064
101. Wang Y, Yin Y, Sun YW, Zhou Y, Chen X, Ding WN, Wang W, Li W, Xu JR, Du YS (2015b) Decreased prefrontal lobe interhemispheric functional connectivity in adolescents with internet gaming disorder: a primary study using resting-state FMRI. PLoS One 10(3):e0118733. doi:10.1371/journal.pone.0118733
102. Weng CB, Qian RB, Fu XM, Lin B, Han XP, Niu CS, Wang YH (2013) Gray matter and white matter abnormalities in online game addiction. Eur J Radiol 82(8):1308–1312. doi:10.1016/j.ejrad.2013.01.031
103. Wilson SJ, Sayette MA, Fiez JA (2004) Prefrontal responses to drug cues: a neurocognitive analysis. Nat Neurosci 7(3):211–214
104. Xu J, Li Y, Lin H, Sinha R, Potenza MN (2013) Body mass index correlates negatively with white matter integrity in the fornix and corpus callosum: a diffusion tensor imaging study. Hum Brain Mapp 34(5):1044–1052. doi:10.1002/hbm.21491
105. Yalachkov Y, Kaiser J, Gorres A, Seehaus A, Naumer MJ (2013) Sensory modality of smoking cues modulates neural cue reactivity. Psychopharmacology 225(2):461–471
106. Yang Z, Xie J, Shao YC, Xie CM, Fu LP, Li DJ, Fan M, Ma L, Li SJ (2009) Dynamic neural responses to cue-reactivity paradigms in heroin-dependent users: an fMRI study. Hum Brain Mapp 30(3):766–775
107. Yuan K, Qin W, Wang G, Zeng F, Zhao L, Yang X, Liu P, Liu J, Sun J, von Deneen KM, Gong Q, Liu Y, Tian J (2011) Microstructure abnormalities in adolescents with internet addiction disorder. PLoS One 6(6):e20708. doi:10.1371/journal.pone.0020708
108. Zhang XC, Salmeron BJ, Ross TJ, Gu H, Geng XJ, Yang YH, Stein EA (2011) Anatomical differences and network characteristics underlying smoking cue reactivity. NeuroImage 54(1):131–141
109. Zhou Y, Lin FC, Du YS, Qin LD, Zhao ZM, Xu JR, Lei H (2011) Gray matter abnormalities in Internet addiction: a voxel-based morphometry study. Eur J Radiol 79(1):92–95. doi:10.1016/j.ejrad.2009.10.025
110. Zijlstra F, Veltman DJ, Booij J, van den Brink W, Franken IH (2009) Neurobiological substrates of cue-elicited craving and anhedonia in recently abstinent opioid-dependent males. Drug Alcohol Depend 99(1–3):183–192. doi:10.1016/j.drugalcdep.2008.07.012

Chapter 6
Similarities and Differences in Psychology

Yu Chen, Yan Sun, Si-Zhi Ai, Jason J. Li, Lin Lu, and Jie Shi

Abstract Addiction is marked by repeating a certain behavior while ignoring the potential physical or mental consequences. Non-substance addiction provides an ideal model for researching the emergence and development of addiction's basic mechanism. Comparative studies of substance and non-substance addiction are helpful to reveal the common basis of addiction development. This article explores this topic from a psychological angle, touching upon sensation seeking, inhibitory control, attentional bias, intertemporal choice and environment. A review of previous literature urges future research to propose a biopsychosocial model of addiction and consider addiction's effect on basic cognitive function alongside cognitive neuroscience technology.

Keywords Substance addiction • Non-substance addiction • Psychology

---

Y. Chen
Department of Pharmacology, School of Basic Medical Sciences, Peking University Health Science Center, Beijing 100191, China

National Institute on Drug Dependence, Peking University, No. 38, Xueyuan Road, Haidian District, Beijing 100191, China

Y. Sun • S.-Z. Ai • J. Shi (✉)
National Institute on Drug Dependence, Peking University, No. 38, Xueyuan Road, Haidian District, Beijing 100191, China
e-mail: shijie@bjmu.edu.cn

J.J. Li
Program in Human Biology, Stanford University, Stanford, CA 94305, USA

L. Lu
Institute of Mental Health/Peking University Sixth Hospital and National Clinical Research Center for Mental Disorders & Key Laboratory of Mental Health, Peking University, Beijing 100191, China

X. Zhang et al. (eds.), *Substance and Non-substance Addiction*, Advances in Experimental Medicine and Biology 1010, DOI 10.1007/978-981-10-5562-1_6

6.1 Introduction

All kinds of addiction affect brain and body's biochemical processes [7]. However, it is not enough to attribute a certain addiction to biological and genetic predisposition [17]. We also need to take personal (e.g., cognitions, personality, etc.) and social factors into consideration.

The essential feature of substance addiction is a cluster of cognitive, behavioral, and physiological symptoms indicating that an individual continues using the substance despite significant substance-related problems, while similar to substance addiction in physiological and psychological changes, non-substance addiction is triggered by environmental cues, which are both described in the fifth edition of the *Diagnostic and Statistical Manual of Mental Disorders* (DSM-5; [2]). In addition to the above common clinical feature, substance addiction and non-substance addiction have similarities in phenomenon. For example, the abuse of nicotine and alcohol usually begins in adolescence and early adulthood just like some non-substance addictions such as pathological gambling and internet addiction [31]. In another example, female patients with pathological gambling have telescoping phenomenon (although women begin gambling later than men, their development of pathological gambling is faster), which is also found in women with alcohol addiction [54, 73]. In addition, substance addiction and non-substance addiction also showed higher rates of comorbidity. Studies have shown that patients with substance addiction have a higher risk of pathological gambling or shopping addiction [64, 70]; individuals who have pathological gambling or internet addiction are more likely to abuse nicotine, alcohol, and other drugs [42, 79]. Although studies show that there are significant correlations between substance addiction and non-substance addiction, studies must further explore their inner mechanism underlying these similarities and differences to elucidate this relationship.

Therefore, this article compares substance addiction with non-substance addiction through the perspective of psychology, providing future directions for research.

6.2 Similarities and Differences

6.2.1 Sensation Seeking

As a stable personality trait or innate propensity, sensation seeking has been applied especially in relation to addiction [33]. Sensation seeking, as defined by Zuckerman, refers to "the need for varied, novel, and complex sensations and experiences, and the willingness to take physical and social risks for the sake of such experiences," which emphasizes novelty and intensity as the two components of sensation seeking [83]. The person with significant sensation seeking always wants to stay awake and has an innate propensity of pleasure or excitement for novel stimulation or potential reward cues, resulting in individual frequent exploratory activities [15]. To measure sensation seeking, the sensation seeking scale focuses on four dimensions: thrill and adventure seeking (TAS), experience seeking (ES), disinhibition (DIS), boredom susceptibility (BS) [82].

Researches in recent years indicated that the scores of addicts in sensation seeking scale are high both in substance addiction and non-substance addiction, especially in the dimension of disinhibition [34, 37, 56]. Moreover, sensation seeking can predict addiction. Cloninger et al. [16] carried out a follow-up study showing that teenagers develop serious alcohol abuse at age 16 whose scores of sensation seeking scale are high at age 11, and consistent results were also found in hypnotics addiction and stimulants addiction [38]; Lynne-Landsman et al. [48] found that if sensation seeking scores were high, sensation seeking had a stable, positive relationship with aggression, delinquency, and substance abuse regardless of age.

Previous studies indirectly support the genetic basis of non-substance addiction. Hereditary factors have been observed to explain 58% of individual differences in sensation seeking. Consequently, the influence of environmental factors seems relatively insignificant, playing only a minor role in determining sensation seeking [29]. Indeed, a study of fraternal twins showed that sensation seeking traits rely mainly on genetic factors, and the Vietnam Era Twin Registry's data reveal that genetic factors significantly influence pathological gambling, explaining 46–55% of the variance [44, 71].

Though the results of the sensation seeking scale may be the same, substance addiction and non-substance addiction have clear, distinct processes. In other words, sensation seeking in substance addiction is influenced by more innate factors whereas sensation seeking in non-substance addiction is influenced by the interaction of both genetic and environmental factors. Indeed, this conclusion is supported by the fact that the relationship between sensation seeking and internet addiction is inconsistent in different studies: studies show highly positive [67], moderate to weakly positive [76], and negative [57] relationships, indicating its high instability. Thus, further refining of experimental design is necessary to clarify this relationship.

6.2.2 *Inhibitory Control*

The cognitive function to restrain the addictive behavior is impaired in addicts, which may reflect how brain inhibitory mechanisms may be affected. Inhibitory control is the psychological process of reducing the influence of irrelevant stimuli and reducing and repressing dominant behavioral responses [3, 14, 46]. In addition, inhibitory control can be divided into two classes: interference inhibition and response inhibition [3, 10, 11, 47]. The main paradigms consist of Stroop task, Flanker task, Go/No-Go task, oddball paradigm and Stop-signal paradigm and so on.

A large number of studies have shown that long-term drug abuse eventually leads to impaired inhibitory control of addicts [25]. Fishbein et al. tested the cognitive ability of heroin addicts and alcohol addicts, analyzing prefrontal cortex activity through a series of neurocognitive tasks. The results showed that all addicts exhibited impaired cognitive function, with their cognitive flexibility and interference inhibition significantly reduced in Stroop task.

The Go/No-Go task has been widely used for inhibitory control in ERP study. N2 (Negative wave occurs 200 milliseconds after the stimulation) under Nogo conditions composition is the embodiment of inhibitory control. The amplitude of Nogo N2 becomes smaller, suggesting that individuals lack inhibitory control [75]. ERP research of Go/No-Go task show that internet addiction group's NogoN2 amplitude is lower than the control group, indicating that the inhibitory control of internet addicts was impaired [22, 81]; at the same time, the amplitude of NogoP3 increased and latent period prolonged, suggesting internet addicts need more mental resources and low efficiency of inhibitory control [21].

There are also some evidences of neuroimaging. The functional connection between the prefrontal cortex and the anterior cingulate is mainly responsible for inhibitory control [26, 55]. Studies have found that heroin addicts' functional connectivity of prefrontal cortex, the anterior cingulate and orbito-frontal weakened [49]. Non-substance addiction had similar findings. Potenza et al. [66] using MRI (Magnetic Resonance Imaging) compared the 13 pathological gamblers with 11 normal subjects in brain activation condition after stroop task; the results showed that functional connectivity of the anterior cingulate and medial orbitofrontal cortex is impaired in pathological gamblers.

As for the study that focuses on the differences of substance addiction and non-substance addiction is few, indicating that inhibitory control is the psychological mechanism of both.

6.2.3 Attentional Bias

Attentional bias refers to the individual's capacity to redistribute attention to novel rather than neutral stimuli [4, 51, 60]. In recent years, researchers have agreed that attention bias is not only symptomatic of mental disease, but also a reinforcer of mental illness [35, 52, 78].

In the context of substance abuse, attentional bias refers to the tendency of substance related cues to grab the attention of experienced substance users [24]. Studies for alcoholics and smokers have shown that substance-addicts exhibit attentional bias towards addictive substance-related stimuli, with attention automatically given to and maintained in addictive substances [23, 61]. When cocaine addicts are exposed to pictures of cocaine, analysis of brainwave activity revealed a significantly slower positive component, which is used to quantify cravings for heroin [27].

Internet addicts are often easily distracted by external cues yet retain strong and sensitive attention to internet-related stimuli [80]. At the same time, Thalmann, Wolfling and Thalemann et al. [74] found, which similar to substance addicts, online game addicts have a significantly slower positive component when exposed to pictures of online gaming. Metcalf and Metcalf and Pammer [58] testing the Stroop paradigm for online game addicts found that online game addicts have obvious attentional bias to online-gaming related words. Decker and Gay [19] study found a similar conclusion.

In general, attentional biases of addicts are significantly affected. Future research should compare attentional bias in substance addiction and non-substance addiction through its three sub-components: facilitated attention, difficulty in disengaging and attentional avoidance [13].

6.2.4 *Intertemporal Choice*

Intertemporal choice refers to the process of weighing the cost and benefits of different choices in the present and future [28, 45]. One of the most important findings about intertemporal choice is that people consider future benefits with less weight than current or recent gains (or losses); this phenomenon is called delay discounting [32], a measure of the degree of patience. For example, when faced with immediate small rewards and delayed large rewards, people often tend to choose immediate small rewards [62]; future benefits are thus discounted because their perceived value decreases with time. Even if the addictive substances is delayed, people still choose to expose themselves to these substances, indicating that delayed rewards showed a high discount value. In essence, delay discounting refers to the process of recalculating the value of the delayed object while considering the length of delay and the change in value between immediate and delayed benefits. Delay discounting is the sensitive and key indicator of intertemporal choice.

In substance addiction, Madden et al. [53] found that the delay discounting of money in opioids addicts is greater than that of the control group, and theirs delay discounting of heroin is greater than the delay discounting of money. Similar results have been found in heroin addicts [41] and cocaine addicts [9]. One study found that mild opiate withdrawal would add more weight to the delay discounting in heroin and money in opioid addicts [30], but cocaine withdrawal (for at least 30 days) does not affect the delay discounting of cocaine [36], indicating that short-term cocaine abstinence does not influence delay discounting. Alcoholics' delay discounting of alcohol was significantly greater than that of the control group [77] and positively correlated with the degree of alcohol addiction [20, 59], while delay discounting of money was not significantly different than the control group [40]. Similar results were found in drug addiction. Smokers' delay discounting of tobacco was significantly greater than that of the control group; those who quit had no significant differences from the control group [8]. Daily smoking was found to be positively correlated with delay discounting [63].

In non-substance addiction, addicts' delay discounting of money is significantly greater than the normal, including gamblers [65] and internet addict [68]. The more serious pathological gambling is, the higher the delay discounting [1].

In conclusion, addicts showed abnormal delay discounting, but substance addiction and non-substance addiction have respective specificity. Different types of addiction align with different types of delay discounting [50]. Some researchers have compared gamblers, drinkers and cannabis addicts in delay discounting. The results found that gamblers had closer relations with money delay discounting, which indicates the prediction effect of money delay discounting for gambling is significantly better than for substance addiction [72].

6.2.5 Environment

Addiction can be influenced not only by genetic factors but also by environmental factors such as family, peers and social environment. Kendler et al. [39] found that after controlling for individual socio-economic status and confounding familial factors, neighborhood social deprivation prospectively predicted risk of drug abuse. One study found that peer and parental neglect may have caused some students who fell behind in school to have high tobacco use [18]. Non-substance addictions also have similar results. Internet addicts have interpersonal and emotional distress generally [12, 43].

One factor must be taken into consideration in substance addiction: the sociocultural factor. Indeed, cultural and ethnic traditions may influence substance use. For instance, societal ideals of masculinity may dictate that men must be able to consumer large quantities of alcohol, event to the point of unconsciousness. Many young people begin to practice drinking even in childhood. A study by Barlow and Durand [5] asked children aged 3–6 to identify apple juice, coffee, spices, beer, whisky and tobacco through their sense of smell. They found that more than half of the older children are able to distinguish tobacco from alcoholic substances, while 20% of younger children are able to distinguish between them, showing that these children have contact with these substances early on.

6.3 Summary and Prospect

From these above studies, we conclude that substance addicts have high sensation seeking scores, suggesting that personality traits and genetic factors significantly influence the formation of substance addiction; for non-substance addiction, only correlational – not causal – relationships may be drawn, indicating that sensation seeking does not play a major role in the formation of non-substance addiction. With regard to inhibitory control and attentional bias, substance addicts have significantly impaired basic cognitive functions in comparison to non-substance addicts. In non-substance addicts, intertemporal choice can effectively predict the severity of their addiction, suggesting that non-substance addicts' higher cognitive functions are relatively unaffected. Finally, environmental factors significantly influence non-substance addiction relative to substance addiction. Substance addiction and non-substance addiction share common cognition, personality, environmental factors, but the effects caused by substance addiction are profound and often permanent due to brain damage, genetic factors, and decline in cognitive function. Non-substance addiction (e.g. internet addiction) is a result of a variety of factors such as family, personality, environment and development (adolescence and early adulthood), indicating that the formation of substance addiction involves more endogenous factors while non-substance addiction involves more exogenous factors, implicating both nature and nurture.

We need to use a more nuanced cause-effect network system to understand addiction. Three models currently exist: biological model, psychological model and social model in traditional psychology research. The biological model regards addiction as a disease, involving specific reward center in the brain. This model is worth promoting, but ignores individual factors. The psychology model uses personality, self-efficacy and cognitive processes to explain addiction. Research showed that psychological factors is critical susceptible factors and maintain factors in addiction, but it cannot explain biological response of dependence and withdrawal. Social model explains addiction from the environmental stressor, family and peer pressure; though the interdisciplinary nature of this perspective has its strengths, this wide range risks overlooking precise nuances. For this reason, we should adopt the biopsychosocial model of health and disease practiced by George Engel [69]. For instance, in substance addiction, the psychological model bases drug use on positive reinforcement, negative reinforcement, individual agency, and expectations. In contrast, the biological model considers inherited genetic vulnerability while the social model considers social expectations, media exposure, and cultural norms [6].

Therefore, future studies should explore mechanism of interactions among those three models on reward and integrate them into a perfect, specific, and clear theoretical model of addiction and set up a mathematical model, if possible.

References

1. Alessi SM, Petry NM (2003) Pathological gambling severity is associated with impulsivity in a delay discounting procedure. Behav Process 64(3):345–354
2. American Psychiatry Association (2013) Diagnostic and statistical manual of mental disorders-DSM-5. American Psychiatric Publishing, Washington, DC
3. Aron AR, Robbins TW, Poldrack RA (2004) Inhibition and the right inferior frontal cortex. Trends Cogn Sci 8(4):170–177
4. Bar-Haim Y, Lamy D, Pergamin L, Bakermans-Kranenburg MJ, van IJzendoorn MH (2007) Threat-related attentional bias in anxious and nonanxious individuals: a meta-analytic study. Psychol Bull 133(1):1–24
5. Barlow D, Durand V (1995) Abnormal psychology. Cole Publishing Company, London

6. Barlow D, Durand V (2011) Abnormal psychology: an integrative approach. Nelson Education, Toronto
7. Berridge KC, Robinson TE (2003) Parsing reward. Trends Neurosci 26(9):507–513
8. Bickel WK, Odum AL, Madden GJ (1999) Impulsivity and cigarette smoking: delay discounting in current, never, and ex-smokers. Psychopharmacology 146(4):447–454
9. Black AC, Rosen MI (2011) A money management-based substance use treatment increases valuation of future rewards. Addict Behav 36(1–2):125–128
10. Booth JR, Burman DD, Meyer JR, Gitelman DR, Parrish TB, Mesulam MM (2004) Development of brain mechanisms for processing orthographic and phonologic representations. J Cogn Neurosci 16(7):1234–1249
11. Brydges CR, Clunies-Ross K, Clohessy M, Lo ZL, Nguyen A, Rousset C, Whitelaw P, Yeap YJ, Fox AM (2012) Dissociable components of cognitive control: an event-related potential (ERP) study of response inhibition and interference suppression. PLoS One 7(3):e34482
12. Caplan S, Williams D, Yee N (2009) Problematic Internet use and psychosocial well-being among MMO players. Comput Hum Behav 25(6):1312–1319
13. Cisler JM, Koster EHW (2010) Mechanisms of attentional biases towards threat in anxiety disorders: an integrative review. Clin Psychol Rev 30(2):203–216
14. Clark JM (1996) Contributions of inhibitory mechanisms to unified theory in neuroscience and psychology. Brain Cogn 30(1):127–152
15. Cloninger CR (1987) A systematic method for clinical description and classification of personality variants – a proposal. Arch Gen Psychiatry 44(6):573–588
16. Cloninger CR, Sigvardsson S, Bohman M (1988) Childhood personality predicts alcohol-abuse in young-adults. Alcohol Clin Exp Res 12(4):494–505
17. Crabbe JC (2002) Genetic contributions to addiction. Annu Rev Psychol 53:435–462
18. Davidson GC, Neale JM (1996) Abnormal psychology. Wiley publishing Company, New York
19. Decker SA, Gay JN (2011) Cognitive-bias toward gaming-related words and disinhibition in world of warcraft gamers. Comput Hum Behav 27(2):798–810
20. Dom G, D'haene P, Hulstijn W, Sabbe B (2006) Impulsivity in abstinent early- and late-onset alcoholics: differences in self-report measures and a discounting task. Addiction 101(1):50–59
21. Dong GH, Lu QL, Zhou H, Zhao XA (2010) Impulse inhibition in people with Internet addiction disorder: electrophysiological evidence from a Go/NoGo study. Neurosci Lett 485(2):138–142
22. Dong GH, Zhou H, Zhao X (2011) Male Internet addicts show impaired executive control ability: evidence from a color-word Stroop task. Neurosci Lett 499(2):114–118
23. Field M, Cox WM (2008) Attentional bias in addictive behaviors: a review of its development, causes, and consequences. Drug Alcohol Depend 97(1):1–20
24. Field M, Munafo MR, Franken IHA (2009) A meta-analytic investigation of the relationship between attentional bias and subjective craving in substance abuse. Psychol Bull 135(4):589–607
25. Fishbein DH, Krupitsky E, Flannery BA, Langevin DJ, Bobashev G, Verbitskaya E, Augustine CB, Bolla KI, Zvartau E, Schech B, Egorova V, Bushara N, Tsoy M (2007) Neurocognitive characterizations of Russian heroin addicts without a significant history of other drug use. Drug Alcohol Depend 90(1):25–38
26. Fox MD, Raichle ME (2007) Spontaneous fluctuations in brain activity observed with functional magnetic resonance imaging. Nat Rev Neurosci 8(9):700–711
27. Franken IHA, Hulstijn KP, Stam CJ, Hendriks VM, van den Brink W (2004) Two new neurophysiological indices of cocaine craving: evoked brain potentials and cue modulated startle reflex. J Psychopharmacol 18(4):544–552
28. Frederick S, Loewenstein G, O'Donoghue T (2002) Time discounting and time preference: a critical review. J Econ Lit 40(2):351–401
29. Fulker DW, Eysenck SB, Zuckerman M (1980) A genetic and environmental analysis of sensation seeking. J Res Personal 14(2):261–281

30. Giordano LA, Bickel WK, Loewenstein G, Jacobs EA, Marsch L, Badger GJ (2002) Mild opioid deprivation increases the degree that opioid-dependent outpatients discount delayed heroin and money. Psychopharmacology 163(2):174–182
31. Grant JE, Brewer JA, Potenza MN (2006) The neurobiology of substance and behavioral addictions. CNS Spectr 11(12):924–930
32. Green L, Myerson J (2004) A discounting framework for choice with delayed and probabilistic rewards. Psychol Bull 130(5):769–792
33. Griffiths M (2009) The psychology of addictive behaviour, Psychology for A2 Level. Harper Collins, London, pp 436–471
34. Hansen EB, Breivik G (2001) Sensation seeking as a predictor of positive and negative risk behaviour among adolescents. Personal Individ Differ 30(4):627–640
35. Hayes S, Hirsch CR, Mathews A (2010) Facilitating a benign attentional bias reduces negative thought intrusions. J Abnorm Psychol 119(1):235–240
36. Heil SH, Johnson MW, Higgins ST, Bickel WK (2006) Delay discounting in currently using and currently abstinent cocaine-dependent outpatients and non-drug-using matched controls. Addict Behav 31(7):1290–1294
37. Hittner JB, Swickert R (2006) Sensation seeking and alcohol use: a meta-analytic review. Addict Behav 31(8):1383–1401
38. Kelly TH, Delzer TA, Martin CA, Harrington NG, Hays LR, Bardo MT (2009) Performance and subjective effects of diazepam and d-amphetamine in high and low sensation seekers. Behav Pharmacol 20(5–6):505–517
39. Kendler KS, Ohlsson H, Sundquist K, Sundquist J (2014) The causal nature of the association between neighborhood deprivation and drug abuse: a prospective national Swedish co-relative control study. Psychol Med 44(12):2537–2546
40. Kirby KN, Petry NM (2004) Heroin and cocaine abusers have higher discount rates for delayed rewards than alcoholics or non-drug-using controls. Addiction 99(4):461–471
41. Kirby KN, Petry NM, Bickel WK (1999) Heroin addicts have higher discount rates for delayed rewards than non-drug-using controls. J Exp Psychol Gen 128(1):78–87
42. Ko CH, Yen JY, Chen CC, Chen SH, Wu KY, Yen CF (2006) Tridimensional personality of adolescents with Internet addiction and substance use experience. Can J Psychiatr 51(14):887–894
43. Ko CH, Yen JY, Yen CF, Chen CS, Chen CC (2012) The association between Internet addiction and psychiatric disorder: a review of the literature. Eur Psychiatry 27(1):1–8
44. Lin SA, Lyons MJ, Scherrer JF, Griffith K, True WR, Goldberg J, Tsuang MT (1998) Familial influences on gambling behavior: an analysis of 3359 twin pairs. Addiction 93(9):1375–1384
45. Loewenstein GF (1988) Frames of mind in intertemporal choice. Manag Sci 34(2):200–214
46. Logan GD, Schachar RJ, Tannock R (1997) Impulsivity and inhibitory control. Psychol Sci 8(1):60–64
47. Lubman DI, Yucel M, Pantelis C (2004) Addiction, a condition of compulsive behaviour? Neuroimaging and neuropsychological evidence of inhibitory dysregulation. Addiction 99(12):1491–1502
48. Lynne-Landsman SD, Graber JA, Nichols TR, Botvin GJ (2011) Is sensation seeking a stable trait or does it change over time? J Youth Adolesc 40(1):48–58
49. Ma N, Liu Y, Li N, Wang CX, Zhang H, Jiang XF, Xu HS, Fu XM, Hu X, Zhang DR (2010) Addiction related alteration in resting-state brain connectivity. NeuroImage 49(1):738–744
50. MacKillop J, Amlung MT, Few LR, Ray LA, Sweet LH, Munafo MR (2011) Delayed reward discounting and addictive behavior: a meta-analysis. Psychopharmacology 216(3):305–321
51. Macleod C, Mathews A, Tata P (1986) Attentional bias in emotional disorders. J Abnorm Psychol 95(1):15–20
52. MacLeod C, Rutherford E, Campbell L, Ebsworthy G, Holker L (2002) Selective attention and emotional vulnerability: assessing the causal basis of their association through the experimental manipulation of attentional bias. J Abnorm Psychol 111(1):107–123
53. Madden GJ, Petry NM, Badger GJ, Bickel WK (1997) Impulsive and self-control choices in opioid-dependent patients and non-drug-using control participants: drug and monetary rewards. Exp Clin Psychopharmacol 5(3):256–262

54. Mann K, Ackermann K, Croissant B, Mundle G, Nakovics H, Diehl A (2005) Neuroimaging of gender differences in alcohol dependence: are women more vulnerable? Alcohol Clin Exp Res 29(5):896–901
55. Margulies DS, Kelly AMC, Uddin LQ, Biswal BB, Castellanos FX, Milham MP (2007) Mapping the functional connectivity of anterior cingulate cortex. NeuroImage 37(2):579–588
56. Mehroof M, Griffiths MD (2010) Online gaming addiction: the role of sensation seeking, self-control, neuroticism, aggression, state anxiety, and trait anxiety. Cyberpsych Behav Soc N 13(3):313–316
57. Mesgarani M, Shafiee S, Ahmadi E, Zare F (2013) Studying the relationship between internet addiction and emotional intelligence, sensation seeking and metacognition among those who referred to cafes. Int Res J Appl Basic Sci 4(4):889–893
58. Metcalf O, Pammer K (2011) Attentional bias in excessive massively multiplayer online role-playing gamers using a modified stroop task. Comput Hum Behav 27(5):1942–1947
59. Mitchell JM, Fields HL, D'Esposito M, Boettiger CA (2005) Impulsive responding in alcoholics. Alcohol Clin Exp Res 29(12):2158–2169
60. Mogg K, Bradley BP, Hyare H, Lee S (1998) Selective attention to food-related stimuli in hunger: are attentional biases specific to emotional and psychopathological states, or are they also found in normal drive states? Behav Res Ther 36(10):227–237
61. Montgomery C, Field M, Atkinson AM, Cole JC, Goudie AJ, Sumnall HR (2010) Effects of alcohol preload on attentional bias towards cocaine-related cues. Psychopharmacology 210(3):365–375
62. O'Donoghue T, Rabin M (1999) Doing it now or later. Am Econ Rev 89(1):103–124
63. Ohmura Y, Takahashi T, Kitamura N (2005) Discounting delayed and probabilistic monetary gains and losses by smokers of cigarettes. Psychopharmacology 182(4):508–515
64. Petry NM (2006) Should the scope of addictive behaviors be broadened to include pathological gambling? Addiction 101(s1):152–160
65. Petry NM, Casarella T (1999) Excessive discounting of delayed rewards in substance abusers with gambling problems. Drug Alcohol Depend 56(1):25–32
66. Potenza MN, Steinberg MA, Skudlarski P, Fulbright RK, Lacadie CM, Wilber MK, Rounsaville BJ, Gore MC, Wexler BE (2003) Gambling urges in pathological gambling – a functional magnetic resonance imaging study. Arch Gen Psychiatry 60(8):828–836
67. Rahmani S, Lavasani MG (2011) The relationship between internet dependency with sensation seeking and personality. Proc Soc Behav 30:272–277
68. Saville BK, Gisbert A, Kopp J, Telesco C (2010) Internet addiction and delay discounting in college students. Psychol Rec 60(2):273–286
69. Schaufeli WB, Bakker AB (2004) Job demands, job resources, and their relationship with burnout and engagement: a multi-sample study. J Organ Behav 25(3):293–315
70. Schlosser S, Black DW, Repertinger S, Freet D (1994) Compulsive buying – demography, phenomenology, and comorbidity in 46 subjects. Gen Hosp Psychiatry 16(3):205–212
71. Slutske WS, Eisen S, True WR, Lyons MJ, Goldberg J, Tsuang M (2000) Common genetic vulnerability for pathological gambling and alcohol dependence in men. Arch Gen Psychiatry 57(7):666–673
72. Stea JN, Hodgins DC, Lambert MJ (2011) Relations between delay discounting and low to moderate gambling, cannabis, and alcohol problems among university students. Behav Process 88(3):202–205
73. Tavares H, Martins SS, Lobo DSS, Silveira CM, Gentil V, Hodgins DC (2003) Factors at play in faster progression for female pathological gamblers: an exploratory analysis. J Clin Psychiatry 64(4):433–438
74. Thalemann R, Wolfling K, Grusser SM (2007) Specific cue reactivity on computer game-related cues in excessive gamers. Behav Neurosci 121(3):614–618
75. van de Laar MC, Licht R, Franken IHA, Hendriks VM (2004) Event-related potentials indicate motivational relevance of cocaine cues in abstinent cocaine addicts. Psychopharmacology 177(1–2):121–129

76. van Holst RJ, van den Brink W, Veltman DJ, Goudriaan AE (2010) Why gamblers fail to win: a review of cognitive and neuroimaging findings in pathological gambling. Neurosci Biobehav R 34(1):87–107
77. Vuchinich RE, Simpson CA (1998) Hyperbolic temporal discounting in social drinkers and problem drinkers. Exp Clin Psychopharmacol 6(3):292–305
78. Waters AM, Valvoi JS (2009) Attentional bias for emotional faces in paediatric anxiety disorders: an investigation using the emotional go/no go task. J Behav Ther Exp Psychiatry 40(2):306–316
79. Yen JY, Yen CF, Chen CC, Chen SH, Ko CH (2007) Family factors of Internet addiction and substance use experience in Taiwanese adolescents. Cyberpsychol Behav 10(3):323–329
80. Zhou ZH, Yuan GZ, Yao JJ (2012) Cognitive biases toward internet game-related pictures and executive deficits in individuals with an internet game addiction. PLoS One 7(11):e48961
81. Zhou ZH, Yuan GZ, Yao JJ, Li C, Cheng ZH (2010) An event-related potential investigation of deficient inhibitory control in individuals with pathological Internet use. Acta Neuropsychiatr 22(5):228–236
82. Zuckerman M (1994) Behavioral expressions and biosocial bases of sensation seeking. Cambridge university press, Cambridge
83. Zuckerman M (2014) Sensation seeking (psychology revivals): beyond the optimal level of arousal. Psychology Press, London

Part III
Comparison Between Substance and Non-substance Addictions in Diagnosis

Chapter 7
Similarities and Differences in Diagnostic Criterion

Zhengde Wei and Xiaochu Zhang

Abstract In this chapter, the main content is to discuss the similarities and differences in diagnostic criteria between substance and non-substance addictions. Firstly, diagnostic criteria of substance addiction were introduced, mainly focused on Diagnostic and Statistical Manual for the Mental Disorders, fifth edition (DSM-5). Then, we described the diagnostic criteria of several non-substance addictions, including gambling disorder, internet addiction, food addiction and hypersexual disorder. Depending on the proof, substance and non-substance addictions have many similarities in symptoms. Though the proposed diagnostic criteria of many non-substance addictions are currently most useful as survey instruments to access the prevalence of the problem, there is little or no validating proof for these diagnostic criteria. Finally, animal model is useful tool for addiction research. But, present animal models for gambling studying do not meet enough diagnostic criteria and could not be regarded as gambling disorder. By introducing the animal models evolved to resemble the diagnostic criteria of substance addiction and two classical paradigms for substance addiction, self-administration and conditioned place preference, we hope it is helpful to improve the validation of animal model of gambling disorder.

Keywords Substance addiction • Non-substance addiction • Diagnostic criteria

Z. Wei
Key Laboratory of Brain Function and Disease, Chinese Academy of Sciences, School of Life Sciences, University of Science & Technology of China, Hefei, Anhui 230027, China

X. Zhang (✉)
Key Laboratory of Brain Function and Disease, Chinese Academy of Sciences, and School of Life Sciences, University of Science & Technology of China, Hefei, Anhui 230027, China

School of Humanities & Social Science, University of Science & Technology of China, Hefei, Anhui 230026, China

Centers for Biomedical Engineering, University of Science & Technology of China, Hefei, Anhui 230027, China
e-mail: zxcustc@ustc.edu.cn

X. Zhang et al. (eds.), *Substance and Non-substance Addiction*, Advances in Experimental Medicine and Biology 1010, DOI 10.1007/978-981-10-5562-1_7

7.1 Diagnostic Criteria of Addiction

7.1.1 Diagnostic Criteria of Substance Addiction

In an effort to give a theoretical framework to clinical observations of substance use, Goodman [39] proposed the term of "addictive disorder" to define addiction (see Table 7.1) [39]. Addictive disorder is characterized by two patterns: (1) recurrent failure to control the behavior and (2) continuation of the behavior despite the knowledge of significant harmful consequences. According to Goodman, addictive disorder is not a collection of distinct disorders, but a group with a similar underlying process.

According to latest Diagnostic and Statistical Manual for the Mental Disorders (DSM-5) [10], the substance addiction encompasses 10 separate classes of drug: alcohol; caffeine; cannabis; hallucinogens; inhalants; opioids; sedatives, hypnotics, and anxiolytics; stimulants; tobacco and other or unknown substances . These 10 classes have common direct brain reward system activation, which is involved in the production of memories and the reinforcement of behaviors. All these substance addictions are not fully distinct. So all these substance addiction can be diagnosed by a common diagnosis (see Table 7.2).

Table 7.1 Goodman's addictive disorder criteria

1. Recurrent failure to resist impulses to engage in a specified behavior
2. Increasing sense of tension immediately prior to the initiation of behavior
3. Pleasure or relief at the time of engaging in the behavior
4. A feeling of a lack of control while engaging in the behavior
5. At least five of the following:
(a) Frequent preoccupation with the behavior or preparatory activities
(b) Frequent engagement in the behavior to a greater extent or over a longer period than intended
(c) Repeated efforts to reduce, control or stop the behaviour
(d) A great deal of time spent in activities necessary for the behaviour, engaging in the behaviour or recovering from its effects
(e) Frequent engaging in the behaviour when expected to fulfill occupational, academic, domestic or social obligations
(f) Important social, occupational or recreational activities given up or reduced because of the behaviour
(g) Continuation of the behaviour despite knowledge of having a persistent or recurrent social, financial, psychological or physical problem that is caused or exacerbated by the behaviour
(h) Tolerance: need to increase the intensity or frequency of the behaviour in order to achieve the desired effect or diminished effect with continued behaviour of the same intensity
(i) Restlessness or irritability if unable to engage in the behaviour
6. Some symptoms of the disturbance have persisted for at least 1 month or have occurred repeatedly over a longer period of time

Table 7.2 DSM-V diagnostic criteria of substance addiction

1.	The substance is often taken in larger amounts or over a longer period than was intended
2.	There is a persistent desire or unsuccessful efforts to cut down or control use of the substance
3.	A great deal of time is spent in activities necessary to obtain the substance, use the substance, or recover from its effects
4.	Craving, or a strong desire or urge to use the substance
5.	Recurrent use of the substance resulting in a failure to fulfill major role obligations at work, school, or home
6.	Continued use of the substance despite having persistent or recurrent social or interpersonal problems caused or exacerbated by the effects of its use
7.	Important social, occupational, or recreational activities are given up or reduced because of use of the substance
8.	Recurrent use of the substance in situations in which it is physically hazardous
9.	Use of the substance is continued despite knowledge of having a persistent or recurrent physical or psychological problem that is likely to have been caused or exacerbated by the substance
10.	Tolerance, as defined by either of the following:
	(a) A need for markedly increased amounts of the substance to achieve intoxication or desired effect
	(b) A markedly diminished effect with continued use of the same amount of the substance
11.	Withdrawal, as manifested by either of the following:
	(a) The characteristic withdrawal syndrome for other (or unknown) substance
	(b) The substance (or a closely related substance) is taken to relieve or avoid withdrawal symptoms

These 11 criteria can fit within the overall grouping of impaired control, social impairment, risky use, and pharmacological criteria. Impaired control over substance use includes criteria 1–4. Social impairment includes criteria 5–7. Risky use of the substance includes criteria 8–9. Criteria 10–11 belong to pharmacological criteria. The presence of two or three criteria suggest a mild substance addiction, four or five suggest moderate, six or more suggest severe addiction.

Compared with DSM-IV [9], the DSM-5 combine substance abuse and substance dependence into substance use disorder, because these two are difficult to distinguish. Mild addiction is more like substance abuse, and moderate or severe is more like substance dependence.

The craving is added in DSM-5 newly. Craving is about an intense desire or urge for the drug. It may occur at any time but is more likely to occur in an environment where the drug was previously obtained or used. Craving has also been shown to be associated with the activation of specific reward regions in the brain. Current craving may be a signal of impending relapse which is usually used as a treatment result measure.

The criteria of substance addictions consist of four groups, including impaired control, social impairment, risky use, and pharmacological criteria; which can apply to each kind of substance addiction diagnosis. Though substance addictions have similarly underlying mechanisms and behavioral symptoms, some substance addictions have their own unique characteristics that must reflect on diagnostic criteria. Withdrawal symptom is an important diagnostic criterion for most of substance addictions but not for hallucinogen use disorder, because withdrawal from phencyclidines has been only reported in animals but not shown in human beings. However, there is evidence of withdrawal symptoms from MDMA (kind of hallucinogen), with two or more withdrawal symptoms revealed in samples of MDMA users. The withdrawal symptoms are also excluded from diagnostic criteria for inhalant use disorder, because the withdrawal symptoms are mild and a few individuals use inhalant avoid withdrawal.

Another well-known handbook for addiction diagnosis is the International Classification of diseases, tenth edition (ICD-10) (Since the ICD-11 is not published formally, we only discuss about ICD-10 in this chapter). It is developed and published by the World Health Organization and used mainly outside the United States (see Table 7.3). ICD-10 criteria for substance addiction are similar with DSM-5, with criteria for tolerance, withdrawal, continued use despite negative problems and various indicators of impaired control. A man who is diagnosed as addict should meet the following character: three or more of these symptoms occurring together for at least 1 month, or if <1 month, occurring together repeatedly within a 12-month period. The main difference between the two diagnoses is that 'difficulties controlling use in terms of onset, termination, or levels of use' is stated explicitly in the ICD-10, whereas DSM-5 does not use the specific language of 'onset' and 'termination'. Despite the difference, if these two criteria sets describe

Table 7.3 ICD-10 diagnostic criteria of substance addiction

1.	Need for significantly increased amounts of substance to achieve intoxication or desired effect or markedly diminished effect with continued use of the same amount of substance	
2.	A physiological withdrawal state of the characteristic withdrawal syndrome for the substance, or use of the substance (or closely related) to relieve or avoid symptoms	
3.	Difficulties controlling use in terms of onset, termination, or levels of use; using in larger amounts or over a longer period than intended; or a persistent desire or unsuccessful efforts to reduce or control use	
4.	Important alternative pleasures or interests given up or reduced because of use	
5.	A great deal of time spent in activities necessary to obtain or use or to recover from its effects	
6.	Persisting with use despite clear evidence and knowledge of harmful physical or psychological consequences	
7.	Strong desire or sense of compulsion to use	
8.	Abuse:	
	(a)	(Harmful use) Clear evidence that substance use contributed to physical or psychological harm, which may lead to disability/adverse consequences
	(b)	The nature of harm should be clearly identifiable (and specified)
	(c)	The pattern of use has persisted for at least 1 month or has occurred repeatedly within a 12-month period
	(d)	Symptoms do not meet criteria for any other mental or behavioral disorder related to substance in the same time period (except for acute intoxication)

the same underlying condition, then small differences between them should not produce large differences in their reliability, validity or concordance.

7.1.2 Diagnostic Criteria of Non-substance Addiction

The main characteristic of non-substance addiction is the failure to resist an impulse or drive to perform a behavior that is harmful to the person or to others. Non-substance addictions discussed in this book involve gambling disorder, internet addiction, food addiction and hypersexual disorder. Each non-substance addiction is characterized by a recurrent pattern of this behavior. The repetitive engagement in these behaviors finally disrupts functioning in other domains. In this respect, the non-substance addictions are similar to substance addictions. Individuals with substance addiction have reported difficulties in resisting the urge to use the drug.

As a consequence of pathological gambling being claimed as formal addiction, it appears reasonably to discuss that other non-substance addictions, such as internet addiction, overeating disorder and hypersexual disorder, may follow suit. In fact, most of these non-substance addictions were diagnosed by proposed criteria based on diagnostic criteria of gambling disorder in DSM-5 or pathological gambling in DSM-IV, e.g., excessive time spent in the behavior, diminished control over the behavior, repeated unsuccessful attempts to cut down or stop the behavior, tolerance, withdrawal, and harmful psychosocial consequences. For instance, Ivan [32] tried to propose diagnostic criteria for food addiction by comparing symptoms of substance addiction with food addiction: (1) tolerance, (2) abstinence symptoms, (3) loss of control, (4) the desire or repeated unsuccessful attempts to stop consumption, (5) lots of time cost in refraining from eating or in the recovery of consumption, (6) waiver of other important activities, and (7) continued use despite physical and psychological problems. Research goes in favor of these symptoms, but some by animal models, some by weak evidence. Just like withdrawal excluded from hallucinogen use disorder because of animal research only, these studies are not enough to make some symptoms included for food addiction.

For many other non-substance addictions which have proposed diagnostic criteria, there is little or no validating proof for these diagnostic criteria. But, these proposed diagnostic criteria are currently most useful as survey instruments to access the prevalence of the problem.

7.1.2.1 Diagnostic Criteria of Gambling Disorder

Gambling disorder, used belongs to impulsive control disorder, is the only non-substance addiction that is now divided into non-substance-related disorder. The essential characteristic of gambling disorder is a persistent and recurrent pathological gambling behavior that disrupts family and personal pursuits. The gambling behaviors activate similar reward systems and produce some similar behavioral symptoms compared with those of the substance use disorders. Therefore the

Table 7.4 DSM-V diagnostic criteria of gambling disorder

Persistent and recurrent problematic gambling behavior leading to clinically significant impairment or distress, as indicated by the individual exhibiting four (or more) of the following in a 12-month period:	
1.	Needs to gamble with increasing amounts of money in order to achieve the desired excitement
2.	Is restless or irritable when attempting to cut down or stop gambling
3.	Has made repeated unsuccessful efforts to control, cut back, or stop gambling
4.	Is often preoccupied with gambling (e.g., having persistent thoughts of reliving past gambling experiences, handicapping or planning the next venture, thinking of ways to get money with which to gamble)
5.	Often gambles when feeling distressed (e.g., helpless, guilty, anxious, depressed)
6.	After losing money gambling, often returns another day to get even ("chasing" one's losses)
7.	Lies to conceal the extent of involvement with gambling
8.	Has jeopardized or lost a significant relationship, job, or educational or career opportunity because of gambling
9.	Relies on others to provide money to relieve desperate financial situations caused by gambling
Specify current severity:	
	Mild: 4–5 criteria met
	Moderate: 6–7 criteria met
	Severe: 8–9 criteria met

diagnostic criteria of gambling disorder (see Table 7.4) are a reference to the diagnostic criteria of substance use disorder, but it also has its own feature. "Chasing one's losses" may be a unique characteristic, with an urgent desire to keep gambling (often with larger bets or greater risks) to undo loss. The individual may try to win back losses all at once without gambling strategy. Lying to family members or others to conceal the extent of gambling is another feature of the gambling disorder. Individuals may also engage in "bailout" behavior, asking family members or others for help with a desperate financial situation that was caused by gambling.

7.1.2.2 Diagnostic Criteria of Internet Addiction

Since problems have arisen with internet use, it has been beneficial for scientific researchers to establish diagnostic criteria. Young has modified the DSM-IV pathological gambling criteria to make them the diagnosis of internet addiction (see Table 7.5), not limited to internet gaming [92]. And her criteria are a significant contribution in offering a concrete basis for establishing internet addiction.

There are however certain problems with Young's diagnosis [13]. One concern is how much of the criteria is depended on self-report and how objective the criteria is. Likewise, the individual's judgment may be impaired and information of self-report may not be accurate. Another issue is that the criteria need to be more specific. There also seems to be an issue of whether or not pathological gambling, which is an impulse control disorder in DSM-IV but now a substance-related and addictive

Table 7.5 Young's criteria of internet addiction

1.	Is preoccupied with the Internet (think about previous online activity or anticipate next online session)
2.	Needs to use the Internet with increased amounts of time in order to achieve satisfaction
3.	Has made unsuccessful efforts to control, cut back, or stop Internet use
4.	Is restless, moody, depressed, or irritable when attempting to cut down or stop Internet use
5.	Has stayed online longer than originally intended
6.	Has jeopardized or risked the loss of a significant relationship, job, educational or career opportunity because of the Internet
7.	Has lied to family members, therapist, or others to conceal the extent of involvement with the Internet
8.	Uses the Internet as a way of escaping from problems or of relieving a dysphoric mood (e.g., feelings of helplessness, guilt, anxiety, depression)
Various subtypes are included:	
1.	Net compulsions (online gambling and online shopping addiction)
2.	Online gaming addiction
3.	Cybersexual addiction
4.	Cyber-related addiction (social working, chat rooms, email and personal messaging addiction)

disorder in DSM-5, is the most accurate criteria to use as a basis for internet addiction.

Ran Tao, director of addiction medicine of the Beijing Military Region Central Hospital, developed a form of diagnostic criteria for identifying internet addiction based on the clinical features of a population of Chinese patients with internet addictions [83]. The proposed internet addiction diagnosis consisted of symptom criterion, clinically significant impairment criterion, course criterion and exclusion criterion (see Table 7.6).

Internet addiction is a compulsive-impulsive spectrum disorder which consists of at least three subtypes: excessive gaming, e-mail/text messaging, and sexual preoccupations. But, only nongambling internet games are included in DSM-V. The diagnostic criteria of internet gaming disorder (see Table 7.7) consist of the following four components: (1) excessive use, often related with a neglect of basic drive or a loss of sense of time, (2) withdrawal, involving feelings of tension, anger, and/or depression when the internet games are inaccessible, (3) tolerance, involving the need for better gaming equipment or more gaming time, and (4) negative repercussions, including lying, poor achievement, arguments, and social isolation.

7.1.2.3 Diagnostic Criteria of Food Addiction

Obesity, which is associated with addictive behavior and often called "food addiction" or "overeating disorder", is a global healthy problem. It is hard to determine whether "food addiction" is an addiction because they lacked psychometrically valid diagnostic

Table 7.6 Proposed internet addiction diagnostic criteria by Tao

(a) Symptom criterion
All the following must be present:
1. Preoccupation with the internet (thinks about previous online activity or anticipates next online session)
2. Withdrawal, as manifested by a dysphoric mood, anxiety, irritability and boredom after several days without internet activity
At least one (or more) of the following:
1. Tolerance, marked increase in internet use required to achieve satisfaction
2. Persistent desire and/or unsuccessful attempts to control, cut back or discontinue internet use
3. Continued excessive use of internet despite knowledge of having a persistent or recurrent physical or psychological problem likely to have been caused or exacerbated by internet use
4. Loss of interests, previous hobbies, entertainment as a direct result of, and with the exception of, internet use
5. Uses the internet to escape or relieve a dysphoric mood (e.g. feelings of helplessness, guilt, anxiety)
(b) Exclusion criterion
Excessive internet use is not better accounted for by psychotic disorders or bipolar I disorder
(c) Clinically significant impairment criterion
Functional impairments (reduced social, academic, working ability), including loss of a significant relationship, job, educational or career opportunities
(d) Course criterion
Duration of internet addiction must have lasted for an excess of 3 months, with at least 6 h of internet usage (non-business/non-academic) per day

criteria or instruments. Researchers in the field of food addiction tried to determine this kind of behavior addiction with different instruments. A structured clinical interview according to the criteria of addiction in DSM-IV was used (see Table 7.8) [21].

The most popularly used scale of food addiction today is standardized Yale Food Addiction Scale (YFAS), which was used almost in all studies related to the concept of food addiction [32]. The YFAS consists of 25 items that can be related with the criteria of substance addiction in DSM-IV. It has shown good convergent validity, good internal reliability and good discriminant validity. By using the YFAS, Gearhardt [37] revealed that the top 3 symptoms of food addiction in adults are:

1. Persistent desire or repeated failures to reduce the amount of food intake.
2. Continued consumption, despite harmful consequences of food.
3. A lot of time spent in trying to reduce the amount of food consumed, as well as a lot of time spent on recovery from overeating.

7.1.2.4 Diagnostic Criteria of Hypersexual Disorder

Hypersexual disorder is proposed as a new psychiatric disorder that would be a part of the Sexual Disorder section for DSM-V. Kafka [45] proposed diagnostic criteria

Table 7.7 DSM-V diagnostic criteria of internet gaming disorder

Persistent and recurrent use of the Internet to engage in games, often with other players, leading to clinically significant impairment or distress as indicated by five (or more) of the following in a 12-month period:	
1.	Preoccupation with Internet games. (The individual thinks about previous gaming activity or anticipates playing the next game; Internet gaming becomes the dominant activity in daily life)
Note: This disorder is distinct from Internet gambling, which is included under gambling disorder	
2.	Withdrawal symptoms when Internet gaming is taken away. (These symptoms are typically described as irritability, anxiety, or sadness, but there are no physical signs of pharmacological withdrawal)
3.	Tolerance—the need to spend increasing amounts of time engaged in Internet games
4.	Unsuccessful attempts to control the participation in Internet games
5.	Loss of interests in previous hobbies and entertainment as a result of, and with the exception of, Internet games
6.	Continued excessive use of Internet games despite knowledge of psychosocial problems
7.	Has deceived family members, therapists, or others regarding the amount of Internet gaming
8.	Use of Internet games to escape or relieve a negative mood (e.g., feelings of helplessness, guilt, anxiety)
9.	Has jeopardized or lost a significant relationship, job, or educational or career opportunity because of participation in Internet games
Note: Only nongambling Internet games are included in this disorder. Use of the Internet for required activities in a business or profession is not included; nor is the disorder intended to include other recreational or social Internet use. Similarly, sexual Internet sites are excluded	

Table 7.8 Modified diagnostic criteria of food addiction based on DSM-IV by Cassin and Ranson

1.	Binge eating larger amount than intended
2.	Continued binge eating despite of knowledge of persistent adverse effects
3.	Persistent desire or unsuccessful efforts to control binge eating
4.	Withdrawal (e.g., restlessness, irritability, headaches)
5.	Great deal of time spent binge eating or recovering from the effects
6.	Tolerance: need to consume more food for desired effect
7.	Important activities given up or reduced because of binge eating

of hypersexual disorder (see Table 7.9) for DSM-5 [45]. The operational criteria for hypersexual was specifically based on elements of two sexual disorders in DSM-IV: Hypoactive sexual Desire Disorder and the Paraphilias. However, hypersexual disorder is defined as a distinct diagnostic category.

Table 7.9 Proposed diagnostic criteria of hypersexual disorder for DSM-V

A.	Over a period of at least 6 months, recurrent and intense sexual fantasies, sexual urges, and sexual behavior in association with four or more of the following five criteria:
1.	Excessive time is consumed by sexual fantasies and urges, and by planning for and engaging in sexual behavior
2.	Repetitively engaging in these sexual fantasies, urges, and behavior in response to dysphoric mood states (e.g., anxiety, depression, boredom, irritability)
3.	Repetitively engaging in sexual fantasies, urges, and behavior in response to stressful life events
4.	Repetitive but unsuccessful efforts to control or significantly reduce these sexual fantasies, urges, and behavior
5.	Repetitively engaging in sexual behavior while disregarding the risk for physical or emotional harm to self or others
B.	There is clinically significant personal distress or impairment in social, occupational or other important areas of functioning associated with the frequency and intensity of these sexual fantasies, urges, and behavior
C.	These sexual fantasies, urges, and behavior are not due to direct physiological effects of exogenous substances (e.g., drugs of abuse or medications) or to Manic Episodes.

7.1.3 Conclusions

Substance and non-substance addictions have many similarities in symptoms including the fact that they may both exhibit chronic, relapsing pattern. As time goes by, both substance and non-substance addictions become less pleasurable (tolerance) and more of a habit or compulsion, or becomes motivated less by positive reinforcement and more by negative reinforcement (e.g., withdrawal). Many individuals with non-substance addictions report a craving state before initiating the behavior, the same as individuals with substance addiction. Additionally, individuals tend to get away from their negative moods by performing addictive behaviors. Many people with non-substance addictions also show a dysphoric state while abstaining from the behaviors, similar to withdrawal of substance addiction. But, there are no reports of physiological prominent or medically serious withdrawals from non-substance addictions. As in substance addiction, financial and marital problems are also common in non-substance addictions. Both substance and non-substance addicts will frequently commit illegal acts, such as embezzlement and theft. All these common features above are critical symptoms that are used to diagnose addictions.

According to the empirical evidence of similarities between substance and non-substance addictions, a syndrome model of addiction that consists of both substance and non-substance addictions was proposed [81]. They suggest that addictions share psychosocial and neurobiological precursors that promote risk for pathogenesis. A premorbid addiction syndrome will develop only if the precursors are accompanied by repeated rewarding engagement in the addictive behavior. Once an addiction becomes manifested, addictions vary in distinctive expression and unique negative consequences. Even though this syndrome model has some natural weakness, it suggests that both substance and non-substance addictions have several similarities.

As noted that different addictions have their own unique feature, which must reflect on the diagnostic criteria, the question arises of whether or not it is suitable to propose a diagnosis of other non-substance addictions based on gambling addiction? Whether there are common diagnostic criteria for all non-substance addictions? What is the difference between substance addictions and non-substance addictions on the diagnosis? Nowadays, proposed diagnostic criteria of behavior addiction have been applied and function well in research. More evidence is needed to support diagnosis.

7.2 Animal Models Evolved to Resemble the Diagnostic Criteria of Addiction

7.2.1 Animal Models Evolved to Resemble the Diagnostic Criteria of Substance Addiction

Thought substance addiction is a human phenomenon, some of the behavioral characteristics of this syndrome can be satisfactorily modeled in animal. In this way, research in animals has contributed to the understanding of the neurobiological basis of substance addiction and of the brain reward system. Simple behavioral models of animals would greatly facilitate the understanding of circuit and molecular mechanisms that promote drug consumption, and could assist in the development of pharmacological therapies to target the compulsive drug intake, which remains intractable to human addiction.

For instance, laboratory models, including animal tissue and live animals, have proven useful for finding molecular targets of the alcohol use disorder and for the discovery of genetic and environmental factors that impact alcohol actions [55]. Evidence from in vivo and in vitro supports that ethanol produces many intoxicating actions by enhancing GABAergic synaptic transmission [40]. Genetic animal models are also important to analyze the genetics of alcohol-related behaviors [60]. High drinkers tend to have low activation of serotonin 5-HT1A or 5-HT2 receptors, which reduces alcohol intake [27, 58]. Blockade of opioid or 5-HT3 receptors also reduces alcohol intake [44, 82].

Substance addiction is clinically defined as substance use disorder by the American Psychiatric Association, in their DSM-5 and dependence syndrome by the World Health Organization in their International Classification of Diseases (ICD-10) [91]. Current clinical definitions of substance addiction are generated by diagnostic criteria applicable to human beings [35, 64]. While most of the symptoms among the diagnostic criteria are psychological, only two are physiological in nature (withdrawal and tolerance). According to current diagnostic criteria, it is impossible to diagnose substance addiction only based on the physiological symptoms while those psychological symptoms are subjective and difficult to measure on animal models. Therefore, argument remains whether animals can be said to be addicted in same way [30, 53]. Similar issues exist for all psychiatric disorders, because of the lack of subjective symptoms [64]. But, larger amount of animal studies related to substance addiction are associated with the behavioral and physiological expression of human addiction. Therefore, animal models of substance addiction diagnosed by current criteria specified applied for animal are meaningful.

7.2.1.1 Appearance of DSM-IV Criteria in Animal Studies of Substance Addiction

Vanderschuren and Ahmed have identified several ways in which those DSM-IV criteria can be researched in animal models, and then described the evidence that these symptoms can be observed in laboratory animals after repeated drug use (see Table 7.10) [85].

In most cases of addiction, the loss of control over drug use is accompanied by a fast increase in drug intake, which is likely to induce neural adaptations into the addicted state [86]. Traditionally, increased drug use over time has been contributed to the occurrence of tolerance or withdrawal symptoms. In contrary, these two symptoms can clearly contribute to the escalation of drug use. In animal research, escalation of drug use has been widely investigated in self-administration settings. A study showed that rats with extended access to cocaine self-administration increased their intake across days gradually, and rats with limited drug access also remained this remarkably stable phenomenon [5], even after several months of testing [6]. Behavioral features of substance addiction, such as increased motivation for the drug [69, 88], and reduced sensitivity to punishment of cocaine seeking [4, 56], have been displayed by rats with a history of escalation of cocaine intake. Escalation of drug self-administration after extended access to drugs has also been showed for other drug, including heroin [7], methylphenidate [57], and methamphetamine [49].

Neurocognitive deficits in substance addiction have been found in a number of studies [14, 36, 38, 70, 78]. These cognitive deficits caused by substance addiction include attention, working memory, planning, impulse control, and decision-making. These deficits contribute to the addiction in some ways. However, it is difficult to distinguish from human studies if these neurocognitive deficits are the cause or consequence of substance addiction. Interestingly, animal studies can make it easy to investigate the relationship between substance addiction and neurocognitive deficits. High impulsivity in rats predicts the vulnerability to nicotine self-administration, alcohol consumption, and cocaine self-administration [15, 29, 31, 71, 73]. Conversely, rats with a period of self-administration of cocaine, metham-

Table 7.10 Appearance of DSM-IV criteria in animal studies of substance addiction

DSM-IV criterion	Behavioral equivalent
1. and 2. Tolerance, withdrawal	Tolerance, escalation of drug use
3. Using more than intended	Impaired control, neurocognitive deficits
4. Difficulty restricting	Resistance to extinction
5. Great deal of time spent	Increased motivation for drug
6. Other activities given up	Drug preference over nondrug rewards
7. Continued use despite problems	Resistance to punishment

phetamine and heroin have been found to have deficits in a variety of cognitive function [19, 28, 59, 80].

Resistance to extinction can be studied in laboratory animals by assessing drug seeking when the drug is not available. Indeed, resistance to extinction has been found in heroin-withdrawal rats with a history of extended access to heroin self-administration [7, 52]. Studies from animals showed that the factors which seem to determine the sensitivity to extinction may be the ways of self-administration and the length of withdrawal [79, 93].

Motivation for drug intake in animals is usually studied by using a progressive ratio schedule of reinforcement, in which animals have to make an increasing number of responses for every subsequent reward [43, 77]. By using this schedule, the motivation of animals' drug intake can be well documented. Animals with a history of escalated cocaine intake were shown to respond more than animals that had limited cocaine access [8, 42, 69, 88]. This effect has subsequently been shown for other substance addictions, including heroin [51] and methamphetamine [87]. Interestingly, a study [66] from animal showed that escalation of cocaine use increased the motivation for cocaine at high unit doses, but reduced the motivation at low doses, suggesting that animals with a history of escalated cocaine use are willing to pay a high price for large amount of the drug. Increases in the motivation for cocaine have also been shown in rats with prolonged cocaine self-administration [16, 30].

As a consequence of drug preference, important social, occupational, or recreational activities are given up or reduced because of drug use, which is one of the core behavioral symptoms of substance addiction. In animal studies, drug preference can be researched by giving other behavioral options during drug access [2, 3]. Chimpanzees preferred morphine over a piece of fresh fruit during drug withdrawal but otherwise preferred fruit [63]. Subsequent studies found that drug preference in animals was dose-dependent [62, 68] and surmountable by increasing the value of the nondrug reward [61]. After extended access to cocaine self-administration, a subgroup of rats continued to prefer drug over sweet water, and when sweet water was the only option available, cocaine-preferring rats drank as fast and as much as nondrug-preferring rats [33]. Interestingly, this subgroup of cocaine-preferring rats continued to take cocaine even when hungry [20]. This subgroup of cocaine-preferring rats may represent severe stage in the transition to substance addiction.

One of the most important features of substance addiction is continued drug intake despite the knowledge of negative consequence. Animal experiment emulated this feature usually use punishment setups, in which seeking or taking drugs was paired with punishment of footshocks or quinine [41, 48]. Rats with limited ethanol experience did reduce their intake of ethanol when quinine was added, and in contrast, rats with extended ethanol intake reduced a little bit or did not reduce their intake [90]. Studies have shown that the intake of drug can become insensitive to punishment after prolonged drug experience.

Based on the DSM-IV diagnostic criteria for substance addiction, some symptoms of substance addiction have been shown to occur in laboratory animals. This data from animals indicates that substance addiction can occur and be researched in laboratory animals. This gives us the opportunity to study the neural and genetic background of substance addiction in animals. We hope these studies of animal would lead to the development of more effective treatments for substance addictions.

There are several questions that need to be addressed in future research. First, the occurrence of only one symptom of substance addiction is not enough for an addiction diagnosis. As a general estimate, the presence of two or more symptoms needs to be met for an addiction based on DSM-5. Thus, it is important to determine how many symptoms should be met together for an animal model of substance addiction. Second, it is important to investigate whether all the symptoms of addiction can occur in animal models or whether some symptoms are specific to humans. Lastly, addictive drugs come from different pharmacological classes and their relative addictive potential varies, and their reinforcing strength depends on environmental factors. Thus, we need to indicate the extent to which the symptoms and their underlying neural, genetic, and environmental factors for substance addiction in general, or drug specific.

7.2.2 Animal Models Evolved to Resemble the Diagnostic Criteria of Gambling Disorder

As we all know, animal models of substance addiction have made a great contribution to the understanding of substance addiction, including symptoms, neural, genetic and environmental factors. Therefore, it comes to us directly that animal models will also contribute to the understanding of non-substance addiction. But, it is still a big challenge to build a valid animal model of non-substance addiction. Substance addictions have a real substance to be addicted to easily for animals. Though substance addiction is unique to humans in nature, we can still build animal models of substance addiction by escalated drug experience. However, it is difficult for non-substance addiction in the same way, because some addictive behaviors are incomprehensible to animals, such as internet surfing. As gambling disorder is the only one formal disorder of non-substance addiction (non-substance-related disorder) involved in DSM-V, we will discuss the application of an animal model to gambling disorder emphatically.

Animal models of gambling disorder are crucial as they allow the dissection of processes and factors related to normal and pathological gambling in a controlled way. What's more, animal models have an alternative perspective because they make it possible to use approaches that are impossible with humans, as in the case of in vivo transgenic ways that allow us to modulate expression of target genes in relevant brain areas.

7.2.2.1 Paradigms Built to Study Gambling Proneness in Animal Models

Many operant paradigms have been built to study gambling proneness in animal models. The rodent Slot Machine Task (rSMT) allows us to assess if the animal discriminates a complete signal from a nearly complete one. By using this task, it has been indicated that rats are susceptible to putative-win signals in non-winning trials [26, 89]. Such a phenomenon is well known as the "near-miss effect", one of the specific consequences of the gambling disorder [76]. The rodent Betting Task (rBT) offers the choice between a small, safe food reward or a larger food reward associated with the possibility of nothing [24]. The Probabilistic-Delivery Task (PDT) is based on a choice between either a small, certain food reward or larger food reward (or not) depending on a given (and progressively decreasing) probability [1]. The rodent Iowa Gambling Task (rIGT) involves the options between a low probability of a large reward and a high probability of a small reward, which is a widely used task for the human gambling behavior [84].

In rSMT, responding on the collect lever resulted in reward delivery only on win trials (3 lights were illuminated); on every other trial type, this response resulted in the punishment of a time out. Animals clearly distinguished win trials from most loss trials, responding on collect lever on 100% of win trials, but only 15–35% of loss trials in which 1 or 0 lights were illuminated. In contrast, animals responded on correct lever on 50–80% of loss trials in which 2 lights were illuminated, suggesting that animals treated these 2-lights trials as more similar to wins than losses [89]. So, this behavioral pattern is extremely similar to the near-miss effects in humans. Data from rSMT, similar to human data, near-misses appeared to promote reward expectancy and game play in rats [46, 76]. Based on human data, we know that the dopamine system has been implicated in the experience of the near-miss effect. Also, investigation of the effect of different dopaminergic agents on performance of the rSMT has been done. The D2-like receptor agonist quinpirole dose-dependent added choice of the collect lever on all loss trials. The D4 receptor agonist PD168077 caused significant deficits in performance, whereas the D4 receptor antagonist L-745,870 caused a parallel improvement in optimal lever choice [26]. These results indicate that reward expectancy on the rSMT is critically depended on the activity of D4 receptors. Temporary inactivation of the ACC using a GABA agonist mixture caused robust deficits in animals' performance, suggesting intact ACC function is vital in this task [25].

The rodent Betting Task was designed to investigate which changes in bet size affect risk tolerance [24]. Altering the amount wagered is a common characteristic of human decision making tasks that purport to assess gambling-like processes. Changing the amount wagered alters willingness to gamble even when the reward remains constant. In the rBT, a choice has to be made between a small, 100% certain reward and a large, 50% uncertain reward. Although the bet size varies between blocks from 1 to 3 pellets, this will not change the relative values of the safe versus uncertain outcomes. Data from this task showed that the bet size had a major impact on the uncertain option, seeing a small but significant decrease in preference for uncertainty as bet size increases [24]. A recent study indicated that rats classified as "wager-sensitive", according to their preference for a larger reward, slightly preferred the uncertain option consistently [24]. Amphetamine increased the choice of the uncertain option only in "wager-sensitive" rats, but not in "wager-insensitive" rats; whereas a D2/3 receptor antagonist decreased selection of an uncertain lever in "wager-insensitive" rats alone. Micro-PET and autoradiography using [^{11}C]raclopride confirmed a significant correlation between high wager sensitivity and low striatal D2/3 receptor density. The authors indicated a clear association between striatal dopaminergic transmission and the sensitivity to bet size, which they posit was linked to human pathological gambling.

7.2.2.2 Proposed Improvements for Animal Models of Gambling Disorder

Though the animal model has facilitated a better understanding of the human gambling behavior, it is important to point out that these animal models have one fundamental issue that has to be solved for future studies [23]. All these tasks only model the gambling behavior, but none of them actually model the gambling disorder. Drug addicted states are produced in animal models by using a paradigm such as self-administration or conditioned place preference. These models need to reach some of the symptoms of substance addiction before being called an addiction model. The most important thing for an animal model of the gambling disorder is to build a valid paradigm based on the diagnostic criteria of the disorder.

Conditioned place preference (CPP) happens when an individual comes to prefer on place more than others due to the preferred place being previously paired with reward. The CPP is widely used to explore the motivational effect of pharmacological stimuli. It plays an important role in the study of the addictive drug. An important promotion to the experimental use of CPP was the observation that rats returned to the chamber where they had received electrical brain stimulation as a reward [65]. Subsequently, CPP became a screening tool for addictive drugs [47]. The most accessible explanation of CPP is the theory of incentive-driven based on Pavlovian conditioning [75]. In this paradigm, the reinforcement of an unconditioned stimulus (US) such as food or drug that the animals "like" or "want" [18], is associated with the stimulus properties of the place, which became conditioned stimuli (CS). Consequently, the CS relates to "incentive value" which leads to the animals to "prefer" them [11, 74]. This is the learning about the association between the rewarding stimulus and the cues in the paired compartment. When an animal prefers a place related to the drug, it may mean that the animal increases its craving and motivation for the drug, and spends a great deal of time craving for the drug, which is a critical symptom of addiction diagnosis based on the DSM-5. Recently, the extinction/reinstatement model has been applied to the conditioned place preference. This model refers to the reinstates drug-paired conditioning stimuli by noncontingent drug exposure after extinction [12]. This phenomenon is associated with drug seeking, involving relapse and craving [22], which are also critical symptoms of the substance addiction diagnosis.

Drug self-administration procedures provide a tool for studying substance addiction under controlled conditions in the laboratory. Under these procedures, an animal performs a response that delivers a dose of drug. These procedures have a high degree of face validity, because they offer the most direct point-to-point correspondence with addictive behavior that happens in nature [67]. The most basic assumption of this paradigm is that drugs, and reinforcers, increase the likelihood of the behavior [34]. Therefore, drug-self-administration is viewed as an operant response produced by classical conditioning. It is assumed that this classical conditioning contributes to substance addiction in two ways [67]. Firstly, stimuli that have been related to the drug can become conditioned reinforcers. Secondly, stimuli associated with the drug can produce conditioned responses that are motivational in nature. Drug self-administration is the most widely used paradigm for addiction, because several diagnostic criteria of substance addiction were met, including escalation of use, difficulty stopping, and continued use despite adverse consequences. With drug self-administration paradigm, rats given extended access to drugs typically increase the amount of drug intake over time [54]. This escalation appears to be associated with a loss of control over intake [50]. Rats given extended access to drugs may take longer to stop drug-seeking when drug delivery is discontinued [7, 72]. Drug addicted rats are also insensitive to aversive consequences and keep responding to the lever when footshock is delivered only [30].

There are two general strategies to design animal models of substance addiction [17]. First, the model has to meet a specific symptom, a neurobiological or psychological characteristic of the substance addiction, such as the escalation of drug intake, resistance to punishment, habitual instrumental performance, high motivation for the drug, impaired cognitive flexibility, or vulnerability to relapse. Indeed, these kinds of models that focus on defined characteristic of substance addictions provide a powerful framework for studying underlying brain mechanisms involved in a specific aspect of the pathology. Secondly, the models try to incorporate several symptoms of substance addiction in humans, providing powerful tools for longitudinal studies or testing pharmacological treatments. In humans, to be diagnosed as addicted, an individual must meet 3 out of 7 diagnostic criteria of substance addiction based on the DSM-IV and 2 out of 11 based on the DSM-5 over the last 12 months. This approach forms the basis of new pre-clinical animal models. This model can capture the multi-symptomatic nature of substance addiction.

7.2.3 Conclusions

New animal models of addiction may promote to identify the neuropharmacologic and molecular mechanisms underlying the addictive-like behavior, developing new affective treatments for addiction by filling the gaps between preclinical and clinical studies.

7.3 Conclusions

Addiction, a neuropsychiatric disorder, is currently identified as syntheses of its behavioral symptoms. Although the addiction diagnosis is according to the summation of symptoms, it is not quite right to think that summing up neurobiological correlates of animal models pertaining to independent symptoms could lead to the understanding of the core pathophysiologic mechanism of addiction. The challenge for future animal models of addiction is to develop a valid approach for symptoms and encapsulate psychological constructs of addiction that can be measured in clinical studies.

Acknowledgments This work was supported by the National Natural Science Foundation of China (31171083, 31230032, 31471071, 31771221), and the Fundamental Research Funds for the Central Universities of China, the National Key Basic Research Program (2016YFA0400900).

Appendices (Tables 7.11, 7.12, 7.13, 7.14, 7.15, 7.16, 7.17, 7.18, and 7.19)

Table 7.11 Diagnostic criteria of alcohol use disorder

A problematic pattern of alcohol use leading to clinically significant impairment or distress, as manifested by at least two of the following, occurring within a 12-month period:		
1.	Alcohol is often taken in larger amounts or over a longer period than was intended	
2.	There is a persistent desire or unsuccessful efforts to cut down or control alcohol use	
3.	A great deal of time is spent in activities necessary to obtain alcohol, use alcohol, or recover from its effects	
4.	Craving, or a strong desire or urge to use alcohol	
5.	Recurrent alcohol use resulting in a failure to fulfill major role obligations at work, school, or home	
6.	Continued alcohol use despite having persistent or recurrent social or interpersonal problems caused or exacerbated by the effects of alcohol	
7.	Important social, occupational, or recreational activities are given up or reduced because of alcohol use	
8.	Recurrent alcohol use in situations in which it is physically hazardous	
9.	Alcohol use is continued despite knowledge of having a persistent or recurrent physical or psychological problem that is likely to have been caused or exacerbated by alcohol	
10.	Tolerance, as defined by either of the following:	
	(a)	A need for markedly increased amounts of alcohol to achieve intoxication or desired effect
	(b)	A markedly diminished effect with continued use of the same amount of alcohol
11.	Withdrawal, as manifested by either of the following:	
	(a)	The characteristic withdrawal syndrome for alcohol
	(b)	Alcohol (or a closely related substance, such as a benzodiazepine) is taken to relieve or avoid withdrawal symptoms

Table 7.12 Diagnostic criteria of cannabis use disorder

A problematic pattern of cannabis use leading to clinically significant impairment or distress, as manifested by at least two of the following, occurring within a 12-month period:	
1.	Cannabis is often taken in larger amounts or over a longer period than was intended
2.	There is a persistent desire or unsuccessful efforts to cut down or control cannabis use
3.	A great deal of time is spent in activities necessary to obtain cannabis, use cannabis, or recover from its effects
4.	Craving, or a strong desire or urge to use cannabis
5.	Recurrent cannabis use resulting in a failure to fulfill major role obligations at work, school, or home
6.	Continued cannabis use despite having persistent or recurrent social or interpersonal problems caused or exacerbated by the effects of cannabis
7.	Important social, occupational, or recreational activities are given up or reduced because of cannabis use
8.	Recurrent cannabis use in situations in which it is physically hazardous
9.	Cannabis use is continued despite knowledge of having a persistent or recurrent physical or psychological problem that is likely to have been caused or exacerbated by cannabis
10.	Tolerance, as defined by either of the following:
	(a) A need for markedly increased amounts of cannabis to achieve intoxication or desired effect
	(b) Markedly diminished effect with continued use of the same amount of cannabis
11.	Withdrawal, as manifested by either of the following:
	(a) The characteristic withdrawal syndrome for cannabis
	(b) Cannabis (or a closely related substance) is taken to relieve or avoid withdrawal symptoms

Table 7.13 Diagnostic criteria of phencyclidine use disorder

A pattern of phencyclidine (or a pharmacologically similar substance) use leading to clinically significant impairment or distress, as manifested by at least two of the following, occurring within a 12-month period:	
1.	Phencyclidine is often taken in larger amounts or over a longer period than was intended
2.	There is a persistent desire or unsuccessful efforts to cut down or control phencyclidine use
3.	A great deal of time is spent in activities necessary to obtain phencyclidine, use the phencyclidine, or recover from its effects
4.	Craving, or a strong desire or urge to use phencyclidine
5.	Recurrent phencyclidine use resulting in a failure to fulfill major role obligations at work, school, or home (e.g., repeated absences from work or poor work performance related to phencyclidine use; phencyclidine-related absences, suspensions, or expulsions from school; neglect of children or household)
6.	Continued phencyclidine use despite having persistent or recurrent social or interpersonal problems caused or exacerbated by the effects of the phencyclidine (e.g., arguments with a spouse about consequences of intoxication; physical fights)
7.	Important social, occupational, or recreational activities are given up or reduced because of phencyclidine use
8.	Recurrent phencyclidine use in situations in which it is physically hazardous (e.g., driving an automobile or operating a machine when impaired by a phencyclidine)

(continued)

Table 7.13 (continued)

9.	Phencyclidine use is continued despite knowledge of having a persistent or recurrent physical or psychological problem that is likely to have been caused or exacerbated by the phencyclidine
10.	Tolerance, as defined by either of the following:
	(a) A need for markedly increased amounts of the phencyclidine to achieve intoxication or desired effect
	(b) A markedly diminished effect with continued use of the same amount of the phencyclidine
Note: Withdrawal symptoms and signs are not established for phencyclidines, and so this criterion does not apply. (Withdrawal from phencyclidines has been reported in animals but not documented in human users)	

Table 7.14 Diagnostic criteria of other hallucinogen use disorder

A problematic pattern of hallucinogen (other than phencyclidine) use leading to clinically significant impairment or distress, as manifested by at least two of the following, occurring within a 12-month period:	
1.	The hallucinogen is often taken in larger amounts or over a longer period than was intended
2.	There is a persistent desire or unsuccessful efforts to cut down or control hallucinogen use
3.	A great deal of time is spent in activities necessary to obtain the hallucinogen, use the hallucinogen, or recover from its effects
4.	Craving, or a strong desire or urge to use the hallucinogen
5.	Recurrent hallucinogen use resulting in a failure to fulfill major role obligations at work, school, or home (e.g., repeated absences from work or poor work performance related to hallucinogen use; hallucinogen-related absences, suspensions, or expulsions from school; neglect of children or household)
6.	Continued hallucinogen use despite having persistent or recurrent social or interpersonal problems caused or exacerbated by the effects of the hallucinogen (e.g., arguments with a spouse about consequences of intoxication; physical fights)
7.	Important social, occupational, or recreational activities are given up or reduced because of hallucinogen use
8.	Recurrent hallucinogen use in situations in which it is physically hazardous (e.g., driving an automobile or operating a machine when impaired by the hallucinogen)
9.	Hallucinogen use is continued despite knowledge of having a persistent or recurrent physical or psychological problem that is likely to have been caused or exacerbated by the hallucinogen
10.	Tolerance, as defined by either of the following:
	(a) A need for markedly increased amounts of the hallucinogen to achieve intoxication or desired effect
	(b) A markedly diminished effect with continued use of the same amount of the hallucinogen
Note: Withdrawal symptoms and signs are not established for hallucinogens, and so this criterion does not apply	

Table 7.15 Diagnostic criteria of Inhalant use disorder

A problematic pattern of use of a hydrocarbon-based inhalant substance leading to clinically significant impairment or distress, as manifested by at least two of the following, occurring within a 12-month period:
1. The inhalant substance is often taken in larger amounts or over a longer period than was intended
2. There is a persistent desire or unsuccessful efforts to cut down or control use of the inhalant substance
3. A great deal of time is spent in activities necessary to obtain the inhalant substance, use it, or recover from its effects
4. Craving, or a strong desire or urge to use the inhalant substance
5. Recurrent use of the inhalant substance resulting in a failure to fulfill major role obligations at work, school, or home
6. Continued use of the inhalant substance despite having persistent or recurrent social or interpersonal problems caused or exacerbated by the effects of its use
7. Important social, occupational, or recreational activities are given up or reduced because of use of the inhalant substance
8. Recurrent use of the inhalant substance in situations in which it is physically hazardous
9. Use of the inhalant substance is continued despite knowledge of having a persistent or recurrent physical or psychological problem that is likely to have been caused or exacerbated by the substance
10. Tolerance, as defined by either of the following:
(a) A need for markedly increased amounts of the inhalant substance to achieve intoxication or desired effect
(b) A markedly diminished effect with continued use of the same amount of the inhalant substance
Specify the particular inhalant: When possible, the particular substance involved should be named (e.g., "solvent use disorder")

Table 7.16 Diagnostic criteria of other Inhalant use disorder

A problematic pattern of opioid use leading to clinically significant impairment or distress, as manifested by at least two of the following, occurring within a 12-month period:
1. Opioids are often taken in larger amounts or over a longer period than as intended
2. There is a persistent desire or unsuccessful efforts to cut down or control opioid use
3. A great deal of time is spent in activities necessary to obtain the opioid, use the opioid, or recover from its effects
4. Craving, or a strong desire or urge to use opioids
5. Recurrent opioid use resulting in a failure to fulfill major role obligations at work, school, or home
6. Continued opioid use despite having persistent or recurrent social or interpersonal problems caused or exacerbated by the effects of opioids
7. Important social, occupational, or recreational activities are given up or reduced because of opioid use
8. Recurrent opioid use in situations in which it is physically hazardous
9. Continued opioid use despite knowledge of having a persistent or recurrent physical or psychological problem that is likely to have been caused or exacerbated by the substance

(continued)

Table 7.16 (continued)

10. Tolerance, as defined by either of the following:	
(a)	A need for markedly increased amounts of opioids to achieve intoxication or desired effect
(b)	A markedly diminished effect with continued use of the same amount of an opioid
Note: This criterion is not considered to be met for those taking opioids solely under appropriate medical supervision	
11. Withdrawal, as manifested by either of the following:	
(a)	The characteristic opioid withdrawal syndrome
(b)	Opioids (or a closely related substance) are taken to relieve or avoid withdrawal symptoms
Note: This criterion is not considered to be met for those individuals taking opioids solely under appropriate medical supervision	

Table 7.17 Diagnostic criteria of sedative, hypnotic, or anxiolytic use disorder

A problematic pattern of sedative, hypnotic, or anxiolytic use leading to clinically significant impairment or distress, as manifested by at least two of the following, occurring within a 12-month period:	
1.	Sedatives, hypnotics, or anxiolytics are often taken in larger amounts or over a longer period than was intended
2.	There is a persistent desire or unsuccessful efforts to cut down or control sedative, hypnotic, or anxiolytic use
3.	A great deal of time is spent in activities necessary to obtain the sedative, hypnotic, or anxiolytic; use the sedative, hypnotic, or anxiolytic; or recover from its effects
4.	Craving, or a strong desire or urge to use the sedative, hypnotic, or anxiolytic
5.	Recurrent sedative, hypnotic, or anxiolytic use resulting in a failure to fulfill major role obligations at work, school, or home (e.g., repeated absences from work or poor work performance related to sedative, hypnotic, or anxiolytic use; sedative-, hypnotic-, or anxiolytic-related absences, suspensions, or expulsions from school; neglect of children or household)
6.	Continued sedative, hypnotic, or anxiolytic use despite having persistent or recurrent social or interpersonal problems caused or exacerbated by the effects of sedatives, hypnotics, or anxiolytics (e.g., arguments with a spouse about consequences of intoxication; physical fights)
7.	Important social, occupational, or recreational activities are given up or reduced because of sedative, hypnotic, or anxiolytic use
8.	Recurrent sedative, hypnotic, or anxiolytic use in situations in which it is physically hazardous (e.g., driving an automobile or operating a machine when impaired by sedative, hypnotic, or anxiolytic use)
9.	Sedative, hypnotic, or anxiolytic use is continued despite knowledge of having a persistent or recurrent physical or psychological problem that is likely to have been caused or exacerbated by the sedative, hypnotic, or anxiolytic
10.	Tolerance, as defined by either of the following:
(a)	A need for markedly increased amounts of the sedative, hypnotic, or anxiolytic to achieve intoxication or desired effect
(b)	A markedly diminished effect with continued use of the same amount of the sedative, hypnotic, or anxiolytic

(continued)

Table 7.17 (continued)

Note: This criterion is not considered to be met for individuals taking sedatives, hypnotics, or anxiolytics under medical supervision	
11.	Withdrawal, as manifested by either of the following:
	(a) The characteristic withdrawal syndrome for sedatives, hypnotics, or anxiolytics
	(b) Sedatives, hypnotics, or anxiolytics (or a closely related substance, such as alcohol) are taken to relieve or avoid withdrawal symptoms
Note: This criterion is not considered to be met for individuals taking sedatives, hypnotics, or anxiolytics under medical supervision	

Table 7.18 Diagnostic criteria of stimulant use disorder

A pattern of amphetamine-type substance, cocaine, or other stimulant use leading to clinically significant impairment or distress, as manifested by at least two of the following, occurring within a 12-month period:	
1.	The stimulant is often taken in larger amounts or over a longer period than was intended
2.	There is a persistent desire or unsuccessful efforts to cut down or control stimulant use
3.	A great deal of time is spent in activities necessary to obtain the stimulant, use the stimulant, or recover from its effects
4.	Craving, or a strong desire or urge to use the stimulant
5.	Recurrent stimulant use resulting in a failure to fulfill major role obligations at work, school, or home
6.	Continued stimulant use despite having persistent or recurrent social or interpersonal problems caused or exacerbated by the effects of the stimulant
7.	Important social, occupational, or recreational activities are given up or reduced because of stimulant use
8.	Recurrent stimulant use in situations in which it is physically hazardous
9.	Stimulant use is continued despite knowledge of having a persistent or recurrent physical or psychological problem that is likely to have been caused or exacerbated by the stimulant
10.	Tolerance, as defined by either of the following:
	(a) A need for markedly increased amounts of the stimulant to achieve intoxication or desired effect
	(b) A markedly diminished effect with continued use of the same amount of the stimulant
Note: This criterion is not considered to be met for those taking stimulant medications solely under appropriate medical supervision, such as medications for attention-deficit/hyperactivity disorder or narcolepsy	
11.	Withdrawal, as manifested by either of the following:
	(a) The characteristic withdrawal syndrome for the stimulant
	(b) The stimulant (or a closely related substance) is taken to relieve or avoid withdrawal symptoms
Note: This criterion is not considered to be met for those taking stimulant medications solely under appropriate medical supervision, such as medications for attention-deficit/hyperactivity disorder or narcolepsy	

Table 7.19 Diagnostic criteria of tobacco use disorder

A problematic pattern of tobacco use leading to clinically significant impairment or distress, as manifested by at least two of the following, occurring within a 12-month period:	
1.	Tobacco is often taken in larger amounts or over a longer period than was intended
2.	There is a persistent desire or unsuccessful efforts to cut down or control tobacco use
3.	A great deal of time is spent in activities necessary to obtain or use tobacco
4.	Craving, or a strong desire or urge to use tobacco
5.	Recurrent tobacco use resulting in a failure to fulfill major role obligations at work, school, or home (e.g., interference with work)
6.	Continued tobacco use despite having persistent or recurrent social or interpersonal problems caused or exacerbated by the effects of tobacco (e.g., arguments with others about tobacco use)
7.	Important social, occupational, or recreational activities are given up or reduced because of tobacco use
8.	Recurrent tobacco use in situations in which it is physically hazardous (e.g., smoking in bed)
9.	Tobacco use is continued despite knowledge of having a persistent or recurrent physical or psychological problem that is likely to have been caused or exacerbated by tobacco
10.	Tolerance, as defined by either of the following:
	(a) A need for markedly increased amounts of tobacco to achieve the desired effect
	(b) A markedly diminished effect with continued use of the same amount of tobacco
11.	Withdrawal, as manifested by either of the following:
	(a) The characteristic withdrawal syndrome for tobacco
	(b) Tobacco (or a closely related substance, such as nicotine) is taken to relieve or avoid withdrawal symptoms

References

1. Adriani W, Laviola G (2006) Delay aversion but preference for large and rare rewards in two choice tasks: implications for the measurement of self-control parameters. BMC Neurosci 7:52
2. Ahmed SH (2005) Imbalance between drug and non-drug reward availability: a major risk factor for addiction. Eur J Pharmacol 526:9–20
3. Ahmed SH (2010) Validation crisis in animal models of drug addiction: beyond non-disordered drug use toward drug addiction. Neurosci Biobehav Rev 35:172–184
4. Ahmed SH, Cador M (2006) Dissociation of psychomotor sensitization from compulsive cocaine consumption. Neuropsychopharmacology 31:563–571
5. Ahmed SH, Koob GF (1998) Transition from moderate to excessive drug intake: change in hedonic set point. Science 282:298–300
6. Ahmed SH, Koob GF (1999) Long-lasting increase in the set point for cocaine self-administration after escalation in rats. Psychopharmacology 146:303–312
7. Ahmed SH, Walker JR, Koob GF (2000) Persistent increase in the motivation to take heroin in rats with a history of drug escalation. Neuropsychopharmacology 22:413–421
8. Allen RM, Dykstra LA, Carelli RM (2007) Continuous exposure to the competitive N-methyl-D: -aspartate receptor antagonist, LY235959, facilitates escalation of cocaine consumption in Sprague-Dawley rats. Psychopharmacology 191:341–351

9. American Psychiatric Association (1994) Diagnostic and statistical manual for mental disorders, 4th edn. APA, Washington
10. American Psychiatric Association (2013) Diagnostic and statistical manual for mental disorders, 5th edn. APA, Washington
11. Bardo MT, Bevins RA (2000) Conditioned place preference: what does it add to our preclinical understanding of drug reward? Psychopharmacology 153:31–43
12. Bardo MT, Miller JS, Neisewander JL (1984) Conditioned place preference with morphine: the effect of extinction training on the reinforcing CR. Pharmacol Biochem Behav 21:545–549
13. Beard KW, Wolf EM (2001) Modification in the proposed diagnostic criteria for Internet addiction. Cyberpsychol Behav 4:377–383
14. Bechara A (2005) Decision making, impulse control and loss of willpower to resist drugs: a neurocognitive perspective. Nat Neurosci 8:1458–1463
15. Belin D, Mar AC, Dalley JW, Robbins TW, Everitt BJ (2008) High impulsivity predicts the switch to compulsive cocaine-taking. Science 320:1352–1355
16. Belin D, Balado E, Piazza PV, Deroche-Gamonet V (2009) Pattern of intake and drug craving predict the development of cocaine addiction-like behavior in rats. Biol Psychiatry 65:863–868
17. Belin-Rauscent A, Belin D (2012) Animal models of drug addiction. InTech, Rijeka
18. Berridge KC (2007) The debate over dopamine's role in reward: the case for incentive salience. Psychopharmacology 191:391–431
19. Briand LA, Flagel SB, Garcia-Fuster MJ, Watson SJ, Akil H, Sarter M, Robinson TE (2008) Persistent alterations in cognitive function and prefrontal dopamine D2 receptors following extended, but not limited, access to self-administered cocaine. Neuropsychopharmacology 33:2969–2980
20. Cantin L, Lenoir M, Augier E, Vanhille N, Dubreucq S, Serre F, Vouillac C, Ahmed SH (2010) Cocaine is low on the value ladder of rats: possible evidence for resilience to addiction. PLoS One 5:e11592
21. Cassin SE, von Ranson KM (2007) Is binge eating experienced as an addiction? Appetite 49:687–690
22. Childress AR, McLellan AT, O'Brien CP (1986) Abstinent opiate abusers exhibit conditioned craving, conditioned withdrawal and reductions in both through extinction. Br J Addict 81:655–660
23. Cocker PJ, Winstanley CA (2015) Irrational beliefs, biases and gambling: exploring the role of animal models in elucidating vulnerabilities for the development of pathological gambling. Behav Brain Res 279:259–273
24. Cocker PJ, Dinelle K, Kornelson R, Sossi V, Winstanley CA (2012) Irrational choice under uncertainty correlates with lower striatal D(2/3) receptor binding in rats. J Neurosci 32:15450–15457
25. Cocker P, Hosking J, Winstanley C (2014a) Dopamine D4 receptors in the anterior cingulate cortex appear to modulate reward expectancy on a rodent slot machine task. Int J Neuropsychopharmacol 17:47–47
26. Cocker PJ, Le Foll B, Rogers RD, Winstanley CA (2014b) A selective role for dopamine D(4) receptors in modulating reward expectancy in a rodent slot machine task. Biol Psychiatry 75:817–824
27. Crabbe JC, Phillips TJ (2004) Pharmacogenetic studies of alcohol self-administration and withdrawal. Psychopharmacology 174:539–560
28. Dalley JW, Laane K, Pena Y, Theobald DE, Everitt BJ, Robbins TW (2005) Attentional and motivational deficits in rats withdrawn from intravenous self-administration of cocaine or heroin. Psychopharmacology 182:579–587
29. Dalley JW, Fryer TD, Brichard L, Robinson ES, Theobald DE, Laane K, Pena Y, Murphy ER, Shah Y, Probst K, Abakumova I, Aigbirhio FI, Richards HK, Hong Y, Baron JC, Everitt BJ, Robbins TW (2007) Nucleus accumbens D2/3 receptors predict trait impulsivity and cocaine reinforcement. Science 315:1267–1270

30. Deroche-Gamonet V, Belin D, Piazza PV (2004) Evidence for addiction-like behavior in the rat. Science 305:1014–1017
31. Diergaarde L, Pattij T, Poortvliet I, Hogenboom F, de Vries W, Schoffelmeer AN, De Vries TJ (2008) Impulsive choice and impulsive action predict vulnerability to distinct stages of nicotine seeking in rats. Biol Psychiatry 63:301–308
32. Dimitrijevic I, Popovic N, Sabljak V, Skodric-Trifunovic V, Dimitrijevic N (2015) Food addiction-diagnosis and treatment. Psychiatr Danub 27:101–106
33. Domingos AI, Vaynshteyn J, Voss HU, Ren X, Gradinaru V, Zang F, Deisseroth K, de Araujo IE, Friedman J (2011) Leptin regulates the reward value of nutrient. Nat Neurosci 14:1562–1568
34. Everitt BJ, Robbins TW (2000) Second-order schedules of drug reinforcement in rats and monkeys: measurement of reinforcing efficacy and drug-seeking behaviour. Psychopharmacology 153:17–30
35. Fernando AB, Robbins TW (2011) Animal models of neuropsychiatric disorders. Annu Rev Clin Psychol 7:39–61
36. Garavan H, Stout JC (2005) Neurocognitive insights into substance abuse. Trends Cogn Sci 9:195–201
37. Gearhardt AN, Corbin WR, Brownell KD (2009) Preliminary validation of the Yale food addiction scale. Appetite 52:430–436
38. Goldstein RZ, Craig AD, Bechara A, Garavan H, Childress AR, Paulus MP, Volkow ND (2009) The neurocircuitry of impaired insight in drug addiction. Trends Cogn Sci 13:372–380
39. Goodman A (1990) Addiction – definition and implications. Br J Addict 85:1403–1408
40. Grobin AC, Matthews DB, Devaud LL, Morrow AL (1998) The role of GABA(A) receptors in the acute and chronic effects of ethanol. Psychopharmacology 139:2–19
41. Grove RN, Schuster CR (1974) Suppression of cocaine self-administration by extinction and punishment. Pharmacol Biochem Behav 2:199–208
42. Hao Y, Martin-Fardon R, Weiss F (2010) Behavioral and functional evidence of metabotropic glutamate receptor 2/3 and metabotropic glutamate receptor 5 dysregulation in cocaine-escalated rats: factor in the transition to dependence. Biol Psychiatry 68:240–248
43. Hodos W (1961) Progressive ratio as a measure of reward strength. Science 134:943–944
44. Johnson BA (2000) Serotonergic agents and alcoholism treatment: rebirth of the subtype concept – an hypothesis. Alcohol Clin Exp Res 24:1597–1601
45. Kafka MP (2010) Hypersexual disorder: a proposed diagnosis for DSM-V. Arch Sex Behav 39:377–400
46. Kassinove JI, Schare ML (2001) Effects of the "near miss" and the "big win" on persistence at slot machine gambling. Psychol Addict Behav 15:155–158
47. Katz RJ, Gormezano G (1979) A rapid and inexpensive technique for assessing the reinforcing effects of opiate drugs. Pharmacol Biochem Behav 11:231–233
48. Kearns DN, Weiss SJ, Panlilio LV (2002) Conditioned suppression of behavior maintained by cocaine self-administration. Drug Alcohol Depend 65:253–261
49. Kitamura O, Wee S, Specio SE, Koob GF, Pulvirenti L (2006) Escalation of methamphetamine self-administration in rats: a dose-effect function. Psychopharmacology 186:48–53
50. Koob GF, Ahmed SH, Boutrel B, Chen SA, Kenny PJ, Markou A, O'Dell LE, Parsons LH, Sanna PP (2004) Neurobiological mechanisms in the transition from drug use to drug dependence. Neurosci Biobehav Rev 27:739–749
51. Lenoir M, Ahmed SH (2007) Heroin-induced reinstatement is specific to compulsive heroin use and dissociable from heroin reward and sensitization. Neuropsychopharmacology 32:616–624
52. Lenoir M, Serre F, Cantin L, Ahmed SH (2007) Intense sweetness surpasses cocaine reward. PLoS One 2:e698
53. Lester D, Freed EX (1973) Criteria for an animal model of alcoholism. Pharmacol Biochem Behav 1:103–107
54. Liu Y, Roberts DC, Morgan D (2005) Effects of extended-access self-administration and deprivation on breakpoints maintained by cocaine in rats. Psychopharmacology 179:644–651

55. Lovinger DM, Crabbe JC (2005) Laboratory models of alcoholism: treatment target identification and insight into mechanisms. Nat Neurosci 8:1471–1480
56. Mantsch JR, Yuferov V, Mathieu-Kia AM, Ho A, Kreek MJ (2004) Effects of extended access to high versus low cocaine doses on self-administration, cocaine-induced reinstatement and brain mRNA levels in rats. Psychopharmacology 175:26–36
57. Marusich JA, Beckmann JS, Gipson CD, Bardo MT (2010) Methylphenidate as a reinforcer for rats: contingent delivery and intake escalation. Exp Clin Psychopharmacol 18:257–266
58. McBride WJ, Li TK (1998) Animal models of alcoholism: neurobiology of high alcohol-drinking behavior in rodents. Crit Rev Neurobiol 12:339–369
59. Mendez IA, Simon NW, Hart N, Mitchell MR, Nation JR, Wellman PJ, Setlow B (2010) Self-administered cocaine causes long-lasting increases in impulsive choice in a delay discounting task. Behav Neurosci 124:470–477
60. Murphy JM, Stewart RB, Bell RL, Badia-Elder NE, Carr LG, McBride WJ, Lumeng L, Li TK (2002) Phenotypic and genotypic characterization of the Indiana University rat lines selectively bred for high and low alcohol preference. Behav Genet 32:363–388
61. Nader MA, Woolverton WL (1991) Effects of increasing the magnitude of an alternative reinforcer on drug choice in a discrete-trials choice procedure. Psychopharmacology 105:169–174
62. Negus SS (2003) Rapid assessment of choice between cocaine and food in rhesus monkeys: effects of environmental manipulations and treatment with d-amphetamine and flupenthixol. Neuropsychopharmacology 28:919–931
63. Negus SS (2006) Choice between heroin and food in nondependent and heroin-dependent rhesus monkeys: effects of naloxone, buprenorphine, and methadone. J Pharmacol Exp Ther 317:711–723
64. Nestler EJ, Hyman SE (2010) Animal models of neuropsychiatric disorders. Nat Neurosci 13:1161–1169
65. Olds J, Milner P (1954) Positive reinforcement produced by electrical stimulation of septal area and other regions of rat brain. J Comp Physiol Psychol 47:419–427
66. Oleson EB, Roberts DC (2009) Behavioral economic assessment of price and cocaine consumption following self-administration histories that produce escalation of either final ratios or intake. Neuropsychopharmacology 34:796–804
67. Panlilio LV, Goldberg SR (2007) Self-administration of drugs in animals and humans as a model and an investigative tool. Addiction 102:1863–1870
68. Paronis CA, Gasior M, Bergman J (2002) Effects of cocaine under concurrent fixed ratio schedules of food and IV drug availability: a novel choice procedure in monkeys. Psychopharmacology 163:283–291
69. Paterson NE, Markou A (2003) Increased motivation for self-administered cocaine after escalated cocaine intake. Neuroreport 14:2229–2232
70. Paulus MP (2007) Decision-making dysfunctions in psychiatry--altered homeostatic processing? Science 318:602–606
71. Perry JL, Larson EB, German JP, Madden GJ, Carroll ME (2005) Impulsivity (delay discounting) as a predictor of acquisition of IV cocaine self-administration in female rats. Psychopharmacology 178:193–201
72. Perry JL, Morgan AD, Anker JJ, Dess NK, Carroll ME (2006) Escalation of i.v. cocaine self-administration and reinstatement of cocaine-seeking behavior in rats bred for high and low saccharin intake. Psychopharmacology 186:235–245
73. Poulos CX, Le AD, Parker JL (1995) Impulsivity predicts individual susceptibility to high levels of alcohol self-administration. Behav Pharmacol 6:810–814
74. Prus AJ, James JR, Rosecrans JA (2009) Conditioned place preference
75. Recorla RA, Wagner AR (1972) A theory of Pavlovian conditioning: variations in the effectiveness of reinforcement and nonreinforcement. In: Black AH, Prokasy WF (eds) Classical conditioning II: current research and theory. Appleton- Century-Crofts, New York, pp 64–99
76. Reid RL (1986) The psychology of the near miss. J Gambl Behav 2:32–39

77. Richardson NR, Roberts DC (1996) Progressive ratio schedules in drug self-administration studies in rats: a method to evaluate reinforcing efficacy. J Neurosci Methods 66:1–11
78. Robbins TW, Ersche KD, Everitt BJ (2008) Drug addiction and the memory systems of the brain. Ann N Y Acad Sci 1141:1–21
79. Rogers JL, De Santis S, See RE (2008) Extended methamphetamine self-administration enhances reinstatement of drug seeking and impairs novel object recognition in rats. Psychopharmacology 199:615–624
80. Schenk S, Harper DN, Do J (2011) Novel object recognition memory: measurement issues and effects of MDMA self-administration following short inter-trial intervals. J Psychopharmacol 25:1043–1052
81. Shaffer HJ, LaPlante DA, LaBrie RA, Kidman RC, Donato AN, Stanton MV (2004) Toward a syndrome model of addiction: multiple expressions, common etiology. Harv Rev Psychiatry 12:367–374
82. Soyka M, Chick J (2003) Use of acamprosate and opioid antagonists in the treatment of alcohol dependence: a European perspective. Am J Addict 12(Suppl 1):S69–S80
83. Tao R, Huang X, Wang J, Zhang H, Zhang Y, Li M (2010) Proposed diagnostic criteria for internet addiction. Addiction 105:556–564
84. Van den Bos R, Lasthuis W, Den Heijer E, Van der Harst J, Spruijt B (2006) Toward a rodent model of the Iowa gambling task. Behav Res Methods 38:470–478
85. Vanderschuren LJ, Ahmed SH (2013) Animal studies of addictive behavior. Cold Spring Harb Perspect Med 3:a011932
86. Vanderschuren LJ, Everitt BJ (2005) Behavioral and neural mechanisms of compulsive drug seeking. Eur J Pharmacol 526:77–88
87. Wee S, Wang Z, Woolverton WL, Pulvirenti L, Koob GF (2007) Effect of aripiprazole, a partial dopamine D2 receptor agonist, on increased rate of methamphetamine self-administration in rats with prolonged session duration. Neuropsychopharmacology 32:2238–2247
88. Wee S, Mandyam CD, Lekic DM, Koob GF (2008) Alpha 1-noradrenergic system role in increased motivation for cocaine intake in rats with prolonged access. Eur Neuropsychopharmacol 18:303–311
89. Winstanley CA, Cocker PJ, Rogers RD (2011) Dopamine modulates reward expectancy during performance of a slot machine task in rats: evidence for a 'near-miss' effect. Neuropsychopharmacology 36:913–925
90. Wolffgramm J (1991) An ethopharmacological approach to the development of drug addiction. Neurosci Biobehav Rev 15:515–519
91. World Health Organisation (1992) International classification of diseases. 10th Revision. Mental and behavioural disorders. World Health Organisation, Geneva
92. Young KS (1998) Caught in the net: understanding Internet addiction. Wiley, New York
93. Zhou W, Zhang F, Liu H, Tang S, Lai M, Zhu H, Kalivas PW (2009) Effects of training and withdrawal periods on heroin seeking induced by conditioned cue in an animal of model of relapse. Psychopharmacology 203:677–684

Chapter 8
Similarities and Differences in Diagnostic Scales

Bin Xuan, Peng Li, Liping Yang, Mingzhu Li, and Jing Zhou

Abstract A scale plays an important role as a diagnostic tool in discriminating between addicts and non-addicts. At the beginning of this chapter, we have briefly introduced the development of substance and non-substance addiction scales, which not only include alcohol addiction, nicotine addiction and pathological gambling, but also the disputed exercise and sex addiction. While it was found that almost all addiction scales contain items relating to social impairment, preoccupation, withdrawal, and tolerance, the variability is more pronounced with non-substance addiction scales. The comparison and trends of addiction scales in the future are discussed in relation to the concept of addiction, development of assessment theory, cross-cultural applicability, and cross-sample applicability.

Keywords Substance addiction • Non-substance addiction scales • Reliability • Validity

8.1 Introduction

What is the nature of addiction? Is it a personal lifestyle choice or a biological vulnerability? The direct activation of reward systems seemed to be the common characteristic in substance addiction [7, 64, 122]. Thus, both pathological gambling and Internet gaming disorder were included in the chapter of "Substance-Related and Addictive Disorders" in the newest version of the Diagnostic and Statistical Manual of Mental Disorders (DSM-V) for its activation of reward systems is similar to illicit drugs. However, other repetitive behaviors were excluded in the DSM-V because of insufficient peer-reviewed evidence to establish the diagnostic criteria. Some subcategories of behavioral addictions or non-substance addictions exist in the literature, such as sex addiction, exercise addiction, and shopping addiction. However, certain researchers believe some behavioral addictions should be treated as impulse-control

B. Xuan (✉) • P. Li • L. Yang • M. Li • J. Zhou
School of Educational Science, Anhui Normal University, Wuhu 241000, China
e-mail: xuanbin@ahnu.edu.cn

X. Zhang et al. (eds.), *Substance and Non-substance Addiction*, Advances in Experimental Medicine and Biology 1010, DOI 10.1007/978-981-10-5562-1_8

disorders [49, 60, 91]. In this chapter, the term "non-substance addictions" will be used because it may cover more in combination with "substance addictions."

Generally, the psychometric properties of scales contain reliability and validity. Reliability concerns internal consistency reliability and test-retest reliability, while validity contains, among other things, construct validity, distinction validity, and criterion validity. In this chapter, we will briefly introduce some substance and non-substance addiction scales used in research and clinical fields, including the psychometric properties and application of them. The similarities and differences of the scales will also be discussed based on the comparison within and between substance scales and non-substance scales.

8.2 Development of Substance Addiction Scales

Tobacco use and the harmful use of alcohol were two out of four risk factors for developing non-communicable diseases according to World Health Organization [132]. Approximately 5.9 % (3.3 million) of all deaths were caused by harmful use of alcohol globally in 2012, while this ratio was 13.3 % in Europe [133]. The number of people using an illicit drug reached 246 million in 2013, with 10 % suffering from drug use disorders or drug dependence and 12.19 million injecting drugs [120]. It is necessary to develop diagnostic tools of substance use and addiction in clinical, social work, and research fields owing to the great influence of abusing of alcohol, tobacco, and illicit drugs. The concept and diagnostic criteria of substance abuse are always changing [4–7, 26, 53, 84]. Some diagnostic instruments are changing with it while other scales are kept stable and remain in widespread use. The diagnostic criteria of substance use disorders include impaired control, social impairment, risky use, and pharmacological criteria in the DSM-V, which were more or less reflected in the diagnostic tools.

8.2.1 Scales of Common Substance Addiction

Alcohol addiction was distinguished from drug addiction when the DSM-I was published, although addiction and personality disorders were contained in the same chapter [4, 84]. Alcoholism, drug dependence, and personality disorders remained in one chapter of the DSM-II, where drug addiction did not include alcohol, tobacco, and caffeine [5]. In the DSM-III disorders caused by tobacco and caffeine were discussed [6]. The variability of the DSM reveals the variability of the addiction concept, which initially is divisive and incomplete, with extension and integration of the concepts accompanied by the development of research on addiction and diagnostic tools (see Table 8.1).

Table 8.1 The format, context, and application of scales of alcohol and nicotine

		MAST	AUDIT	CAGE	FTQ	FTND	HONC	CDS-12	WISDM	NDSS
Format	I/S[a]	I/S	I/S	I/S	S	S	I	S	S	S
	Brief version	SMAST	AUDIT-C					CDS-5	BWISDM	
			AUDIT-3							
Context[b]	Quantity and frequency		Y		Y			Y		
	Purpose						Y		Y	Y
	Consequence[c]	Y	Y	Y	Y	Y	Y	Y	Y	Y
	Guilty	Y	Y	Y						
	Help-seeking	Y								
	Illegal behavior	Y								
	Psychiatric symptom[d]	Y								
	Physiological symptom[e]	Y	Y							
Application[f]	Cross-culture	Y	Y	Y	Y	Y	Y	Y	Y	Y
	Legal system	Y	Y	Y						
	Adolescent	Y	Y	Y		Y	Y		Y	Y

[a] *I* interview, *S* self-report,
[b] Y = the scales contained the context
[c] Consequence contained impaired control and social impairment
[d] Psychiatric symptoms contained blackouts and flashbacks
[e] Serious physiological symptoms contained memory loss, hepatitis, convulsions, and bleeding
[f] Y = the scale was used in the field, OY = the scale was only used in the field

8.2.1.1 Alcohol Addiction Scales

The Michigan Alcoholism Screening Test (MAST) is a reliable, quantifiable, structured interview instrument to detect alcoholism developed by Selzer [106], which could be rapidly administered by nonprofessionals. The scale consists of 25 questions, four of which are scored 0 or 1, 17 scored 0 or 2, three scored 0 or 5, and one scored 0. Based on the scores of all questions, the participants can be divided into nonalcoholic, suggestive of alcoholism and alcoholic. Excluding the item scored 0, Selzer et al. [107] used the interview instrument as a self-report scale. It is reported that the scale for adults yielded high internal consistency, good validity, and a small influence of social desirability bias. To simplify the scoring system, Selzer et al. [107] introduced the Shorter Michigan Alcoholism Screening Test in the same paper. This test consisted of thirteen items scored 0 or 1, and the validity coefficients of it were slightly better than the MAST.

The CAGE scored 0 or 1 was published in 1970. It contains only four items (Cutting down, Annoyance by criticism, Guilty feeling, and Eye-openers). However, more recent literature on the scale could be found than for the MAST. Because of the conciseness, the CAGE is popular in many fields, especially in clinics. However, if a person reaches the cut-off score, which was 1 or 2 for different groups or purpose, it only means that there is high likelihood of the presence of alcoholism [37].

Scholars developed the Alcohol Use Disorders Identification Test (AUDIT), which was funded by the WHO and designed to detect early hazardous and harmful alcohol use for cross-culture environment [102, 103]. The test contains three items on alcohol consumption, three on drinking behavior, two on adverse psychological reactions, and two on alcohol-related problems. All the items are scored from 0 to 4. According to WHO guidelines [131], total scores between 8 and 15 indicate that simple advice is most appropriate, scores between 16 and 19 suggest brief counseling and continued monitoring, and scores of 20 or above clearly warrant further diagnostic evaluation for alcohol dependence or referral to specialist for diagnostic evaluation and treatment.

Generally, the MAST and the CAGE are used widely for their conciseness and sensitivity. However, some scholars have revealed the reliability of the MAST to be low. In particular, a high false positive ratio is seen in common samples, although it is sensitive to detecting alcoholism [46, 76, 119]. At the same time, the performance of the CAGE with some samples is also suboptimal. Dhalla and Kopec [32] highlighted that the CAGE was not effective for white women, pregnant women, college students, and those who drunk less severely. Messiah et al. [80] thought it would be more sensitive when using the CAGE for assessing community temperance level. Although the WHO [131] indicated that the AUDIT is appropriate for medical patients, driving-while-intoxicated offenders, violent criminals, enlisted men, officers and so on, it seemed to preform badly in an Asian population. Leung and Arthur [73] added 8 items to the initial 10 items to improve its cultural sensitivity. When revising the AUDIT for Hong Kong, Kim et al. [65] suggested a score of twelve as the cut-off in a Korean sample, while Kim et al. [66] suggested eight as the cut-off to detect alcohol use disorders for Korean American men. Kawada et al. [63]

revealed that the internal consistency of the Japanese version of the AUDIT was low (0.67) for Japanese adults.

8.2.1.2 Nicotine Addiction Scales

Initially several methods, such as theoretical models and biochemistry markers (e.g. carbon monoxide level) were used for investigating cigarette consumption [40]. Fagerstrom [39] developed the first nicotine addiction scale called the Fagerstrom Tolerance Questionnaire (FTQ). Thereafter, Heatherton et al. [55] revised the FTQ to the Fagerstrom Test for Nicotine Dependence (FTND), which is mainly used for descriptive studies in clinical fields. The FTND consists of six items, two of which are scored from 0 to 3, while the remaining four are scored 0 or 1. Because the FTND corrected some questions of the FTQ on psychology and concepts, its coefficient of internal consistency reached an acceptable level and its score was closely related to smoking severity.

DiFranza et al. [33] developed the Hooked on Nicotine Checklist (HONC) based on the assumption that nicotine dependence is caused by the loss of autonomy. This scale consists of ten questions scored 0 or 1. O'Loughlin et al. [85] found that the test-retest reliability of every item ranged from 0.61 to 0.93, and that of the whole scale was 0.61. The internal consistency reliability of the HONC reached 0.94, while the correlation between the test score and the maximum amount smoked ($r = 0.65$) as well as the maximum smoking frequency ($r = 0.78$) was significant [33].

The Cigarette Dependence Scales (CDS-12/CDS-5) developed by Etter et al. [36] were based on the definition of cigarette dependence from the DSM-IV and the International Statistical Classification of Diseases (ICD-10). The CDS-12 contains twelve items and the CDS-5 contains five. The test-retest reliability and internal consistency reliability of the two scales were both above 0.83, and it is sensitive to change over time.

The Wisconsin Inventory of Smoking Dependence Motives (WISDM) was developed by Piper et al. [88] and was based on theoretically grounded motives for drug use. The scale consists of 13 subscales reflecting different characteristics of smoking and is made up of 68 items scored from 1 (not true of me at all) to 7 (extremely true of me). Although all the subscales of the WISDM are significantly correlated to the severity of smoking, the criteria of dependence and recrudescence in the DSM-IV, and the test-retest reliability reached 0.88, it was too long to use in clinical settings. Smith et al. [114] developed the Brief WISDM, consisting of 37 items and 11 subscales.

Based on the concept of alcohol dependence, Shiffman et al. [108] developed the Nicotine Dependence Syndrome Scale (NDSS) that consists of drive, priority, continuity, stereotypy, and tolerance subscales. The NDSS contains 19 items scored from 1 (not at all true) to 5 (extremely true).

The FTND does not meet the psychometric criteria, and its operational definition is not complete according to the definition of the DSM-IV, although it is prevalent in clinical practice. Etter [35] highlighted that the FTND omits some important components of dependence listed in the DSM-IV and the ICD-10 such as smoking more than expected, tolerance, unsuccessful effort to withdraw, and smoking despite knowledge of having physical problems. Piper et al. [87] thought the FTND should be viewed as a descriptive or clinical tool for its weakness. The subsequent scales, such as the HONC or the CDS-12/CDS-5, both have good psychometric properties. Compared with other scales, HONC has a more reasonable cut-off point, more sensitivity and interpretability to addiction occurrence, and low dependence. The HONC may be more sensitive to studying populations with infrequent smoking [95], while the CDS-12/CDS-5 are more suitable for epidemiological study because of its conciseness.

8.2.2 *Scales of Other Substance Addiction*

Scales of other substance addiction (see Table 8.2) will discuss below. Skinner [111] designed the Drug Abuse Screening Test (DAST) for clinical use, which consists of 28 items scored 0 or 1. Thereafter the DAST-20 was created by excluding eight

Table 8.2 The format, context, and application field of the scales for other substance addiction

		DAST DAST-20	SDS	SMAST-AID	CAGE-AID	SDSS	ASSIST v1.0	ASSIST v3.1
Format	I/S[a]	S	S	I/S	I/S	I	I/S	I/S
	Brief version	DAST-10						
Context[a]	Quantity and frequency						Y	Y
	Purpose							
	Consequence[c]	Y	Y	Y	Y	Y	Y	Y
	Guilt	Y		Y	Y			
	Help-seeking			Y			Y	
	Illegal behavior	Y		Y			Y	Y
	Psychiatric symptom[d]	Y						
	Physiological symptom[e]	Y						
Application[f]	Cross-culture	Y	Y		Y	Y	Y	Y
	Legal system	Y						Y
	Adolescent	Y	Y		Y		Y	Y

[a]*I* interview, *S* self-report
[b]Y = the scales contained the context
[c]Consequence contained impaired control and social impairment
[d]Serious psychiatric symptoms contained blackouts and flashbacks
[e]Serious physiological symptoms contained memory loss, hepatitis, convulsions, bleeding
[f]Y = the scale was used in the field, OY = the scale was only used in the field

items only related to alcohol. The DAST focused on the symptoms caused by the drug. The correlation between the DAST and the DAST-20 is 0.99, and the internal consistency reliability of the DAST is 0.92 compared to the DAST-20's 0.95. Skinner and Goldberg [112] reported the internal consistency reliability of the DAST-20 in narcotic abusers was 0.79, and the cut-off value suggested was 6. The scale could be divided into five subscales including dependence, social problems, medical problems, polydrug abuse, and previous treatment.

To detect the degree of dependence experienced by users, Gossop et al. [48] developed the Severity of Dependence Scale (SDS). This scale focuses on impaired control, preoccupation, and anxieties when using drugs but does not assess tolerance, withdrawal, or reinstatement. The scale contains five items scored from 0 to 3, four of which relate to the frequency of control (0 = never/almost never; 1 = sometimes; 2 = often; 3 = always/nearly always), and the other relates to difficulty of control (0 = not difficult; 1 = quite difficult; 2 = very difficult; 3 = impossible).

The Substance Dependence Severity Scale (SDSS) developed by Miele et al. [81] was to assess the degree of substance addiction. It is appropriate for many kinds of substances including alcohol, cocaine, heroin, and other drugs. The SDSS is a semi-structured interview instrument, which contains six questions rated in symptom severity and symptom frequency with the last 30-day time frame. Test–retest reliability was good for alcohol, cocaine, heroin (0.54–0.88) but not for cannabis (0.41–0.62).

The Alcohol, Smoking, and Substance Involvement Screening Test (ASSIST) was completed by the WHO Assist Working Group [128] to solve the problem of cross-culture and insincere answer in Australia, Brazil, Ireland, India, Israel, the Palestinian Territories, Puerto Rico, the United Kingdom, and Zimbabwe. The first version contained twelve items and the subsequent versions had only eight questions. In version 3.1, the first item is an initial screening item that asks about lifetime substance use and if the respondent reports no lifetime substance use the interview is terminated. The subsequent items are about the frequency of substance use, dependence symptoms, substance-related problems, and injection drug use. Four items are scored from 0 (never) to 4 (daily or almost daily) and the remaining items are scored from 0 (no, never) to 2 (yes, in the past 3 months).

The DAST and SDS both specifically target drug use, but as researchers found more and more similarities within substance addiction, new scales, such as the SDSS and the ASSIST, appeared targeting cross-substance. Meanwhile, some scales developed for alcohol addiction were adapted to include drugs, such as the Short Michigan Alcoholism Screening Test Adapted to Include Drugs (SMAST-AID) and the CAGE Adapted to Include Drugs (CAGE-AID).

Although only the ASSIST was designed for the cross-culture application, the DAST and SDS were also used in other countries. However, the DAST did not perform well in Asia. Carey et al. [15] reported 59 % (16 out of 27) interviewees did not meet the diagnostic criteria of addiction when the cut-off in the DAST-10 was 3. However, the internal consistency reliability was 0.94 in India, and the sensitivity was 0.66 in Korea according to the diagnosis in DSM-III-TR when the cut-off was 2 [67], and 0.79 in Hong Kong when the cut-off was 1 [69]. Compared with the DAST, the SDS performed much better in Asia. Chen et al. [21] revealed that the internal consistency reliability (0.75) and the test-retest reliability after a week (0.85) of the Chinese version of the SDS were good. Moreover, the correlation between the scores and the diagnosis based on the DSM-IV was significant ($r = 0.54$, $p < 0.001$). Besides the research in Asia, the performance of the German version of the SDS for cannabis was acceptable, in terms of internal consistency reliability (0.80), sensitivity (0.94), and specificity (0.74) according to the DSM-IV diagnosis.

8.2.3 Analyses of Substance Addiction Scales

8.2.3.1 Comparison Among Substance Scales

More scales than those listed above have been created for use in research, clinical work, or the justice system. The scales, developed by different researchers, vary in format, context, and application fields. However, this variety is only theoretical and their performance has been supported by a large amount of research and practice.

The scales assessed here could be either in self-report or interview format. However, the variability caused by format was small. The only difference between the self-report version and the interview version of the MAST was that the self-report version did not include the item scored zero [107]. Aertgeerts et al. [1] found no difference between the oral and the written versions of the CAGE, and there was no significant difference between the groups with or without an open-ended question about alcohol. With the widespread use of computers, computer-administered modality has also been explored. Wolff and Shi [130] found no difference between the computer-administered modality and the interviewer-administered modality when applying the ASSIST. Similarly, McNeely et al. [79] found that participant's performance did not differ between an audio computer-assisted self-interview and a traditional interview.

Almost all the scales regarding alcohol or drugs have a self-report version, although some initially only had an interview version. In contrast, few nicotine addiction scales have an interview version, except for the HONC. It is likely that most nicotine abusers can complete the scales by themselves, but that it is more difficult for the severe alcohol or drug abuser to answer the self-report version. Besides the difference between self-report interview versions, the application of brief versions in clinics is another factor, for example the CAGE, the DAST-10, the SMAST, and the brief versions of the AUDIT. In fact, the CAGE is brief, but may still be too long for use in clinics owing to the time limitation. Here, the AUDIT-C can be used which consists of three items on alcohol consumption, or the AUDIT-3 where the third question is: "How often do you have six or more drinks on one occasion?" [3, 24, 100].

As for the context, every aspect relating to the substance should be contained in the scales, such as the quantity, frequency, purpose, and consequence of substance use (e.g. guilt and help-seeking behavior). Alcohol and drug scales specifically ask about the consequences of substance use. This could include feeling guilty, potential unlawful acts (driving or fighting under the influence), serious psychiatric symptoms (blackouts or flashbacks), and serious physiological symptoms (memory loss, hepatitis, convulsions, or bleeding). In contract, the purpose of substance use (concentration, losing weight, or joy) is often found in nicotine addiction scales.

The application of these scales can be divided into three fields depending on culture, law, or age, which are also barriers to widespread use. Culture is an important barrier in all the scales, but this can be avoided. Another barrier is the law concerning the balance between individual rights and public power. In most cases, the drugs are illegal and addiction to them might relate to the measurement of penalty. Therefore new scales were employed without thorough verification. Gavin et al. [45] indicated that the DAST should be applied cautiously in the justice system. Moreover, a report from Saltstone et al. [101] revealed that the dimensionality of the DAST was different for female offenders in jail compared to other groups although the internal consistency reliability was good. Thus, scales developed for the justice system were needed; for example, the UNCOPE designed by Hoffmann et al. [57]. However, little research focused on nicotine addiction in the justice system, just as there were few items related to illegal nicotine scales. The age of participants was another important factor concerned with applicability of scales. Animal models of addiction reveal differences between adults and adolescents [56, 68, 92, 105, 109, 113]. Conrod and Nikolaou [23] thought that the developmental model of addiction is suitable for adolescents. According to White and Labouvie [127], the usual measures of frequency, quantity, and variability of alcohol use are not sufficient to accurately diagnose the problem status of adolescents. Moreover, the progressive nature of the disease, medical complications, physical dependence, and other chronic symptoms are less clearly associated with adolescent alcohol problems. Thus, some new scales were developed for adolescents, such as the Rutgers Alcohol Problems Index (RAPI) and the Drug Use Screening Inventory (DUSI). Some existing scales were revised; for example, the DAST-a and the ASSIST-y were revised from the DAST and the ASSIST respectively. College students are important participants in research as they are in the transition between adolescence and adulthood, and therefore, some scales, such as the Young Adult Alcohol Problems Screening Test and CRAFFT (the combination of the six questions related to "CAR, RELAX, ALONE, FORGET, FAMILY or FRIENDS, TROUBLE") were developed for them.

8.2.3.2 Comparison Between Substance Scales and the Diagnostic Criteria in the DSM-V

In the DSM-V, diagnosis of a substance use disorder can be grouped within several categories, namely: impaired control, social impairment, risky use, and pharmacological criteria. Each of these groups consists of two, three, or four criteria, which can be summarized as follows: (1) more than intended, (2) unsuccessful efforts, (3) preoccupation, (4) craving, (5) failure in fulfilling obligations, (6) ignoring social or interpersonal problems, (7) giving up or reducing social activities, (8) physically hazardous using method, (9) ignoring physical or psychological problems, (10) tolerance, and (11) withdrawal.

Most of the scales assessed the consequences of substance addiction, but the DSM-V highlighted another factor, namely continuing substance use despite having persistent or recurrent problems caused or exacerbated by the effects of the substance (see Tables 8.3 and 8.4). This means that in the DSM-V the consequences of substance use were not only the effects caused by the substance but also knowing the harm of substance. After the publication of the DSM-III, a theoretical, descriptive diagnosis was advocated because of the physiological basis of addiction. This is because almost all addictions, including drugs, pathological gambling, and Internet gaming disorder, show the activation of the brain's reward system [84]. This was also consistent with the DSM-V. In addition, the independence of addiction diagnosis and personality disorder diagnosis meant the separation of addiction and morality to a certain extent. However, this separation was not reflected in the scales, as the items on guilt were still included in most addiction scales.

In the ASSIST, alcohol, smoking, and substance (drug) use are paratactic, and the parataxis means there is some difference among them, which is reflected in the DSM-V too. Since alcohol-related scales have fewer items than others, the items in alcohol scales are related to impaired control (craving or unsuccessful efforts) and social impairment. In nicotine scales the focus is on impaired control (craving) and withdrawal, while in other substance addiction scales the focus is on impaired control (unsuccessful efforts) and social impairment (failure in fulfilling obligations). Craving in drug addiction is so intense that it is meaningless to evaluate it, but the emphasis on unsuccessful efforts to stop using the drug and failure to fulfill obligations may be consequences of using the drug. This is less strong in nicotine addiction. In addition, the social function of people who smoking may be the main reason that no scales for nicotine are concerned with social impairment.

The pharmacological criteria of the DSM-V are rarely present in drug addiction scales. On one hand, neither tolerance nor withdrawal is necessary for a diagnosis of a substance use disorder, and only laboratory tests are effective in detecting tolerance. On the other hand, significant withdrawal has not been documented in humans after repeated use of phencyclidine, other hallucinogens, and inhalants. Thus, there are rare the items related to pharmacological criteria on drugs.

Table 8.3 Comparison of alcohol, nicotine, and their diagnostic criteria in the DSM-V

DSM-V	MAST	SMAST	AUDIT	CAGE	FTQ/FTND	HONC	CDS-12	CDS-5	WISDM/BWISDM	NDSS
1			√						√	√
2	√	√				√				√
3							√		√	
4			√	√		√	√	√	√	√
5	√	√								
6	√									
7			√							√
8										
9					√		√			
10										√
11					√	√	√		√	√

(1) 1–11 is the 11 criteria in DSM-V; (2) √ means the scale contained the corresponding criterion in DSM-V

Table 8.4 Comparison of other substance addiction and diagnostic criteria in the DSM-V

DSM-V	DAST/ DAST-20	DUSI	SDS	SMAST-AID	CAGE-AID	SDSS	ASSIST v1.0	ASSIST v3.1
1							√	
2	√	√	√	√			√	√
3							√	
4		√			√			√
5	√	√		√			√	√
6	√							
7		√				√		
8							√	√
9								
10		√						
11	√							

(1) 1–11 is the 11 criteria in DSM-V; (2) √ means the scale contained the corresponding criterion in DSM-V

8.3 Development of Non-substance Addiction Scales

Does non-substance addiction exist? Scholars have different views on it. Pathological gambling was originally classed as "disorders of impulse control that are not classified" in the DSM-III, and it was classed as non-substance-related disorders in the DSM-V. However, irresistible, repetitive, and harmful behaviors, such as out-of-control gambling, eating, shopping, sexuality, exercise, work, and Internet use are also considered as non-substance addiction. Non-substance addiction has a negative impact on many factors such as marriage, property, emotion, profession, education, and even threatens to life, therefore, the measurement and identification is very important. However, scholars have compiled a wide variety of scales with different uses based on various definitions and requirements. The lack of diagnostic criteria for non-substance addiction has been a barrier to adequate research.

Pathological gambling was the first non-substance addiction behavior to be studied. Gamblers Anonymous Twenty Questions (GA20) is the earliest pathological gambling scale developed by Custer and Custer [27]. The measurement research on non-substance addiction began with pathological gambling and gradually extended to pathological Internet use and Internet gaming addiction. Many scholars have begun to focus on the measurement of a range of non-substance addictions such as exercise addiction, compulsive shopping, work addiction, food addiction, sex addiction, mobile phone addiction etc.

8.3.1 Scales of Common Non-substance Addictions

In this section, each non-substance addiction will be introduced and discussed in terms of available measurement scales and their performance and applicability.

8.3.1.1 Pathological Gambling

The Gambling Treatment Team at South Oaks Hospital developed the South Oaks Gambling Screen (SOGS) under the diagnostic criteria of pathological gambling in the DSM-III and the DSM-III-R. The purpose is to assess the impact of gambling in multiple dimensions including emotional, family/society, occupation, education, money, and so on [72]. The scale focuses mainly on lifetime gambling activities and related behaviors. It includes 16 questions and 35 response items, 20 of which were scored 1 (yes) or 0 (no). A summed score of 0 means "no problem," 1–4 means "some problem," 5 or more means "probable" pathological gambling. The other non-scored items investigated the type of gambling, the maximum amount of money spent on gambling in one day, the gambler's family background, and whether gamblers quarreled with important people because of money. The psychometric properties of the scale were good, and it is the most widely used scale for pathological gambling to date.

However, wide application was not equal to perfection. The SOGS does not contain all the features about gambling and is not suitable for teenagers and cross-cultural participants. Moreover, it is too long to apply in some settings and therefore, development of new scales was very necessary. The SOGS-Revised for Adolescents (SOGS-RA) is targeted at teenagers. It contains two parts, non-scored items to investigate gambling activities and scored items to evaluate the severity of the problem gambling. The scale consists of 12 scored items, some of which are scored 0 or 1, and others that have 4 options [129]. The sum score divides participants into three groups: no gambling problems (0–1), at risk gamblers (2–3), and problem gamblers (5 or more). The reliability of the scale in a male sample was 0.80, and it was correlated with gambling activity count (r = 0.39), aggregate gambling frequency (r = 0.54) and amount of money gambled in the past year (r = 0.42).

For quick identification of pathological gamblers, Johnson et al. [59] used two items, item 3 (have you ever felt the need to bet more and more money?) and item 6 (have you ever had to lie to people important to you about how much you gambled?), from a 12-statement questionnaire based on the diagnostic criteria of pathological gambling in the DSM-IV. These two questions (scored 0 or 1) made up the Lie/Bet Questionnaire, and a total score of more than 0 indicates a participant at risk of pathological gambling. The scale has a high consistency (Kappa = 0.811), sensitivity (0.99), and specificity (0.91).

Ferris and Wynne [41] developed the Canadian Problem Gambling Index (CPGI), which contains four parts: (1) a detailed measurement of respondents' involvement in various forms of gambling; (2) the assessment of problem gambling; (3) an evaluation of correlates of problem gambling (e.g. family history, alcohol or drug use); and (4) demographic variables. The scale contains 31 items and only 9 items in the second part reflect the incidence of pathological gambling, scored on a 4-point scale (0 = never; 1 = sometimes; 2 = most of the time; 3 = always). Many studies have used only these nine items and named it the Problem Gambling Score Index (PGSI). Participants can be classified into one of five categories based on their total score: non-gambling (never gambled at all in the past 12 months); non-problem gambling (0); low risk gambling (1–2.5); moderate risk gambling (3–7.5); and problem gambling (8–27). The nine items belong to one dimension, of which the internal consistency reliability and the retest reliability was 0.84 and 0.78, respectively. The CPGI scores have a high correlation ($r = 0.83$) with the DSM and the SOGS, and a moderate correlation ($r = 0.48$) with the clinical interview. Various applications have proven that the performance of CPGI is superior to that of the Victorian Gambling Screen (VGS) and the SOGS.

The VGS, which is based on Australian culture and the definition of harmful consequences of pathological gambling, was developed for the cross-cultural application by Wenzel et al. [126] and funded by the Commission of the Victorian Casino and Gaming Authority (VCGA). This scale has three factors and 21 items: harm to self (15 items), harm to partner (3 items), and enjoyment of gambling (3 items). Only the first factor is scored and is scored on a 5-point scale (0 = never; 1 = rarely; 2 = sometimes; 3 = often; 4 = always). The total score range is between 0 and 60. Participants are classified into categories based on their total score: borderline problem gamblers (9–20), and problem gamblers (21 or more). The VGS is superior to the SOGS, but is not as good as the CPGI in various cases.

We summarized these scales of pathological gambling above in Table 8.5.

Table 8.5 The format, context, and application of scales of pathological gambling

		SOGS	SOGS-RA	Lie/Bet	CPGI	VGS
Format	I/S[a]	S	S	S	S	S
	Brief version				PGSI	
Context[b]	Frequency	Y	Y		Y	
	Purpose					Y
	Consequence[c]	Y	Y		Y	Y
	Guilty	Y	Y		Y	Y
	Help-seeking					
	Illegal behavior	Y	Y		Y	
	Lie	Y	Y	Y	Y	Y
	Finance	Y	Y	Y	Y	Y
Application[d]	Cross-culture	Y	Y	Y	Y	
	Legal system	Y			Y	
	Adolescent		OY			Y

[a]*I* interview, *S* self-report
[b]Y = the scales contained the context
[c]Consequence contained impaired control and social impairment
[d]Y = the scale was used in the field, OY = the scale was only used in the field

8.3.1.2 Pathological Internet Use

Young [135] developed Young's Diagnostic Questionnaire (YDQ) based on the diagnostic criteria of pathological gambling in the DSM-IV. The questionnaire consists of eight items with the response options "yes" or "no." If a subject answers "yes" to at least five items, he/she is considered "Internet dependent." In the same year, Young [134] developed the Internet Addiction Test (IAT) to assess the influence of excessive Internet use on one's life. The IAT is an extension of the YDQ and is a self-report questionnaire including 20 questions scored on a 5-point scale ranging from 1 (rarely) to 5 (always). The total score is between 20 and 100, and a higher score means a more serious problem caused by Internet addiction. Young suggested that a score of 20–39 points was an average online user who had complete control over his/her usage; a score of 40–69 signified frequent problems due to Internet usage; and a score of 70–100 meant that Internet use was causing significant problems. The IAT was used far more widely than the YDQ and was evaluated positively with good psychometric properties, and it is currently the most widely used questionnaire to screen for Internet addiction.

Chen et al. [20] put forward that the concept of Internet addiction models should include, among others, the following aspects: tolerance of Internet addiction, compulsive Internet use, withdrawal from Internet use, and related problems of Internet addiction. This concept emphasizes the psychological behavior level and combines the diagnostic criteria of addiction in the DSM-IV, clinical cases, conversations with heavy Internet users, and the traditional addiction diagnostic model. Chen et al. developed the Chinese Internet Addiction Scale (CIAS) and revised it as CIAS-R. The revised version has a better factor structure than the original version. The CIAS-R contains 26 questions and two subscales: core systems of Internet addiction and related problems of Internet addiction. These questions are scored from 1 (very inconsistent) to 4 (very consistent), the total score ranges from 26 to 104, and a higher score indicates a more serious level of Internet addiction.

In addition to the two frequently used scales introduced above, many other scales are not popular for various reasons. Some scales take longer to administer, have better psychometric properties, have an acceptable theoretical model, and so on. Moreover, the development of many scales is not always based on the diagnostic criteria of the DSM-IV. For example, the Online Cognitive Scale (OCS) developed by Davis et al. [28] is based on cognitive-behavioral model of pathological Internet use (PIU).

8.3.1.3 Internet Gaming Addiction

In order to facilitate research on gaming addiction, Lemmens et al. [71] developed the Game Addiction Scale for Adolescents(GAS)based on the seven criteria of pathological gambling in the DSM-IV: salience, tolerance, mood modification, relapse, withdrawal, conflict, and problems. Each criterion contains 3 questions and thus the GAS contains 21 questions scored from 1 (never) to 5 (very often). Participants are asked to indicate the frequency with which they have experienced each of the described situations over the past 6 months. A monothetic approach is used where a rating of three or above on all items indicates gaming addiction. This scale was a second-order factor model; its reliability in two samples was 0.94 and 0.92 respectively. The authors also developed a simplified version containing seven items according to the scale, and each criterion corresponded to one question. The reliability of the simplified version in two samples was 0.86 and 0.81 respectively. Both GAS-21 and GAS-7 were significantly correlated with time spent playing games, and with some psychological variables, such as loneliness, life satisfaction, social competence, and aggression. The correlations between the two scales and these variables had no significant difference, which means that the two scales were equally effective.

Pontes et al. [90] developed Internet Gaming Disorder Test (IGD-20) according to IGD diagnostic criteria in DSM-V. The IGD-20 has 20 items, which not only reflects the nine diagnostic criteria of IGD listed in the DSM-V, but also conforms to the theoretical framework of the components model of addiction. The IGD-20 examines both online and offline gaming activities over a 12-month period as the IGD criteria of DSM-V are based on persistent and recurrent gaming. Participants rate all items of this test on a 5-point Likert scale: 1 (strongly disagree), 2 (disagree), 3 (neither agree nor disagree), 4 (agree), or 5 (strongly agree).

In order to evaluate whether the IGD diagnostic criteria in the DSM-V could be a foundation for developing new scales, and to explore its psychometric properties, Pontes and Griffiths [89] developed the Internet Gaming Disorder Scale–Short-Form (IGDS9–SF). This scale examined both online and offline gaming activities over a 12-month period. The nine questions comprising the IGDS9-SF were scored from 1 (never) to 5 (very often), the total score ranges from 9 to 45. A higher score indicates a more serious level of gaming addiction. Participants can be divided into disordered gamers (36–45) and non-disordered gamers (9–36).

In fact, there are no less than 20 gaming addiction scales, but the applicability and psychometric properties of many scales still need further testing. Along with the development, normalization, theorization, and better psychometric properties the trends of the gaming addiction scales should be considered.

We summarized these scales of pathological Internet use and Internet gaming addiction above in Table 8.6.

8.3.2 *Scales of Other Non-substance Addiction*

The DSM-V currently only includes pathological gambling in the non-substance addiction. Because peer-reviewed evidence is not enough to establish diagnostic criteria and course description of behavioral disorder identification, other non-substance addiction types, such as sexual addiction, sport addiction, and shopping addiction are not included in the DSM-V. These non-substance addiction studies are not all carried out within the framework of addiction. Along with the findings of

Table 8.6 The format, context, and application of scales of pathological Internet use and Internet gaming addiction

		IAT	CIAS-R	GAS-21	IGD-20	IGDS9 – SF
Format	I/S[a]	S	S	S	S	S
	Brief version			GAS-7		
Context[b]	Frequency	Y	Y	Y	Y	
	Purpose			Y	Y	
	Consequence[c]	Y	Y	Y	Y	Y
	Guilty					
	Help-seeking					
	Illegal behavior					
	Lie			Y	Y	Y
	Finance					
Application[d]	Cross-culture	Y		Y	Y	Y
	Legal system					
	Adolescent	Y	Y	OY	Y	Y

[a]*I* interview, *S* self-report
[b]Y = the scales contained the context
[c]Consequence contained impaired control and social impairment
[d]Y = the scale was used in the field, OY = the scale was only used in the field

neurobiology research, researchers have begun to consider other addictions from the perspective of addiction. Thus, the resulting research is becoming richer and includes a variety of other non-substance addiction diagnostic criteria. Here we give a brief introduction to the scales that are widely used in non-substance addiction and have good psychometric properties.

8.3.2.1 Exercise Addiction

In 1987, de Coverley Veale [29] proposed "exercise dependence" and its diagnostic criteria. Exercise dependence is also known as "negative addiction," "exercise addiction," "obligatory exercising," "exercise abuse," "excessive exercise," etc. The term "exercise addiction" is not often distinguished from "exercise dependence" and, in order to remain consistent with the information presented earlier, here we use the term "exercise addiction."

Godin and Shephard [47] attempted to develop an easy-to-use, easy-to-score, and reliable scale for measuring sporting behavior—the Godin Leisure Time Exercise Questionnaire (GLTEQ). The scale has two questions and test-retest reliability after 2 weeks was 0.74 and 0.80. A higher score indicates a higher level of exercise. The GLTEQ is widely used because it has only two questions and a simple score calculation, but studies have generally only used three questions of its first part, and rarely report psychological characteristics of its reliability, validity, etc. [82, 123, 124].

Exercise addiction scales represented by the GLTEQ are mostly single-dimensional scales that assess only a single aspect of exercise addiction. Hausenblas and Downs [54] developed the Exercise Dependence Scale (EDS) according to the criteria of substance addiction in the DSM-IV. The EDS includes 28 items where participants respond on a 5-point Likert scale ranging from 1 (never) to 5 (always). It has been revised to the Exercise Dependence Scale-Revised (EDS-R) including 21 items scored from 1 (never) to 6 (always). There are seven factors in the EDS and EDS-R, the reliability of each subscale is between 0.55 and 0.95, and the reliability of the total scale is between 0.87 and 0.97. The score of each factor and total score of EDS-R were significantly correlated with exercise frequency. It can distinguish between individuals with different levels of exercise, and thus, can be widely used in exercise addiction research, from regular exercisers (in the general population) to professional athletes. It can also be applied in intercultural environments [11, 25].

For ease of use, Terry et al. [118] proposed the Exercise Addiction Inventory (EAI) with a total of six questions, each on a 5-point scale ranging from 1 (strongly disagree) to 5 (strongly agree). A score in the range of 0–12 means that there is no addiction, 13–23 is indicative of a symptomatic individual, and 24 or more indicates someone at risk of developing exercise addiction. The score on the EAI was significantly correlated with the Obligatory Exercise Questionnaire (OEQ) and EDS. It has been used in Spain, Denmark, England, France, Italy, and other countries [51, 74, 75, 110].

8.3.2.2 Sex Addiction

Excessive sexual behavior may have adverse effects; in particular, inappropriate sexual activity increases individual's risk of contracting AIDS and other diseases [61, 83]. This has led to researchers' interest in sexual addiction. In order to distinguish between sex addicts and non-addicts, researchers have developed and continued to improve related measurement tools as their understanding of sexual addiction deepens. Carnes [16], the earliest proposer of sexual addiction, developed the Sexual Addiction Screening Test (SAST) for screening addicts. It is widely used in clinical screening of sexual addiction. The scale contains 25 items scored 0 (no) or 1 (yes), and the total number of positive response items is the individual's score. A person with a score equal to 13 or higher can be diagnosed as a sex addict. Self-reported sex addicts in the questionnaire scored significantly higher than non-sex addicts did. Delmonico et al. [31] showed that the SAST was significantly correlated with other measures of addiction, such as the Sexual Dependency Inventory-Revised and the Garos Sexual Behavior Index [44]. Although the SAST is easy to use and showed good psychometric properties, it was originally only applied to heterosexual males. Subsequent studies developed the W-SAST and G-SAST to assess female and homosexual sex addiction [22].

In order to assess the level of individual sexual addiction conveniently, Carnes et al. [17] developed a revised version of the SAST (SAST-R) with only 20 items. The SAST-R showed good reliability and validity similar to the original study.

Carnes et al. [18] extracted six items from the original SAST to form a brief screening application (PATHOS) for assessing sexual addiction, and confirmed the reliability and validity with inpatients and undergraduate students.

For distinguishing different categories of sexual addiction, Carnes et al. developed a sexual dependency inventory, and updated it in 1996, 2011, and 2015; the latest version is the SDI-4.0 [50]. The item numbers are varied in each version, and subjects need to respond to each item with frequency (how often does the idea/behavior/fantasy/feeling appear) and power (the influence of this frequency) as indicators.

With the advent of the Internet, researchers began to pay attention to Internet sexual addiction. Delmonico and Miller [30] developed a 25-item Internet Sex Screening Test (ISST) that measures online sexual behavior problems. The ISST includes five subscales of online sexual compulsivity, online social sexual behavior, online isolated sexual behavior, online sexual spending and interest in online sexual behavior. The total number of positive responses is the total score, and a higher score indicates a higher level of sexual addiction. The ISST is the only scale to measure if individual's Internet sexual behavior has become clinically problematic; however, there is little evidence for its reliability and validity.

8.3.2.3 Work Addiction

In 1971, Oates first defined workaholism as a compulsive, uncontrollable need for continuous working, and emphasized that its cognitive behavior patterns are similar to substance addiction. Subsequent research tends to view workaholics as an addictive behavior. The first tool to quantify the level of work addiction was the Work Addiction Risk Test (WART) by Robinson [96]. He defines workaholism as excessive attention and indulgence in work with impairment to the worker's health, intimacy, and children's education. The WART includes 25 items scored from 1 (never true) to 4 (always true). The questionnaire consisted of five sub-questionnaires initially. The internal consistency reliability alpha was 0.88 [97], the test-retest reliability with 151 subjects was 0.83 after 2 weeks, the internal consistency coefficient was 0.85 [99], the split-half reliability was 0.85, and the average correlation coefficient between projects was 0.26. The items can measure different aspects of the work addiction structure [98]. The WART is a good way to identify non-workaholics, but cannot identify workaholics well. Only 5 % of the control group was considered workaholics, while 43 % of the workaholics were not considered so [42].

In order to distinguish between workaholics, work enthusiasts, and other types of workers, Spence and Robbins [115] developed the Workaholism Battery (WorkBAT). This scale includes 25 items, which can be divided into 3 subscales: work involvement (8 questions), drive (7 questions), and enjoyment of work (10 questions; [58]). Participants are asked to indicate their responses on a 0–4 or 0–6 scale. According to the respondents' scores on these three subscales, they can be divided into six categories: workaholic, work enthusiasts, enthusiastic workaholic, unengaged worker, relaxed worker, and disenchanted worker [115]. However, studies with a

Japanese sample showed that the work involvement dimension overlaps with the drive dimension and cannot be independently dimensioned [62], The WorkBAT is a two-factor scale, and that the work involvement dimension should be discarded [78].

Andreassen et al. [9] argued that the previous studies considered workaholism to be an addictive behavior, but did not developed scales based on addiction. Following this, they developed the Bergen Work Addiction Scale (BWAS). There are seven items in the BWAS based seven core components of addiction (salience, mood modification, tolerance, withdrawal, conflict, relapse, and problems). Participants were asked to response on a 5-point Likert scale (never, rarely, sometimes, often, always) and were considered workaholics if they indicated "always" on four or more questions. The internal consistency reliability was 0.84 in the workaholic group and 0.80 in the control group. Except for the work enjoyment subscale, the BWAS score was significantly correlated with scores of other measures of work addiction. The BWAS had good convergent validity and discriminative validity, which can distinguish workaholics well [9]. As a relatively new scale, the reliability and validity of BWAS has a good reliability and validity, and has been applied in other countries [8, 86], however, the validity of the cross-cultural applications need to be verified further.

8.3.2.4 Shopping Addiction

Some researchers believe that shopping addiction is a kind of addiction, while others classify it as an obsessive-compulsive disorder. Thus, researchers have developed scales to measure shopping addiction based on different theoretical frameworks. Valence et al. [121] argued that obsessive shopping should be separated from other repetitive behaviors, distinguishing compulsive buying from impulsive buying, and thus developed the first Compulsive Buying Measurement Scale (CBMS). This scale has only one dimension, with 13 items scored on a 5-point scale, its internal consistency coefficient is 0.88 and can effectively identify compulsive buyers. CBMS is also widely used in other cultures [10], but some items are still controversial. In order to make up for the scale's shortcomings, Faber and O'Guinn [38] developed a 7-item, 5-point Compulsive Buying Scale (CBS) to evaluate the participant's thinking, mood, and behavior (including general shopping behavior). The CBS is a single-factor questionnaire, with a load of 0.69 and above on all seven items, and an internal consistency coefficient of 0.95. It showed a good external validity. Self-reported addicts and questionnaire-screened addicts had similar scores on all the relevant factors, but had a significant difference from non-addicted shoppers. It should be noted that the CBS includes some culture-specific items, making it somewhat difficult to adapt for different cultures [77].

Compulsive shopping is not only an obsessive-compulsive behavior, but also an impulse-control disorders behavior. Based on this, Ridgway, Kukar-Kinney, and Monroe [94] developed the Richmond Compulsive Buying Scale (RCBS). There are six items with two dimensions: obsessive-compulsive buying and impulsive buying. Participants are asked to respond on a 7-point scale. Individuals scoring more than 24 points are considered as compulsive buyers. The internal consistency coefficient of the scale is good (0.84), and the consistency coefficients of the two dimensions are 0.77 and 0.78, respectively.

Andreassen et al. [10] developed the Bergen Shopping Addiction Scale (BSAS) to measure addiction according to the relevant criteria of addiction, which is the first scale in a completely addiction-based paradigm. The questionnaire is primarily based on the seven addictive criteria (salience, mood modification, conflict, tolerance, withdrawal, relapse, and problems) and each criterion has only one related question. The BSAS is a single-dimension questionnaire, with 5-point scaling ranging from complete disagreement (0 point) to complete agreement (4 points). A higher total score indicates a higher level of addiction. Its internal consistency coefficient is 0.87, it has a good factor structure, and the correlation coefficient between the scores of BSAS and CBMS is 0.80. The scale's validity, reliability, and applicability in more settings still need to be verified.

8.3.3 *Analyses of Non-substance Addiction Scales*

8.3.3.1 Comparison of Non-substance Addiction Scales

First, we will discuss the theoretical basis of scale development. Non-substance addiction scales, such as the SOGS, the IAT, and the IGD often depend on the case studies and literature on the diagnostic criteria of substance addiction. We found that scales on pathological gambling are primarily based on the criteria of substance addiction, and that later non-substance addiction scales are often based on the diagnostic criteria of substance addiction or pathological gambling. This is because no non-substance addiction scales were available for reference before the pathological gambling scale. Moreover, there are some common characteristics between pathological gambling and substance abuse; for example, they can both activate the reward system. After the gambling scales included diagnostic criteria, the subsequent non-substance could reference them.

The contents of the scales will be discussed here. On the one hand, most non-substance addiction contains some common aspects, such as tolerance, unsuccessful withdrawal, preoccupation, intention to engage in the behavior, lying, and social impairment. On the other hand, the diagnostic criteria of non-substance addiction kept changing to adapt to different non-substance categories. This is evident in the changes from the DSM-III to the DSM-V as in the diagnostic criteria for pathological gambling and IGD. For example, the diagnostic criteria for pathological gambling include "chasing" one's losses and financial difficulty, but there are not included in IGD diagnosis in the DSM-V. IGD criteria includes "loss of interests in previous hobbies and entertainment as a result of, and with the exception of, Internet games" and "continued excessive use of Internet games despite knowledge of psychosocial problems," which is reflected in the non-substance addiction scales. From the perspective of compulsive behavior and addiction, different scales focus on different aspects. In shopping addiction studies, researchers emphasized the compulsivity in repetitive shopping behavior and this is reflected in the scale development [34, 38]. Ridgway et al. [94] believed that compulsive shopping is not only a compulsive behavior, but also a behavior of lacking impulsive control. They developed the RCBS, which includes two dimensions of obsessive-compulsive buying and impulsive buying. With the deepening of the concept of addiction, Andreassen et al. [10] developed the BSAS to measure the addiction related characteristics. Similarly, Robinson [96, 97] indicated that workaholism concentrated too much on work and the impairment of other functions, and hence developed the WART to focus on both compulsive tendencies and the impairment of work involvement and feelings about the self. Spence and Robbins [115] believed that workaholism is a kind of addiction and emphasized the imbalance between job involvement and happiness. Different scales are based on different theoretical bases and most of the scales showed good stability. There are strong correlations between the same types of non-substance addiction scales as well as between these scales and scales based on the diagnostic criteria for addiction. For example, the correlation coefficient of the CBMS developed in 1988 and the BSAS based on the criteria for addiction reached 0.80 [10].

As for the usage of the scales, the most frequently used scales have been widely used in the studies in a range of countries. Most of them are aimed at adults, but there are also some scales for adolescents, such as the SOGS-R. Because the majority of Internet users and game players are adolescents, the GAS and the IGD are often used with them. Many scales are also applied in cross-cultural studies with good psychometric properties.

8.3.3.2 Comparison of Non-substance Addiction Scales and DSM-V

Non-substance addiction scales have shown high consistency with the diagnostic criteria or have been developed according to the DSM criteria directly. However, the DSM criteria are mainly used in clinical environments, and non-substance addiction scales are mainly used in general population surveys or scientific research. Therefore, they do not always match completely. Some criteria of the DSM are not found in diagnostic scales, and research scales may include some items that do not belong to the criteria. For example, the pathological gambling scales do not include withdrawal, but do include self-awareness.

Comparison of Pathological Gambling Scales, Pathological Internet Use Scales and the Diagnostic Criteria of Pathological Gambling in DSM-V

Frequently used pathological Internet use scales, such as the IAT and CIAS-R, are at least partly developed based on the criteria of pathological gambling in the DSM-IV. Moreover, there are no official diagnostic criteria for pathological Internet use. Therefore, both pathological gambling and pathological Internet use are

Table 8.7 Comparisons of pathological gambling scales, pathological Internet use scales, and DSM-V

DSM-V	SOGS	SOGS-RA	Lie/Bet questionnaire	CPGI	VGS	IAT	CIAS-R
1			√	√		√	√
2						√	√
3	√	√			√	√	√
4	√				√	√	√
5					√	√	
6	√	√		√	√		
7	√	√	√		√	√	
8	√	√		√	√	√	√
9	√	√		√			

(1) 1–9 indicate the nine criteria of DSM-V; (2) √ indicates the items of scales are consistent with DSM-V

compared to the diagnostic criteria of pathological gambling in the DSM-V (see Table 8.7).

The DSM-V defined gambling disorder as "persistent and recurrent problematic gambling behavior leading to clinically significant impairment or distress," and put forward nine diagnostic criteria. Subjects need to respond according to their experience from the past 12 months. If they meet four to five criteria they are considered as having "mild gambling disorder," if they meet six to seven criteria they are considered as having "moderate gambling disorder," and meeting eight to nine criteria means they are considered as having "severe gambling disorder."

In order to compare and analyze the scales conveniently, we simplified each diagnostic criterion into a phrase. The nine criteria correspond to the following phrases: (1) increasing need, (2) withdrawal symptoms, (3) unsuccessful efforts to stop, (4) preoccupation, (5) intention, (6) "chasing" one's losses, (7) lie, (8) social impairment, and (9) financial difficulties. Table 8.7 shows a comparison between pathological gambling, pathological Internet, use and the DSM-V.

From the above table, we see that the scales of pathological gambling and pathological Internet use do not cover all diagnostic criteria and contain some measurements that are not included in the DSM-V, such as "have you ever felt guilty about the way you gamble or what happens when you gamble?" In other words, the scales of pathological gambling and pathological Internet use are not completely consistent with the diagnostic criteria of the DSM-V.

Comparison of Internet Gaming Addiction Scales, Pathological Internet Use Scales and the Diagnostic Criteria of Internet Gaming Disorder in DSM-V

DSM-V defined Internet gaming disorder as "persistent and recurrent use of the Internet to engage in games, often with other players, leading to clinically significant impairment or distress," and put forward nine diagnostic criteria. Subjects were required to answer the questions according to their situation in the past 12 months.

Table 8.8 Comparison of internet gaming addiction scales, pathological Internet use scales, and DSM-V

DSM-V	GAS	IGD-20 & IGDS-SF9	IAT	CIAS-R
1	√	√	√	√
2	√	√	√	√
3	√	√	√	√
4	√	√	√	√
5		√		
6	√	√		
7	√	√	√	
8	√	√	√	
9	√	√	√	√

(1) 1–9 indicate the nine criteria of DSM-V; (2) √ indicates the items of scales are consistent with DSM-V

If the answers meet or exceed five criteria, subjects are considered gaming addicts. There is a difference between mild, moderate, and severe levels of damage due to daily activities. The more serious the gaming addiction, the more time the individual spends on the game, and the more damage to the social function is present.

In order to compare the DSM-V with frequently used Internet gaming addiction scales, we simplified each diagnostic criterion into a phrase, so the nine criteria correspond to the following: (1) preoccupation, (2) withdrawal symptoms, (3) tolerance, (4) unsuccessful attempts to stop, (5) loss of interest, (6) excessive use, (7) lying, (8) intention to game, and (9) social impairment. Pathological Internet use and Internet gaming addiction are both based on the Internet and, consider this, Internet gaming addiction may be a special form of pathological Internet use. Thus, it is necessary to compare pathological Internet use scales with the DSM-V. Table 8.8 shows a comparison between Internet gaming addiction scales, pathological Internet use scales, and the DSM-V

Comparison of Other Non-substance Addiction Scales and the DSM-V

Compared to the addiction diagnostic criteria of the DSM-V, various non-substance addiction scales showed large differences. Besides pathological gambling, pathological Internet use, and Internet gaming addiction, other non-substance addiction scales focused more on negative emotions caused by repetitive behavior. For example, shopping addicts and workaholics showed anxiety and other negative emotions if they could not shop or work. In addition, researchers paid more attention to the intrinsic driving force or the compulsivity in these non-substance addiction scales, and the DSM-V emphasized that individual's is inability to control the addiction. There are diagnostic criteria for impairment of the individual's physical health in the DSM-V, but not in non-substance addiction scales. Both the DSM-V and non-substance addiction scales are concerned with the addiction's effects on the individual's social functioning. Most non-substance addiction scales only mention a certain aspect of social functions, such as family or social activities. In addition, items relating to tolerance were not included in the non-substance addiction scales.

8.4 Comparisons of Substance Addiction and Non-substance Addiction Scales

8.4.1 Comparisons of Addiction Scales Based on the DSM-V and Other Criteria

According to the DSM-V, the diagnostic criteria of substance addiction and non-substance addiction both include unsuccessful efforts to stop, preoccupation, withdrawal, tolerance, and social impairment. This is reflected in the related scales. Almost all of the scales include an item related to unsuccessful efforts to stop, and most of them include an item relating to social impairment. Social impairment and related features in addiction usually involves craving, consequences (cognitive and social impairment), tolerance, occupancy time, and so on, although the pattern of time occupation is not exactly the same. In addition, the purpose of using the substance is not involved in the diagnosis of substance addiction, but some substance addiction scales do enquire about it (such as the DUSI, HONC, WISDM, and NDSS). Questions around the purpose also appear in scales about pathological gambling and Internet gaming disorders.

Social impairment is a diagnostic criterion for both substance and non-substance addiction and may be the biggest and most immediate impact on the addiction. The impairment is partly caused by addicts' inability to continue their social responsibilities (such as family identity or work). On one hand, addiction brings about a loss of interest as addicts find it difficult to obtain satisfaction from daily life, as their reward system is abnormal. On the other hand, addictive substances or behaviors take time, including the time it takes to engage in the addictive behavior and the symptoms that follow. For example, alcohol abuse may lead to confusion. However, not all the substance addiction scales include items about social impairment. As described in the first section of this chapter, smoking is a social activity, and thus few scales of tobacco addiction include social impairment. In contrast, almost all scales about alcohol, substance, and non-substance addiction include items about social impairment.

Preoccupation, withdrawal, and tolerance are diagnostic criteria shared by both substance and non-substance addiction. Slight differences exist between scales, however. Most nicotine addiction scales include items about withdrawal and preoccupation, while in substance addiction scales including alcohol, drugs, and other substances, only the DAST/DAST-20 contain withdrawal. The ASSIST v1.0 and the UNCOPE both contain preoccupation and tolerance related items. Tolerance related items are rarely found in substance addiction scales (tolerance related items are only found in the DUST for drug addiction and NDSS for nicotine addiction). In contrast, withdrawal is not included in the frequently used pathological gambling scales. Only the Lie/Bet Questionnaire and CPGI contain tolerance related questions and the SOGS and VGS contain items related to preoccupation. Two frequently used Internet gaming disorder scales contain more than one item of preoccupation, tolerance, and withdrawal, and items about tolerance and withdrawal are also found in the other non-substance addiction scales. It should be noted that the indexes of non-substance addiction include absolute time spent on the activity, relative time spent on the activity (the ratio of occupation), and frequency. Besides the temporal indexes, absolute amount used and daily amount used are also included in substance addiction scales.

Diagnostic criteria for pathological gambling and Internet gaming disorder both include lying as a criterion, which does not appear in substance addiction. Pathological gambling also contains "chasing" one's losses and financial difficulties. IGD includes three specific criteria: loss of interest, excessive, use and mood modification. This indicates that there is more variability in the diagnostic criteria of non-substance addiction and this is reflected in the specificity of scale items. For the other non-substance addictions that are not included in the DSM-V, the impairment of social function may be the most common part of all of addiction scales. These scales are more concerned about the compulsivity brought about by repetitive behaviors, such as negative emotional experience, and less concerned about the addiction's negative influence on the individual's health and tolerance.

An isolated study cannot present useful information, and other information is needed as reference or basis for any scale study. The diagnostic criteria of the DSM and ICD can serve as accepted references in substance addiction; however, frequency is often taken as a criterion since the diagnosis system is not perfect in non-substance addiction. For example, Ridgway et al. [94] collected the income, self-reported shopping frequency, and expenditure in the study of the RCBS. The amount of substance used, frequency, and blood indicators (including metabolites) are taken as criteria for substance addiction. Substance addiction scales usually involve the investigation of substance dose. This is not the case in non-substance addiction except for food addiction. In the study of food addiction, body mass index (BMI) is also collected.

Although there is no addictive personality, scholars are still interested in the relationship between addiction and personality. Franken et al. [43] believe that the biopsychological theory of personality and, specifically, the Behavioral Approach System (BAS) is relevant to substance addiction. Stautz and Cooper [117] showed that impulsivity related personality traits are associated with juvenile alcohol abuse. Andreassen et al. [10] indicated that shopping addiction is positively correlated with extroversion and neuroticism, and negatively with conscientiousness, agreeableness, and intellect/imagination. In contrast with personality traits, SriRajaskanthan and Preedy [116] believe that blood alcohol, mean corpuscular cell volume, gamma-glutamyl transferase, serum aminotransferases, and carbohydrate deficient transferrin are the biochemical markers of alcohol addiction. Aflatoonian et al. [2] found that opioid use is associated with blood group type. Since there is less research on addictive biochemical markers, physiological research is rarely found.

In addition to personality, the effects of gender and family history on addiction are also discussed more in relation to substance addiction [12, 14, 52, 70, 136]. In contrast, little literature on family history was found in non-substance addiction, and discussions on gender difference were only found for sexual addiction [13, 125].

8.4.2 Applicability of Substance and Non-substance Scales

In the clinical field, especially in emergency situations, the number of questions may be an important factor to affect the scale's application. Patients may struggle to complete the scale if it is too long and difficult. Therefore, there are some simplified versions for use in clinical field as well as some original scales that are short enough. This is more common in substance addiction scales, but is also seen in non-substance addiction scales, such as the Lie/Bet Questionnaire for pathological gambling and Problematic and Risky Internet Use Screening Scale (PRIUSS-3) for Internet gaming disorder.

In substance addiction, the difference between adults and adolescents is large enough to develop scales with different items. This may be due to the differences in the nervous system of developing adolescents and mature adults and the related physiological responses and symptoms. While in non-substance addiction the studies of pathological gambling are mainly aimed at adults, the average age of the subjects suffering from Internet gaming addiction ranges from adolescence to early adulthood (although the subjects in some studies were 58 years old, the average age is 26.14). This suggests that age related problems have assessed in non-substance addiction scales. On one hand, this may be due to insufficient exploration within non-substance addiction. On the other hand, at present non-substance addiction studies focus on certain group to reduce the influence of age (e.g. Internet gaming addicts are generally adolescents, workaholics are generally adults). A future direction is to refine non-substance addiction scales and extend these to different age ranges.

The secondary consequences of substance use are more obvious, especially the effects of neuroactive substances under different levels of control (such as alcohol and drugs) on the nervous system. Therefore, the substance addiction scales also consider severe addicts' physiological and psychiatric symptoms (such as blackouts, flashbacks, memory loss, hepatitis, convulsions, bleeding, etc.) which are influenced by neuroactive substances and subsequent illegal behaviors. However, no similar items are found in non-substance addiction scales.

Despite all scales being faced with the issue of cross-cultural applicability, non-substance addiction scales may encounter more challenges, and are shown to be completely different within varied cultural environments. Chen and Nath [19] found that the psychological measurement structures of Internet addiction was different within cross-cultural environments. Schaufeli et al. [104] found that work is considered to be more important and the average amount of hours worked in a week are longer in Japanese than in other regions in a study with workaholics. Raylu and Oei [93] believed that there is a higher incidence of pathological gambling in China. Culture related items of the Compulsive Buying Scale (CBS) are also difficult to adapt for diverse cultures [77].

8.4.3 *Trends in Substance and Non-substance Addiction Scales*

Alcohol and drug addiction are both in the addiction section of the DSM-I. In the DSM-V alcohol, caffeine, cannabis, hallucinogens, inhalants, opioids, sedatives/hypnotics/anxiolytics, stimulants, tobacco, and others (or unknown) are all in the substance use disorder section. This structural evolution suggests the trend of the integration of substance addiction. Alcohol addiction has been studied thoroughly and presented an example for other substance addictions. For example, the items of the DAST are consistent with the items of MAST. The SMAST and CAGE can also be used for substance addiction, although these scales were originally used for alcohol addiction. The earlier substance addiction scale was considered unfit for alcohol addiction since there is difference between alcohol and drugs. With the deepening of the understanding of addiction, researchers began to use the scales in alcohol addiction and found similar performance to substance addiction. After the year 2000, new substance addiction scales (such as the SDSS and ASSIST) will no longer limited the range to a single drug, but includes almost all the related substance, which highlights the integration trend of substance addiction scales.

Pathological gambling, as a typical non-substance addiction, is classified as a "disorder of impulse control that not classified" found in the DSM-III. It is the only non-substance related disorder that is viewed as a substance-related and addictive disorder in the DSM-V. Internet gaming addiction is incorporated into the DSM-V as having "conditions for further study," but its position is uncertain. Other excessively repetitive behavior is not accepted as an addiction diagnosis because of its lack of peer-reviewed evidence of diagnostic criteria. Some types of non-substance addiction may be more suitable to be classified as impulse-control disorders. Although most studies on non-substance addiction were based on the diagnostic criteria of pathological gambling, the specific addiction type made the classification more and more refined. The focus of each non-substance addiction is completely different, and thus non-substance addiction scales need a long time to be integrated so that they can be defined and classified in the ICD and the DSM more clearly.

Besides the trend of integration, substance addiction scales tended to be more concerned with psychometric properties and theoretical models. From the view of diagnostic requirements, discriminant validity was emphasized in the early stages of scale development. The subsequent research is to extend the range, to explore other measurement indexes, and to develop new scales with more comprehensive and effective indexes. The earlier studies with the MAST and CAGE only reported the validity results of sensitivity and specificity, and then added the reliability results in subsequent studies. In addition, substance addiction also showed a trend of more dependence on theoretical frameworks. We can found a separation between adult versions and adolescent versions in new scales based on addiction theories, which are also found in non-substance addiction scales. With the development of measurement techniques, we paid more attention to psychometric indexes instead of only face validity, and took these as an important standard of scale quality.

At the same time, cross-cultural communication is becoming increasingly popular along with socioeconomic development. Thus, cross-cultural applicability becomes an important feature and future direction in the development of the concept of addiction. The AUDIT and ASSIST, developed by the WHO, were used to face global health and cross-cultural problems. Other scales have more or less application in cross-culture environments. In the scales developed by the WHO, the usage amount/frequency are selected as research indexes and these are seen less in other scales. This indicates that they have wider value in a cross-cultural context even though the usage amount/frequency and other indexes are not sensitive in a specific cultural background.

8.5 Conclusion

The scales play an important role in discriminating between addicts and healthy participants, but they are meaningless without the necessary psychometric properties. The reliability and validity answer two important questions of the scales: one is what they really measure, and the other is whether the measurement is consistent. In this chapter, we have reviewed many scales targeting different conditions, and the variability on their reliability and validity in different fields indicates their applicability. There is no universal scale, and thus, it is necessary to select a scale or scales depending on the application.

Although there are many differences between substance addiction and non-substance addiction, the direct activation of reward systems is the shared mechanism of addiction. It is difficult to measure the physiological features of addiction, but the consequences can be measured. This is reflected in the scale items of unsuccessful efforts, social impairment, preoccupation, withdrawal, and tolerance. Moreover, there is a trend of integrating substance addiction scales and non-substance addiction scales. Besides the integration of addiction scales, cross-cultural application is another trend for worldwide cross-cultural communication. However, only the projects provided by the WHO address the problem, and the items regarding the usage amount and frequency are less seen in other scales.

Compared to the problem of cross-types (substance and non-substance) and cross-culture, the variability of the addiction concept, the specific need (such as the emergency or the legal system) and measurement theories and techniques are more difficult to deal with. The integration trend of addiction types and cultures indicates the feasibility of developing a universal scale; however, the variability of other factors makes it impossible. Thus, the balance between universality and specificity will run through the development of addiction scales.

Overall, the important factors can be summarized as four points throughout the development of the modern addiction scales: changes in the addiction concept, development of measurement theories and techniques, cross-cultural applicability, and sample applicability. Besides the developing trend of a focus on psychometric properties, the application range and method develop over time to a certain extent. These all influence the design and development of substance and non-substance addiction scales by expanding the groups studied, adding computer-aided investigation, widening the cross-cultural application, and considering the arguments between the developmental and mature models in theory and practice.

Acknowledgments The work was supported by National Natural Science Foundation of China 31171076. We thank Dr. Xiaochu Zhang for making suggestions about the manuscript.

References

1. Aertgeerts B, Buntinx F, Fevery J, Ansoms S (2000) Is there a difference between CAGE interviews and written CAGE questionnaires? Alcoholism Clin Exp Res 24(5):733–736
2. Aflatoonian MR, Meymandi MS, Divsalar K, Mahmoudi M, Heravi G (2011) Possible association between human blood types and opioid addiction. Am J Addict 20(6):581–584. doi:10.1111/j.1521-0391.2011.00170.x
3 Allen JP, Reinert DF, Volk RJ (2001) The alcohol use disorders identification test: an aid to recognition of alcohol problems in primary care patients. Prev Med 33(5):428–433. doi:10.1006/pmed.2001.0910
4 American Psychiatric Association (1952) Diagnostic and statistical manual of mental disorders, 1st edn. American Psychiatric Pub, Washington, DC
5 American Psychiatric Association (1968) Diagnostic and statistical manual of mental disorders, 2nd edn. American Psychiatric Pub, Washington, DC
6 American Psychiatric Association (1980) Diagnostic and statistical manual of mental disorders, 3rd edn. American Psychiatric Pub, Washington, DC
7 American Psychiatric Association (2013) Diagnostic and statistical manual of mental disorders, 5th edn. American Psychiatric Pub, Washington, DC
8 Andreassen CS, Griffiths MD, Hetland J, Kravina L, Jensen F, Pallesen S (2014) The prevalence of workaholism: a survey study in a nationally representative sample of Norwegian employees. PLoS One 9(8):e102446. doi:10.1371/journal.pone.0102446
9 Andreassen CS, Griffiths MD, Hetland J, Pallesen S (2012) Development of a work addiction scale. Scand J Psychol 53(3):265–272. doi:10.1111/j.1467-9450.2012.00947.x
10 Andreassen CS, Griffiths MD, Pallesen S, Bilder RM, Torsheim T, Aboujaoude E (2015) The Bergen shopping addiction scale: reliability and validity of a brief screening test. Front Psychol 6:1374. doi:10.3389/fpsyg.2015.01374
11 Antunes HKM, Andersen ML, Tufik S, Mello D, M. T. l. (2006) Physical stress and physical exercise dependence. Rev Bras Med Esporte 12(5):234–238
12 Bobzean SA, DeNobrega AK, Perrotti LI (2014) Sex differences in the neurobiology of drug addiction. Exp Neurol 259:64–74. doi:10.1016/j.expneurol.2014.01.022
13 Brand M, Laier C, Pekal J (2015) Comparison of cyber sex addiction in men and women. Z Psychosom Med Psychother 61(1):49–49
14 Cappella JN, Lerman C, Romantan A, Baruh L (2005) News about genetics and smoking – priming, family smoking history, and news story believability on inferences of genetic susceptibility to tobacco addiction. Commun Res 32(4):478–502. doi:10.1177/0093650205277320

15 Carey KB, Carey MP, Chandra PS (2003) Psychometric evaluation of the alcohol use disorders identification test and short drug abuse screening test with psychiatric patients in India. J Clin Psychiatry 64(7):767–774
16 Carnes P (1983) The sexual addiction. CompCare Publication, MN
17 Carnes P, Green B, Carnes S (2010) The same yet different: refocusing the Sexual Addiction Screening Test (SAST) to reflect orientation and gender. Sex Addict Compulsivity J Treat Prev 17(1):7–30
18 Carnes PJ, Green BA, Merlo LJ, Polles A, Carnes S, Gold MS (2012) PATHOS: a brief screening application for assessing sexual addiction. J Addict Med 6(1):29–34. doi:10.1097/ADM.0b013e3182251a28
19 Chen LD, Nath R (2016) Understanding the underlying factors of Internet addiction across cultures: a comparison study. Electron Commer Res Appl 17:38–48. doi:10.1016/j.elerap.2016.02.003
20 Chen SH, Weng LJ, Su YJ, Wu HM, Yang PF (2003) Development of Chinese internet addiction scale and its psychometric study. Chin J Psychol 45:279–C294
21 Chen VC, Chen H, Lin TY, Chou HH, Lai TJ, Ferri CP, Gossop M (2008) Severity of heroin dependence in Taiwan: reliability and validity of the Chinese version of the Severity of Dependence Scale (SDS[Ch]). Addict Behav 33(12):1590–1593. doi:10.1016/j.addbeh.2008.06.001
22 Coleman-Kennedy C (2002) Assessment and diagnosis of sexual addiction. J Am Psychiatr Nurses Assoc 8(5):143–151. doi:10.1067/mpn.2002.128827
23 Conrod PJ, Nikolaou K (2016) Annual research review: on the developmental neuropsychology of substance use disorders. J Child Psychol Psychiatry 57(3):371–394. doi:10.1111/jcpp.12516
24 Cortes-Tomas MT, Gimenez-Costa JA, Motos-Selles P, Sancerni-Beitia MD (2016) Different versions of the Alcohol Use Disorders Identification Test (AUDIT) as screening instruments for underage binge drinking. Drug Alcohol Depend 158:52–59. doi:10.1016/j.drugalcdep.2015.10.033
25 Costa S, Cuzzocrea F, Hausenblas HA, Larcan R, Oliva P (2012) Psychometric examination and factorial validity of the exercise dependence scale-revised in Italian exercisers. J Behav Addict 1(4):186–190. doi:10.1556/JBA.1.2012.009
26 Crocq MA (2007) Historical and cultural aspects of man's relationship with addictive drugs. Dialogues Clin Neurosci 9(4):355–361
27 Custer RL, Custer LF (1978) Characteristics of the recovering compulsive gambler: a survey of 150 members of Gamblers Anonymous
28 Davis RA, Flett GL, Besser A (2002) Validation of a new scale for measuring problematic internet use: implications for pre-employment screening. Cyber Psychol Behav 5(4):331–345
29 de Coverley Veale DM (1987) Exercise dependence. Br J Addict 82(7):735–740. doi:10.1111/j.1360-0443.1987.tb01539.x
30 Delmonico D, Miller J (2003) The internet sex screening test: a comparison of sexual compulsives versus non-sexual compulsives. Sex Relatsh Ther 18(3):261–276. doi:10.1080/1468199031000153900
31 Delmonico DLSS, Bubenzer DLSS, West JDSS (1998) Assessing sexual aAddiction with the sexual dependency inventory-revised. Sexual Addiction and Compulsivity (No.3), 179
32 Dhalla S, Kopec JA (2007) The CAGE questionnaire for alcohol misuse: a review of reliability and validity studies. Clin Invest Med 30(1):33–41
33 DiFranza JR, Savageau JA, Fletcher K, Ockene JK, Rigotti NA, McNeill AD et al (2002) Measuring the loss of autonomy over nicotine use in adolescents – the DANDY (development and assessment of nicotine dependence in youths) study. Arch Pediatr Adolesc Med 156(4):397–403
34 Edwards EA (1993) Development of a new scale for measuring compulsive buying behavior. J Financ Couns Plan
35 Etter JF (2009) Dependence on the nicotine gum in former smokers. Addict Behav 34(3):246–251. doi:10.1016/j.addbeh.2008.10.018

36 Etter JF, Le Houezec J, Perneger TV (2003) A self-administered questionnaire to measure dependence on cigarettes: the cigarette dependence scale. Neuropsychopharmacology 28(2):359–370. doi:10.1038/sj.npp.1300030
37 Ewing JA (1984) Detecting alcoholism the CAGE questionnaire. JAMA 252(14):1905–1907
38 Faber RJ, O'Guinn TC (1992) A clinical screener for compulsive buying. J Consum Res 19(3):459. doi:10.1086/209315
39 Fagerstrom KO (1978) Measuring degree of physical dependence to tobacco smoking with reference to individualization of treatment. Addict Behav 3(3–4):235–241
40 Fagerstrom KO, Schneider NG (1989) Measuring nicotine dependence: a review of the fagerstrom tolerance questionnaire. J Behav Med 12(2):159–182
41 Ferris J, Wynne H (2001) The Canadian problem gambling index: final report. Can Centre Subst Abuse
42 Flowers CP, Robinson B (2002) A structural and discriminant analysis of the work addiction risk test. Educ Psychol Meas 62(3):517–526. doi:10.1177/00164402062003008
43 Franken IH, Muris P, Georgieva I (2006) Gray's model of personality and addiction. Addict Behav 31(3):399–403. doi:10.1016/j.addbeh.2005.05.022
44 Garos S, Stock WA (1998) Measuring disorders of sexual frequency and control: the garos sexual behavior index. Sex Addict Compulsivity J Treat Prev 5(3):159–177
45 Gavin DR, Ross HE, Skinner HA (1989) Diagnostic validity of the drug abuse screening test in the assessment of DSM-III drug disorders. Br J Addict 84(3):301–307
46 Gibbs LE (1983) Validity and reliability of the Michigan alcoholism screening test: a review. Drug Alcohol Depend 12(3):279–285. doi:10.1016/0376-8716(83)90071-6
47 Godin G, Shephard RJ (1985) A simple method to assess exercise behavior in the community. Can J Appl Sport Sci 10(3):141–146
48 Gossop M, Darke S, Griffiths P, Hando J, Powis B, Hall W, Strang J (1995) The Severity of Dependence Scale (SDS): psychometric properties of the SDS in English and Australian samples of heroin, cocaine and amphetamine users. Addiction 90(5):607–614
49 Grant JE, Potenza MN, Weinstein A, Gorelick DA (2010) Introduction to behavioral addictions. Am J Drug Alcohol Abuse 36(5):233–241. doi:10.3109/00952990.2010.491884
50 Green BA, Arnau RC, Carnes PJ, Carnes S, Hopkins TA (2015) Structural congruence of the sexual dependency inventory—4th edition. Sex Addict Compulsivity 22(2):126–153. doi:10.1080/10720162.2015.1023386
51 Griffiths MD, Urbán R, Demetrovics Z, Lichtenstein MB, de la Vega R, Kun B et al (2015) A cross-cultural re-evaluation of the Exercise Addiction Inventory (EAI) in five countries. Sports Med Open 1(1):5. doi:10.1186/s40798-014-0005-5
52 Gungor B, Gulseren S, Dalmis A, Zorlu N (2013) The relation of alcohol addiction onset and family history with impulsivity and compulsivity. Anatol J Psychiatry 1. doi:10.5455/apd.38074
53 Hasin DS (2012) Combining abuse and dependence in DSM-5. J Stud Alcohol Drugs 73(4):702–704
54 Hausenblas HA, Downs DS (2002) How much is too much? The development and validation of the exercise dependence scale. Psychol Health 17(4):387–404
55 Heatherton TF, Kozlowski LT, Frecker RC, Fagerstrom KO (1991) The fagerstrom test for nicotine dependence: a revision of the fagerstrom tolerance questionnaire. Br J Addict 86(9):1119–1127
56 Hodgson SR, Hofford RS, Wellman PJ, Eitan S (2009) Different affective response to opioid withdrawal in adolescent and adult mice. Life Sci 84(1–2):52–60. doi:10.1016/j.lfs.2008.11.002
57 Hoffmann NG, Hunt DE, Rhodes WM, Riley KJ (2003) UNCOPE: a brief substance dependence screen for use with arrestees. J Drug Issues 33(1):29–44
58 Huang JC, Hu C, Wu TC (2010) Psychometric properties of the Chinese version of the workaholism battery. J Psychol 144(2):163–183. doi:10.1080/00223980903472219
59 Johnson EE, Hamer R, Nora RM, Tan B, Eisenstein N, Engelhart C (1997) The Lie/Bet questionnaire for screening pathological gamblers. Psychol Rep 80(1):83–88. doi:10.2466/pr0.1997.80.1.83

60 Jorgenson AG, Hsiao RC, Yen CF (2016) Internet addiction and other behavioral addictions. Child Adolesc Psychiatr Clin N Am 25(3):509–520. doi:10.1016/j.chc.2016.03.004
61 Kalichman SC, Johnson JR, Adair V, Rompa D, Multhauf K, Kelly JA (1994) Sexual sensation seeking: scale development and predicting AIDS-risk behavior among homosexually active men. J Pers Assess 62(3):385–397
62 Kanai A, Wakabayashi M, Fling S (1996) Workaholism among employees in Japanese corporations: an examination based on the Japanese version of the workaholism scales. Jpn Psychol Res 38(38):192–203
63 Kawada T, Inagaki H, Kuratomi Y (2011) The alcohol use disorders identification test: reliability study of the Japanese version. Alcohol 45(3):205–207. doi:10.1016/j.alcohol.2010.08.012
64 Keiflin R, Janak PH (2015) Dopamine prediction errors in reward learning and addiction: from theory to neural circuitry. Neuron 88(2):247–263. doi:10.1016/j.neuron.2015.08.037
65 Kim JS, Oh MK, Park BK, Lee MK, Kim GJ (1999) Screening criteria of alcoholism by alcohol use disorders identification test(AUDIT) in Korea
66 Kim SS, Gulick EE, Nam KA, Kim SH (2008) Psychometric properties of the alcohol use disorders identification test: a Korean version. Arch Psychiatr Nurs 22(4):190–199. doi:10.1016/j.apnu.2007.07.005
67 Kim Y (2014) Validating a Korean version of the drug abuse screening Test-10 (DAST-10). J Soc Serv Res 40(2):232–241. doi:10.1080/01488376.2013.875096
68 Kota D, Martin BR, Robinson SE, Damaj MI (2007) Nicotine dependence and reward differ between adolescent and adult male mice. J Pharmacol Exp Ther 322(1):399–407. doi:10.1124/jpet.107.121616
69 Lam LP, Leung WC, Ip P, Chow CB, Chan MF, Ng JW et al (2015) Validation of the Drug Abuse Screening Test (DAST-10): a study on illicit drug use among Chinese pregnant women. Sci Rep 5:11420. doi:10.1038/srep11420
70 Leite P, Range B, Kukar-Kiney M, Ridgway N, Monroe K, Ribas Junior R et al (2013) Cross-cultural adaptation, validation and reliability of the brazilian version of the Richmond Compulsive Buying Scale. Rev Bras Psiquiatr 35(1):38–43. doi:10.1016/j.rbp.2012.10.004
71 Lemmens JS, Valkenburg PM, Peter J (2009) Development and validation of a game addiction scale for adolescents. Media Psychol 12(1):77–95. doi:10.1080/15213260802669458
72 Lesieur HR, Blume SB (1987) The South Oaks Gambling Screen (SOGS): a new instrument for the identification of pathological gamblers. Am J Psychiatry 144(9):1184–1188. doi:10.1176/ajp.144.9.1184
73 Leung SF, Arthur D (2000) The alcohol use disorders identification test (AUDIT): validation of an instrument for enhancing nursing practice in Hong Kong. Int J Nurs Stud 37(1):57–64. doi:10.1016/S0020-7489(99)00052-8
74 Lichtenstein MB, Christiansen E, Bilenberg N, Stoving RK (2014) Validation of the exercise addiction inventory in a Danish sport context. Scand J Med Sci Sports 24(2):447–453. doi:10.1111/j.1600-0838.2012.01515.x
75 Lichtenstein MB, Jensen TT (2016) Exercise addiction in CrossFit: prevalence and psychometric properties of the exercise addiction inventory. Addict Behav Rep 3:33–37. doi:10.1016/j.abrep.2016.02.002
76 Magruder-Habib K, Stevens HA, Alling WC (1993) Relative performance of the MAST, VAST, and CAGE versus DSM-III-R criteria for alcohol dependence. J Clin Epidemiol 46(5):435–441. doi:10.1016/0895-4356(93)90020-2
77 Maraz A, Eisinger A, Hende B, Urban R, Paksi B, Kun B et al (2015) Measuring compulsive buying behaviour: psychometric validity of three different scales and prevalence in the general population and in shopping centres. Psychiatry Res 225(3):326–334. doi:10.1016/j.psychres.2014.11.080
78 McMillan LHW, Brady EC, O'Driscoll MP, Marsh NV (2002) A multifaceted validation study of Spence and Robbins' (1992) workaholism battery. J Occup Organ Psychol 75(3):357–368
79 McNeely J, Strauss SM, Rotrosen J, Ramautar A, Gourevitch MN (2016) Validation of an audio computer-assisted self-interview (ACASI) version of the alcohol, smoking and sub-

stance involvement screening test (ASSIST) in primary care patients. Addiction 111(2):233–244. doi:10.1111/add.13165
80 Messiah A, Encrenaz G, Sapinho D, Gilbert F, Carmona E, Kovess-Masfety V (2008) Paradoxical increase of positive answers to the Cut-down, Annoyed, Guilt, Eye-opener (CAGE) questionnaire during a period of decreasing alcohol consumption: results from two population-based surveys in Ile-de-France, 1991 and 2005. Addiction 103(4):598–603. doi:10.1111/j.1360-0443.2007.02120.x
81 Miele GM, Carpenter KM, Cockerham MS, Trautman KD, Blaine J, Hasin DS (2000) Substance Dependence Severity Scale (SDSS): reliability and validity of a clinician-administered interview for DSM-IV substance use disorders. Drug Alcohol Depend 59(1):63–75. doi:10.1016/S0376-8716(99)00111-8
82 Motl RW, McAuley E, Klaren R (2014) Reliability of physical-activity measures over six months in adults with multiple sclerosis: implications for designing behavioral interventions. Behav Med 40(1):29–33. doi:10.1080/08964289.2013.821966
83 Muench F, Parsons JT (2004) Sexual compulsivity and HIV: identification and treatment. Focus 19(6):1–5
84 Nathan PE, Conrad M, Skinstad AH (2016) History of the concept of addiction. Annu Rev Clin Psychol 12:29–51. doi:10.1146/annurev-clinpsy-021815-093546
85 O'Loughlin J, Tarasuk J, Difranza J, Paradis G (2002) Reliability of selected measures of nicotine dependence among adolescents. Ann Epidemiol 12(5):353–362. doi:10.1016/S1047-2797(01)00312-X
86 Orosz GB, Dombi E, Andreassen CS, Griffiths MD, Demetrovics Z (2015) Analyzing models of work addiction: single factor and Bi-Factor models of the Bergen work addiction scale. Int J Ment Health Addict, 1–10
87 Piper ME, McCarthy DE, Baker TB (2006) Assessing tobacco dependence: a guide to measure evaluation and selection. Nicotine Tob Res 8(3):339–351. doi:10.1080/14622200600672765
88 Piper ME, Piasecki TM, Federman EB, Bolt DM, Smith SS, Fiore MC, Baker TB (2004) A multiple motives approach to tobacco dependence: the Wisconsin Inventory of Smoking Dependence Motives (WISDM-68). J Consult Clin Psychol 72(2):139–154. doi:10.1037/0022-006X.72.2.139
89 Pontes HM, Griffiths MD (2015) Measuring DSM-5 internet gaming disorder: development and validation of a short psychometric scale. Comput Hum Behav 45:137–143. doi:10.1016/j.chb.2014.12.006
90 Pontes HM, Kiraly O, Demetrovics Z, Griffiths MD (2014) The conceptualisation and measurement of DSM-5 internet gaming disorder: the development of the IGD-20 test. PLoS One 9(10):e110137. doi:10.1371/journal.pone.0110137
91 Potenza MN (2006) Should addictive disorders include non-substance-related conditions? Addiction 1(Suppl 1):142–151
92 Radke AK, Gewirtz JC, Carroll ME (2015) Effects of age, but not sex, on elevated startle during withdrawal from acute morphine in adolescent and adult rats. Behav Pharmacol 26(5):485–488. doi:10.1097/FBP.0000000000000151
93 Raylu N, Oei TP (2004) Role of culture in gambling and problem gambling. Clin Psychol Rev 23(8):1087–1114. doi:10.1016/j.cpr.2003.09.005
94 Ridgway NM, Kukar-Kinney M, Monroe KB (2008) An expanded conceptualization and a new measure of compulsive buying. J Consum Res 35(4):622–639. doi:10.1086/591108
95 Robert JW, Judith AS, Sameer G, Noelle S, Karen F, Jennifer H, Joseph RD (2006) A comparison of the hooked on nicotine checklist and the fagerstrom test for nicotine dependence in adult smokers. Nicotine Tob Res 8(4):575–580
96 Robinson BE (1989) Work addiction: hidden legacies of adult children. Health Communications
97 Robinson BE (1999) The work addiction risk test: development of a tentative measure of workaholism. Percept Mot Skills 88(1):199–210. doi:10.2466/pms.1999.88.1.199
98 Robinson BE, Post P (1994) Validity of the work addiction risk test. Percept Mot Skills 78(1):337–338. doi:10.2466/pms.1994.78.1.337

99 Robinson BE, Post P, Khakee JF (1992) Test-retest reliability of the work addiction risk test. Percept Mot Skills 74(3 Pt 1):926. doi:10.2466/pms.1992.74.3.926
100 Rubinsky AD, Dawson DA, Williams EC, Kivlahan DR, Bradley KA (2013) AUDIT-C scores as a scaled marker of mean daily drinking, alcohol use disorder severity, and probability of alcohol dependence in a U.S. general population sample of drinkers. Alcohol Clin Exp Res 37(8):1380–1390. doi:10.1111/acer.12092
101 Saltstone R, Halliwell S, Hayslip MA (1994) A multivariate evaluation of the Michigan alcoholism screening test and the drug abuse screening test in a female offender population. Addict Behav 19(5):455–462
102 Saunders JB, Aasland OG, Amundsen A, Grant M (1993) Alcohol consumption and related problems among primary health care patients: WHO collaborative project on early detection of persons with harmful alcohol consumption-I. Addiction 88(3):349–362. doi:10.1111/j.1360-0443.1993.tb00822.x
103 Saunders JB, Aasland OG, Babor TF, de la Fuente JR, Grant M (1993) Development of the Alcohol Use Disorders Identification Test (AUDIT): WHO collaborative project on early detection of persons with harmful alcohol consumption--II. Addiction 88(6):791–804
104 Schaufeli WB, Shimazu A, Taris TW (2009) Being driven to work excessively hard the evaluation of a two-factor measure of workaholism in The Netherlands and Japan. Cross-Cult Res 43(4):320–348. doi:10.1177/1069397109337239
105 Schramm-Sapyta NL, Walker QD, Caster JM, Levin ED, Kuhn CM (2009) Are adolescents more vulnerable to drug addiction than adults? Evidence from animal models. Psychopharmacology 206(1):1–21. doi:10.1007/s00213-009-1585-5
106 Selzer ML (1971) The Michigan alcoholism screening test: the quest for a new diagnostic instrument. Am J Psychiatry 127(12):1653–1658. doi:10.1176/ajp.127.12.1653
107 Selzer ML, Vinokur A, van Rooijen L (1975) A self-administered Short Michigan Alcoholism Screening Test (SMAST). J Stud Alcohol 36(1):117–126
108 Shiffman S, Waters A, Hickcox M (2004) The nicotine dependence syndrome scale: a multidimensional measure of nicotine dependence. Nicotine Tob Res 6(2):327–348. doi:10.1080/1462220042000202481
109 Shram MJ, Funk D, Li Z, Le AD (2008) Nicotine self-administration, extinction responding and reinstatement in adolescent and adult male rats: evidence against a biological vulnerability to nicotine addiction during adolescence. Neuropsychopharmacology 33(4):739–748. doi:10.1038/sj.npp.1301454
110 Sicilia A, Alias-Garcia A, Ferriz R, Moreno-Murcia JA (2013) Spanish adaptation and validation of the Exercise Addiction Inventory (EAI). Psicothema 25(3):377–383. doi:10.7334/psicothema2013.21
111 Skinner HA (1982) The drug abuse screening test. Addict Behav 7(4):363–371
112 Skinner HA, Goldberg AE (1986) Evidence for a drug dependence syndrome among narcotic users. Addiction 81(4):479–484. doi:10.1111/j.1360-0443.1986.tb00359.x
113 Slotkin TA, Stadler A, Skavicus S, Seidler FJ (2016) Adolescents and adults differ in the immediate and long-term impact of nicotine administration and withdrawal on cardiac norepinephrine. Brain Res Bull 122:71–75. doi:10.1016/j.brainresbull.2016.03.006
114 Smith SS, Piper ME, Bolt DM, Fiore MC, Wetter DW, Cinciripini PM, Baker TB (2010) Development of the brief wisconsin inventory of smoking dependence motives. Nicotine Tob Res 12(5):489–499. doi:10.1093/ntr/ntq032
115 Spence JT, Robbins AS (1992) Workaholism: definition, measurement, and preliminary results. J Pers Assess 58(1):160–178. doi:10.1207/s15327752jpa5801_15
116 SriRajaskanthan R, Preedy V (2006) Biochemical markers of alcoholism and their clinical effectiveness. Clin Eff Nurs 9:e280–e285. doi:10.1016/j.cein.2006.09.004
117 Stautz K, Cooper A (2013) Impulsivity-related personality traits and adolescent alcohol use: a meta-analytic review. Clin Psychol Rev 33(4):574–592. doi:10.1016/j.cpr.2013.03.003
118 Terry A, Szabo A, Griffiths M (2004) The exercise addiction inventory: a new brief screening tool. Addict Res Theory 12(5):489–499. doi:10.1080/16066350310001637363

119 Tulevski IG (1989) Michigan Alcoholism Screening Test (MAST) — its possibilities and shortcomings as a screening device in a pre-selected non-clinical population. Drug Alcohol Depend 24(3):255–260. doi:10.1016/0376-8716(89)90064-1
120 U. N. Office on Drugs Crime (2015) World drug report 2015. United Nations Office on Drugs and Crime, Vienna
121 Valence G, D'Astous A, Fortier L (1988) Compulsive buying: concept and measurement. J Consum Policy 11(4):419–433
122 Volkow ND, Morales M (2015) The brain on drugs: from reward to addiction. Cell 162(4):712–725. doi:10.1016/j.cell.2015.07.046
123 Weikert M, Motl RW, Suh Y, McAuley E, Wynn D (2010) Accelerometry in persons with multiple sclerosis: measurement of physical activity or walking mobility? J Neurol Sci 290(1–2):6–11. doi:10.1016/j.jns.2009.12.021
124 Weikert M, Suh Y, Lane A, Sandroff B, Dlugonski D, Fernhall B, Motl RW (2012) Accelerometry is associated with walking mobility, not physical activity, in persons with multiple sclerosis. Med Eng Phys 34(5):590–597. doi:10.1016/j.medengphy.2011.09.005
125 Weinstein A, Zolek R, Babkin A, Lejoyeux M (2016) Gender differences in sex addiction. J Behav Addict 51:45–45
126 Wenzel M, McMillen J, Marshall D, Ahmed E (2004) Validation of the Victorian Gambling Screen. Gambling Research Panel, Melbourne
127 White HR, Labouvie EW (1989) Towards the assessment of adolescent problem drinking. J Stud Alcohol 50(1):30–37. doi:10.15288/jsa.1989.50.30
128 WHO Assist Working Group (2002) The Alcohol, Smoking and Substance Involvement Screening Test (ASSIST): development, reliability and feasibility. Addiction 97(9):1183–1194. doi:10.1046/j.1360-0443.2002.00185.x
129 Winters KC, Stinchfield RD, Fulkerson J (1993) Toward the development of an adolescent gambling problem severity scale. J Gambl Stud 9(1):63–84
130 Wolff N, Shi J (2015) Screening for substance use disorder among incarcerated men with the Alcohol, Smoking, Substance Involvement Screening Test (ASSIST): a comparative analysis of computer-administered and Interviewer-administered modalities. J Subst Abus Treat 53:22–32. doi:10.1016/j.jsat.2015.01.006
131 World Health Organization (2001) The alcohol Use disorders identification test. Guidelines for use in primary care. World Health Organization, Geneva
132 World Health Organization (2010) Global status report on noncommunicable diseases 2010. Retrieved from
133 World Health Organization (2014) Global status report on alcohol and health, 2014
134 Young KS (1998) Caught in the net: how to recognize the signs of internet addiction--and a winning strategy for recovery. Wiley
135 Young KS (1998) Internet addiction: the emergence of a new clinical disorder. Cyberpsychol Behav 1(3):237–244. doi:10.1089/cpb.1998.1.237
136 Zhou Y, Zhao M, Zhou C, Li R (2016) Sex differences in drug addiction and response to exercise intervention: from human to animal studies. Front Neuroendocrinol 40:24–41. doi:10.1016/j.yfrne.2015.07.001

Chapter 9
Biochemical Diagnosis in Substance and Non-substance Addiction

Wenwen Shen, Huifeng Liu, Xiaohu Xie, Haixiong Liu, and Wenhua Zhou

Abstract An optimal biochemical marker for addiction would be some easily traced molecules in body specimens, which indicates indulgent addictive behaviors, or susceptibility to certain addictive stimuli. In this chapter, we discussed existing literature about possible biomarkers, and classified them into three categories: origin forms and metabolites of substances, markers from biochemical responses to certain addiction, and genetic and epigenetic biomarkers suggesting susceptibility to addiction. In every category, we examined studies concerning certain type of addiction one by one, with focuses mainly on opiates, psychostimulants, and pathological gambling. Several promising molecules were highlighted, including those of neurotrophic factors, inflammatory factors, and indicators of vascular injury, and genetic and epigenetic biomarkers such as serum miRNAs. DNA methylation signatures and signal nucleotide polymorphism of candidate gene underlying the addiction.

Keywords Substance addiction • Non-substance addiction • Biochemical diagnosis

Addictive disorders are often diagnosed according to the behaviour criteria. The criteria include compulsive and impulsive acts toward addictive stimuli, consequent social impairment, increased tolerance to the stimuli, and physical and psychological uneasiness when withdrawal. Subjective as they are, diagnoses may be varied depending on the integrity and self-awareness of patients, as well as the judgment from psychiatrists. The situation calls for more objective measures of diagnoses for addiction and evaluations for addictive patients.

Biochemical markers are among the most promising objective indices for addictive disorders. Biological specimens such as urine or blood sample, have a long history as a source for measuring health and remains an important tool for clinical

W. Shen • H. Liu • X. Xie • H. Liu • W. Zhou (✉)
Zhejiang Provincial Key Laboratory of Addiction Research, Medical School of Ningbo University, Ningbo 315010, Zhejiang Province, China
e-mail: whzhou@vip.163.com

X. Zhang et al. (eds.), *Substance and Non-substance Addiction*, Advances in Experimental Medicine and Biology 1010, DOI 10.1007/978-981-10-5562-1_9

diagnosis. The specimens not only are easy to collection, but also require minimal cooperation from the patients. In alcohol use disorders, several biomarkers have been used routinely. Such indices include serum gamma-glytamyltransferase (GGT), serum carbohydrate deficient transferrin (CDT), and blood mean corpuscular volume (MCV), *etc.* in monitoring heavy alcohol drinking. It is also believed that other addictive substances can induce biochemical reactions so that certain biomarkers of indulgent use can be found. On the other hand, addictive disorders are considered as diseases of multifactorial inheritance. It is likely that certain inheritable features carry susceptibility to one or more addictive disorders, and affect the responsiveness to certain treatment. These genetic biomarkers, if established, would help clinicians with more personalized therapeutic choices.

According to their origins, we would divide these potential biomarkers into three categories, and discuss them separately:

1. abused substance and its metabolites;
2. biochemical and biomolecular responses to substances and addictive behaviours, which include secretion and modification of small molecules, proteins, etc.
3. epigenetic and genetic biomarkers suggesting susceptibility to addiction.

9.1 Substance and Its Metabolites

Substance in its original or metabolite form can be found in specimens like blood, urine, sweat, saliva, hair, and nail. Their existence confirms use of substance in a recent period. The time window for detection depends on the specimens used. A positive result from hair or nail indicates substance use in recent weeks or months; a positive urine or blood sample suggests substance use in day's range; and a breath with alcohol lasts only in hour's range. We should therefore choose specimens under different purpose. Regular urine/blood samples are used so that we can monitor patient's abstinence in a consecutive way. Convenient methods such as immunogold test paper for urine samples has made urine the most regular specimens in clinical practice. And breath examination of alcohol is usually used to detect dangerous use pattern, for example, drunk driving.

The existence of a certain substance in urine depends on a lot of factors. After a single dose, Substances like heroin or methamphetamine can be found in urine for 2–4 days. But repeated daily use or binge use prolongs their period of detection to 1 week or longer. The pH of the urine, fluid intake and clearance, and other personal differences in metabolism all contribute to the variation of time window. The time of detection of the most common abused substance has been listed in Table 9.1, as reviewed by Moeller et al. [1].

Serum metabolites for alcoholism has also been developed as indicators of drinking. Among them, phosphatidyl ethanol (PEth) is one of the most promising biomarkers. PEth can be detected for 3 weeks after only a few days of moderately heavy drinking (four drinks/day).

Table 9.1 Length of time drugs of abuse can be detected in urine, modified from [1]

Drugs and metabolites		Time window of detection
Alcohol		
	Alcohol as origin form	7–12 h
	Ethyl glucuronide (EtG) (heavy drinking)	1–2 days, binge drinking 5 day
	Ethyl sulfate (EtS)	1–2 day
	phosphatidyl ethanol (PEth)	3 weeks
Psychostimulants		
	Amphetamine	48 h
	Methamphetamine	48 h, binge use >1 week
	Cocaine metabolites	2–4 days
Opioid		
	Codeine	48 h
	Hydromorphone	2–4 days
	Methadone	3 days
	Morphine (main metabolite of heroin)	48–72 h
	Oxycodone	2–4 days
	Propoxyphene	6–48 h
Barbiturate		
	Short-acting (eg, pentobarbital)	24 h
	Long-acting (eg, phenobarbital)	3 weeks
Benzodiazepine		
	Short-acting (eg, lorazepam)	3 days
	Long-acting (eg, diazepam)	30 days
Marijuana		
	Single use	3 days
	Moderate use (4 times/week)	5–7 days
	Daily use	10–15 days
	Long-term heavy smoker	>30 days
Hallucinogen		
	Phencyclidine	8 days

The same abused substances can be detected in non-substance addiction, such as pathological gambling, sex addiction, and internet use disorder. People with non-substance addiction usually comorbid with substance addiction, or use abusive drugs as tools. These drugs either help them maintain energy (psychostimulants and nicotine), stimulate motivation (psychostimulants), or enhance pleasant sensation (psychostimulants, opiates). The first two conditions may suggest more compulsive urges in these patients to carry out the non-substance addictive behaviour. And the latter could be a clue of tolerance to the non-substance addictive stimuli. Therefore, detection of substance use in non-substance addiction often indicate more severe degree of the non-substance addiction. Treatment of such comorbidity should emphasize on the non-substance addiction, even though the problem of substance abuse usually gains more attention in clinical practice.

9.2 Biochemical/Biomolecular Responses to Substances and Addictive Behaviours

Though mechanisms underlying different addictions vary in certain ways, some biochemical reactions of our body to substance or non-substance addiction are similar. They both relate with the brain reward system and hypothalamic-pituitary-adrenal (HPA) axis. They also share similar behavioural patterns, like exhaustive adherence to the addictive stimuli, erratic hours for work and rest, poor hygiene habits, and other behaviours that may increase the risks of infections. It is possible that the brain neurotrophic factors, immune function, and vascular integrity be changed by these behavioural patterns.

We would review the evidence of biochemical changes in different addictive disorders, with particular interest in the following aspects: (1) diagnostic abilities of the indices that are suggestive of indulgence; (2) prognostic abilities that relate the indices to the outcome of an individual; (3) indices of particular damage caused by the addiction.

9.2.1 Developed Biomarker for Alcoholism

Numbers of biochemical indices for alcoholism have been developed (Table 9.2). They include serum gamma glutamyl transferase (GGT), serum aspartate amino transferase (AST), serum alanine amino transferase (ALT), serum carbohydrate-deficient transferrin (CDT), and blood mean corpuscular volume (MCV). Among them, GGT, ALT and AST are liver enzymes that indicate damage of liver cells. After abstinence, heavy drinkers have elevated GGT, ALT and AST levels lasting

Table 9.2 Summary of serum biomarkers for alcohol use

Biomarkers	Sensitivity %	Diagnostic specificity %	Time to return to normal limits after abstinence	Indicated drinking patterns
Gamma-glutamyltransferase (GGT)	34–85	11–95	2–4 weeks	>=5 drinks/day for weeks
Alanine aminotransferase (ALT)	18–58	50–57	2–4 weeks	Heavy drinking for weeks
Aspartate aminotransferase	15–69	47–68	2–4 weeks	Heavy drinking for weeks
Carbohydrate-deficient transferrin (CDT)	34–94	82–100	2–4 weeks	>=5 drinks/day for months
Mean corpuscular volume (MCV)	34–89	26–95	3 months or longer	Heavy drinking
Phosphatidyl ethanol (Peth)			2 weeks	>=3–4 drinks for days

Specificity and sensitivity from http://emedicine.medscape.com/article/285913-workup#c6

for weeks. MCV enlargement is less specific to liver enzymes, and could last several months after abstinence. CDT is a group of isoforms of the iron transport protein transferrin. It is usually measured as the percentage of total transferrin that is carbohydrate deficient. CDT is considered to bear high specificity of heavy drinking, but might be less sensitive in female and in episodic heavy drinking. A combination of these biomarkers would help to enhance the identification of hazardous drinking, and measure the level of daily drinking.

9.2.2 Hypothalamic-Pituitary-Adrenal Axis hormones

Generally, addictive patients have basal levels of serum cortisol and corticotropin comparable to healthy controls. Normal cortisol and corticotropin levels have been reported in alcoholics 2, 3, current opiate users 4, pathological gamblers, and patients with internet use disorder 5. But sporadic reports of raised cortisol or corticotropin levels can be seen. For example, Kiefer et al. reported higher basal serum levels of cortisol in alcoholic patients with 15–25 h' abstinence 6. Sinha et al. also reported higher basal corticotropin levels in alcoholic patients who had been abstinent for 1 month 7.

Meanwhile, it is agreed that the stress coping reactions of the HPA axis hormones are altered, which are demonstrated in experimental settings. Current heroin users fail to manifest an elevation of corticotropin in the metyrapone stimulation test, which drug blocks the production of cortisol in the adrenal glands, and leads to the rise of corticotropin. But the same test lead a higher than normal reaction in patients who had been abstinent from heroin. Abstinent patients with cocaine use disorders may also be over reactive in the metyrapone stimulation test. These results suggest hypersensitivity of the stress system in the withdrawal phase. Similar results can been observed in psychological studies where subjects are exposed to neutral, stressful, and alcohol cues, respectively 7. In this study, healthy individuals exhibited low corticotropin and cortisol response to neutral, relaxing scripts, and significantly elevated responses to stressful ones. In comparison, the alcoholics who had been abstaining for 1 month had higher corticotropin and lower cortisol responses. Notably, the response differences to neutral, stressful, and alcohol cue scripts were largely diminished, indicating a lasting stressful state in abstinent patients. Other study using Trier Social Stimulation Test also found diminished cortisol stress responses measured by the area under curve of serum cortisol 2.

The impairment of stress hormone responses may be predictors of the future relapse. In the study by Sinha et al., shorter time to alcohol relapse during 3 months' follow-up were associated with higher cortisol responses in the stress condition and higher cortisol to corticotropin ratio 7. And lack of cortisol responses in the Trier social stimulation test is a predictor of alcohol relapse 2. In cocaine dependent patients, the serum cortisol and corticotropin response to stress were also positively correlated to the average cocaine use per occasion, but not time to relapse 8. Preliminary study of non-substance addiction also suggest a negative correlation between serum cortisol levels and the severity of pathological gambing 5.

In conclusion, hyperactivities of corticotropin may be common in abstinent subjects with addiction. It may cause hypersensitivities even in neutral situation, and a seemingly blunted response under stress. Although complicated to carry out, the cortisol responses to stress might be a promising predictor of future relapse.

9.2.3 Neurotrophic Factors

Brain-derived neurotrophic factor (BDNF) is a kind of neurotrophins, which help the growth, survival and differentiation of developing neurons in the central and peripheral nervous system. Abundant in central nervous system, BDNF is also found in human circulation. The source of BDNF in the peripheral blood may include the pituitary gland, platelets, vascular endothelial cells, lymphocytes and monocytes, etc. [9]. During clotting process, activated platelets release BDNF into the blood, resulting an approximately ten-times accumulation of BDNF concentration in the serum 10. The highly concentrated BDNF in the serum may cover up the probable diurnal changes of BDNF plasma levels observed in men, but not observed in women. Demographic factors, such as age and gender, were considered as irrelevant to the serum levels of BDNF.

Studies of BDNF serum levels in patients with substance addiction yields controversial results (see Table 9.3). Some preclinical studies showed no significant difference between alcoholics in early withdrawal (day 1) and healthy controls, and no significant time effect in early withdrawal days (day 1, 7, 14) [3, 11]. On the other side, one study in Taiwan reported significantly reduced BDNF serum levels in alcoholics, especially those with delirium tremens. And the serum levels of BDNF were raised in withdrawal day 7, compared to the baseline [12]. In heroin users, BDNF serum levels were increased in both current users and in those of early withdrawal (<7 day), and remain in high level after 1 month abstinence [13, 14]. But negative result was also reported in patients with heroin use disorder [15]. In patients with psychostimulant use disorders, reduction of serum BDNF levels in early withdrawal (day 1–21) were reported [16, 17], and its levels were increasing during early abstinence. One study examined the cocaine users after 3 weeks withdrawal, and found increased levels of serum BDNF, with higher levels of BDNF predicting quicker relapse [18]. However, a group in China reported the opposite trend, with higher levels of serum BDNF in methamphetamine users in early withdrawal (<7 days), and reduced to normal after 1 month abstinence [19].

Fewer studies were done concerning non-substance addiction. One preliminary study showed increased serum levels of BDNF in 14 male patients with pathological gambling. There were no significant correlations between BDNF serum levels and severity of pathological gambling or clinical and demographic variables [20]. The same investigators failed to confirm the phenomenon in patients with internet use disorders [21].

If we overview of the BDNF serum levels across these researches (Table 9.3), we could find great discrepancies of the values, from 500 to 26,000 pg/ml in the mean levels of healthy subjects. Great differences existed even in the studies using the

Table 9.3 Clinical researches of BDNF as a biomarker for addictive disorders

Abusive substance or behaviour	Sample	Methods and materials	No. of patients (No. of healthy controls)	Time of detection	BDNF(pg/ml) in patients	BDNF(pg/ml) in control	Conclusion	Correlation	Author and year
Alcohol	Serum	ELISA, R&D	52 male [41]	Withdrawal day 1, 7, 14	WD1: 630 ± 501 WD7: 548 ± 421 WD14: 614 ± 418	535 ± 345	No difference	Negatively associated with alcohol withdrawal severity on day 1	Heberlein 2010 [11]
Alcohol	Serum	ELISA, Promega	DT 25, non-DT 40 (39)	Withdrawal day 1, 7	Non-DT 12,300 ± 3300–>13,400 ± 3500; DT 6200 ± 2600–>8900 ± 4400;	14,800 ± 4700	Lower in pts. with DT, increasing from baseline to day 7	None	Huang 2011 [12]
Alcohol	Serum	ELISA, Promega	27 current abstinent		2853 ± 2113	/			Hilburn 2011 [22]
Alcohol	Serum	ELISA, ChemiKine	14 male (10)		12,690 ± 4790–>(post social stress)16390 ± 6220	11,760 ± 5050–>14,780 ± 4650	No difference	None	Meng 2011 [3]
Opiate	Serum	ELISA, R&D	72 (90),	Withdrawal <7 day or 1 month	1565 ± 511–>1454 ± 556	1241 ± 335	Higher. No time effect.		Zhang 2014 [14]

(continued)

Table 9.3 (continued)

Abusive substance or behaviour	Sample	Methods and materials	No. of patients (No. of healthy controls)	Time of detection	BDNF(pg/ml) in patients	BDNF(pg/ml) in control	Conclusion	Correlation	Author and year
Opiate	Serum	ELISA, R&D	27 male (21)		About 870 ± 60->865 ± 60	About 570 ± 50	Higher		Heberlein 2011 [13]
Opiate	Serum	ELISA, R&D	15 (15)		5092 ± 109	5433 ± 101	No difference	None	Angelucci 2007 [15]
Cocaine	Serum	ELISA, Promega	22 current abstinent		3660 ± 2378	/		(Psychostimulant) Negatively correlated with current abstinent days	Hilburn 2011 [22]
Cocaine	Serum	ELISA, R&D	15 (15)		5182 ± 171	5433 ± 101	No difference	None	Angelucci 2007 [15]
Cocaine	Serum	ELISA, R&D	35 (34)	Withdrawal 3 weeks	About 35,000 ± 11,000 (30,000 ± 10,000 for non-relapsers vs 39,000 ± 10,000 for relapsers)	About 26,000 ± 11,000 ng/ml	Higher. Relapsers had higher levels than non-relapsers	None	D'Sa 2011 [18]
Cocaine	Serum	ELISA, Aushon BioSystems	23 (46),	Withdrawal day 1 and 14	51,676 ± 17,505–>60,643 ± 22,607	76,044 ± 32,661	Lower. increasing across 12 days	Baseline and post-detoxification BDNF correlated with baseline caving	Corominas-Roso 2012 [17]
Methamphetamine	Serum	ELISA, Promega	59 (59)	Withdrawal 1–7 days: 32, withdrawal 8-21 days: 27	9840 ± 4850 (9320 ± 4330, 10,450 ± 5410, respectively)	16,260 ± 4720	Lower. No time effect	None	Chen 2014 [16]

Methamphetamine	Serum	Promega	15 current abstinent		3779 ± 2803	/			Hilburn 2011 [22]
Methamphetamine	Serum	ELISA, R&D	179 (90)	Withdrawal <7 days or 1 month	1460 ± 490 (1621 ± 591–>1364 ± 581)	1241 ± 336	Higher. Decreasing over time	None	Ren 2016 [19]
Gambling	Serum	ELISA, Promega Inc.	14 male (13)		4762 ± 1335	3479 ± 1231	Higher	None	Geisel 2012 [20]
Internet use	Serum	ELISA, Promega Inc.	11 male (10)		4095 ± 1196	4556 ± 1538	No difference	None	Geisel 2013 [21]

DT deliriumtremens

same kits and methods. It implicates that methodological improvement are needed to interpret these contradictory results. And the meaning of BDNF in the peripheral blood need further clarification, given its close relationship with the clotting process.

Glial cell line-derived neurotrophic factor (GDNF) is a small protein which may promote the survival and differentiation of dopaminergic neurons and motor neurons. It is not clear of the source of serum GDNF, but its serum levels are significantly higher than those of cerebrospinal fluid [23]. Some preliminary studies showed that serum levels of GDNF were consistently downregulated in alcoholics from withdrawal day 1 to 14 [11],and the serum levels of GDNF in patients with opiate dependence was comparable to healthy controls [13].

S100B belongs to a family of small, acidic proteins of S100. It is a calcium-binding peptide secreted by the astrocytes. It acts as a neurotrophic factor in developing brains, but are elevated and potentially deleterious in adult brains with injuries. The increase of serum S100B levels accurately indicates the presence of acute brain damage or neurodegenerative diseases, which is more sensitive than brain imaging. Serum levels of S100B were decreasing during the withdrawal day 1–5 in alcoholic patients, suggesting a tendency of brain recovery over early withdrawal, though the average levels of serum S100B were within normal range [24].

Galanin is a neuropeptide expressed in central nervous system and in enteric neurons. It is involved in the modulation and inhibition of actin potentials in neurons. It also has roles in development as well as acting as a trophic factor. Galanin serum levels significantly decreased in patients with alcohol dependence in early withdrawal (day 1–14), and were negatively associated with the craving of alcohol [25].

9.2.4 Inflammatory Factors

The behaviours of substance abuse may exert two-faced actions to the immune system. Opiates suppresses immune reaction by inhibiting proliferation of T- and B-cells, suppressing T cell-mediated cytotoxicity, and decreasing the activity of NK cells. Meanwhile, the addictive behaviours, including smoking, injection, and binge use, *etc.* may also increase the incidence of infectious diseases, therefore increase the inflammatory factors in the blood.

Significant elevation was noted in blood levels of C-reactive protein, erythrocyte sedimentation rate, total lymphocyte count, serum globulins and the globulin-albumin ratio in patients with substance addiction, suggesting an enhanced inflammatory status [26]. Another study found an approximately 10% elevation of serum levels of IgG, IgA, and IgM in drug users,as well as increased numbers of monocytes, neutrophils, and eosinophils, and reduced lymphocyte counts peripherally [27]. Meanwhile, decreased function of T-lymphocytes was suggested in heroin users, whose blood cells exert a pattern of less IFN-gamma and more IL-10 secretion after PBS and LPS stimulation [28]. And in patients with methamphetamine dependence, the serum levels of complement factor H increased to 5.8 times to the healthy control, which may reduce the complement activity on pathogenic cells, increasing susceptibility to microbial infections [29].

The idea of using inflammatory factors as biomarkers for alcoholism has been discussed in more details in the review by Achur et al. [30]. Notably, inflammatory status is relevant to the withdrawal process of alcohol. Elevated serum levels of TNF-alpha and IL-6 were found in male alcoholics from withdrawal day 1 to day 14 [6, 31]. While the serum levels of IL-6 decreased across the withdrawal process, the levels of TNF-alpha remains high. Moreover, the serum levels of IL-6 was negatively associated with craving, depression, and trait anxiety, and the serum levels of TNF-alpha were associated with the serum levels of BDNF [31].

On the other side, immune function is related to the liver function of patients with alcoholism. A study examined the function of monocytes in secreting IL-1beta, TNF-alpha, IL-6, and IL-12 in alcoholics with or without liver diseases. It suggested that the patients without liver diseases had higher functioned monocytes compared to healthy controls and the patients with liver cirrhosis. And the monocytes from active-drinking, liver-cirrhotic patients had significantly lower ability to produce IL-1beta and TNF-alpha [32]. Therefore, it might be important to note the liver condition of the patients with alcoholism. An investigation to the alcoholics without liver disease found only elevated serum levels of IL-6, but not IL-8, IL-10, IL-12, or TNF-alpha [33].

To conclude, patients with substance addiction usually suffer from inflammatory status, which also may be related to the liver function. By far, no relevant study has been released about peripheral immune function in pathological gambling or other non-substance addictive disorders.

9.2.5 Indicators of Vascular Injury

Homocysteine is a homologue of the amino acid cysteine. Enhanced serum levels of homocysteine are correlated with higher risks of endothelial injury, which induces inflammation, atherogenesis, and ultimately ischemic injury of the organs. Blood alcohol concentration is closely correlated with total plasma homocysteine ($R^2 = 0.7662$) [34]. Further studies showed elevated serum levels of homocysteine, which gradually declined during early withdrawal [24, 35]. Its serum levels were associated with serum levels of folate and riboflavin, and about half patients had constantly low levels of serum homocysteine [35]. The serum levels of asymemetric di-methylarginine (ADMA) decrease during the withdrawal day 0–3, but return to normal levels compared to the controls. There were no association between serum levels of homocysteine and ADMA [36]. Correspondingly, serum levels of vascular endothelial growth factor A (VEGFA) also increased in patients with alcoholism during withdrawal day 1–14, which declined over time [37].

Serum levels of homocysteine and VEGFA were also investigated in patients with opiate use disorders. The serum levels of homocysteine were significantly higher in 103 opium-addicted patients than 114 controls (11.49 ± 7.45 vs. 8.02 ± 3.87 μmol/l). But the serum levels of VEGF-A did not differ significantly with the controls in the patient with injective diamorphine maintenance [38].

9.2.6 Lipids and Adipocyte-Derived Hormones

It was suggested in cocaine users that higher relapse rates were associated with lower serum levels of total cholesterol [39]. Serum levels of total cholesterol and LDL-c were also negatively related with heroin craving in patients with methadone maintenance therapy [40].

Alcoholics had higher serum levels of leptin during early withdrawal [6]. But the serum levels of leptin and adiponectin were decreased in patients with heroin dependence, while the levels of resistin were increased. Unlike the healthy controls, whose levels of adipocyte-derived hormones were associated with the body mass index (BMI), the serum levels of these hormones were irrelevant to BMI in patients with heroin dependence [41]. The irrelevance may be explained by the close association of adipocyte-derived hormones with inflammatory factors (TNF-alpha) [6]. One year maintenance on methadone normalized the serum levels of leptin, but not adiponectin or resistin [41]. Adiponectin down-regulation was also confirmed in 88 patients with opium dependence, while the serum levels of leptin were comparable to the controls [42].

9.2.7 Other Proteins

Alpha-synuclein is a protein abundant in the presynaptic terminals. It may interact with phospholipids and proteins and segregate abnormally under conditions such as Alzheimer disease and Parkinson disease. Serum levels of alpha-synuclein were significantly higher in alcoholics (14 ± 13 ng/ml vs 6 ± 10 ng/ml), and it is positively related to the alcohol craving measured by obsessive-compulsive drinking scale [43]. The same phenomenon was confirmed in cocaine users. Patients with cocaine dependence has significantly higher levels of alpha-synuclein than controls (35.7 ± 26.6 ng/ml vs. 4.1 ± 2.5 ng/ml), and its levels were positively associated with the scores of Minnesota Cocaine Craving Questionnaire.

Mitochondrial (mt) abnormalities were also associated with opiate addiction, with reduced plasma levels of mtDNA copy number and increased levels of mtDNA damage ratio in heroin users [44]. It suggests an increased oxidant stress, which might be associated with the inflammatory status in substance users. Meanwhile, elevated serum levels of insulin-like growth factor-1(IGF-1) were found in patients with opiate dependence (26.6 ± 10.4 vs. 22.6 ± 9.2 nmol/L), suggesting immune and hepatic activation [45].

A slight elevation of prolactin (9.3 ± 4.1 vs. 7.3 ± 2.9 ng/ml) was observed in patients with cocaine dependence. The levels of PRL were also correlated with addiction severity index of drug use, alcohol use, and psychological problems, but not with the treatment retention or outcome [46].

9.3 Epigenetic and Genetic Biomarkers

There are three major modes of epigenetic regulation, namely histone acetylation and methylation, DNA methylation, and non-coding RNAs. All of them profoundly affect the expression of genes, the translation into proteins, cellular structures and functions, and ultimately the apparent phenotypes. Given technical convenience and affordability, changes of DNA methylation are among the most promising ones to be developed into biomarkers. Signatures of DNA methylation of the peripheral blood cells have been related with exposure of drugs of addiction [47].

9.3.1 Methylation of DNAs

One of the convenient strategies of searching epigenetic biomarker is to focus on genes of interest. Genes that involve production, binding and degeneration of neurotransmitters are given particular interests, followed by genes of neurotrophic factors and other proteins in the neural system. Significant changes in methylation at many loci have been associated with alcoholism, including the OPRM1, the serotonin transporter, the dopamine transporter, DNA methyltransferase 3b, NMDA receptor subtype 2b, monoamine oxidase A, vasopressin and atrial natriuretic peptide, proopiomelanocortin, orexin A, nerve growth factor, MeCP2, leptin, alpha-synuclein, and homocysteine-induced endoplasmic reticulum protein. Nicotine dependence has also been associated with decreased methylation at the Monoamine Oxidase A and B (MAO-A, MAO-B) promoter methylation in peripheral blood, differential patterns of methylation at ras association domain family 1 isoform A (RASSF1A) and catechol-O-methyltransferase (COMT) [47].

OPRM1 encode the opioid receptor μ, which is the receptor for endogenous β-endorphin and enkephalins, as well as for synthetic opioids. Nielsen et al. analysed methylation at sixteen CpG dinucleotides in the promoter of the OPRM1 gene in the peripheral lymphocytes, and found two GpG sites associated with methadone maintaining patients who were heroin dependent [48]. The results were then extended ethnically from Caucasians to African Americans and Hispanics, although with ethnic differences. Overall hypermethylation of OPRM1 CpG sites was suggested in these patients with opioid dependence, as compared to healthy controls [49]. And these methylation was suggested with the reduction of OPRM1 expression in the lymphocytes. These findings were confirmed by Chorbov et al. [50], who also showed an increased CpG methylation at the OPRM1 promoter in the blood and sperm in patients with opioid dependence. They suggest that the hypermethylation may block the binding of transcription activators such as Sp1, leading to silencing of the gene. Epigenetic heritability is implicated as the modification can be observed in the sperm.

CB1 is the gene for Cannabinoid receptor 1. A recent study found hypomethylation of CB1 promotor was associated with clinical measurements of cannabis abusing, including craving. The levels of methylation was also negatively related with CB1 gene expression [51]. Further studies are needed to verify the result.

Elevated methylation of BDNF was also suggested to be associated with drug dependence. Methylation of the promotor CpG5 in BDNF in peripheral blood was correlated with opioid addiction and increased negative mood during abstinence [52]. Meanwhile, hypomethylation at BDNF gene was suggested in the whole blood of the adolescent offspring who had contacted nicotine via maternal smoking [53].

With the development of array-based platforms, several loci have emerged as definite indicators of smoking. The two most prominent locus were coagulation factor II (thrombin) receptor-like 3 (F2RL3) and the aryl hydrocarbon receptor repressor (AHRR) [47]. F2RL3 is a member of the proteinase-activated receptor family, locating on chromosome 19p13.11. There is robust evidence that hypomethylation at F2RL3 locus is strongly associated with smoking across ages, genders, and ethnicities. Given its evidence of being a candidate biomarker for heart disease, the F2RL3 locus seem to bridge the well-recognized association between smoking and cardiovascular mortality. AHRR is a key regulator of the aryl hydrocarbon receptor (AHR) pathway which is responsible for the detoxification of toxins locating on chromosome 19p13.11. hypomethylation of AHRR locus is strongly associated with smoking, and this association is also valid across ages, genders, and ethnicities. It is considered a mature biomarker that is ready for clinical use.

In brief, several DNA methylation signatures have been developed as potential biomarkers for addictive disorders, especially for alcoholism and nicotine dependence. Investigations of epigenetic biomarkers for less popular substances, for example heroin and methamphetamine, are still insufficient, albeit they are clinically more important.

9.3.2 Serum microRNAs

MicroRNAs (miRNAs) are a class of short noncoding RNA that can regulate the expression of large numbers of protein-coding mRNA transcripts. By binding to the 3′ untranslated region (3′ UTR) of target transcripts, they block their translation into the encoded protein, or trigger their destabilization and degradation. MicroRNAs have potential as useful biomarkers for clinical use because of their stability and ease of detection in many tissues, especially in blood. They are present in the serum and plasma of humans, and the levels of miRNAs in serum are stable, reproducible, and consistent among individuals of the same species [54]. More recently, miRNA in the adult nervous system has been revealed to play a role in neuronal plasticity, including the regulation of synaptic protein synthesis, dendritic spine morphogenesis, and plasticity-related diseases. Evidence is also emerging for miRNA involvement in physiological higher-order brain functions, such as learning, memory, emotion, and mental illness [55, 56].

Abnormal patterns of miRNA expression have been found in drug addictions [57]. In central nerve system, specific miRNAs whose expression is altered are identified regulating the motivational effects of addictive drugs [58, 59]. Drug abuse also impair immune system function, as heroin impact the expression of miRNAs related to immunity and virus infection [60, 61]. For example, the levels of miRNA-582-5p and miRNA-590-5p in the peripheral monocytes of heroin abusers were significantly decreased, as compared with those in healthy controls [62]. Heroin use also dysregulated a panel of plasma miRNAs [61]. Correlation of the miRNA with psychiatric measures of addiction should be emphasized as a strategy to find potential biomarkers.

9.3.3 Genetic Markers

Genetic factors contribute to the formation of addictive disorders. Sibling and twin studies found the heritability of addictive disorders around 30–70 %. The genetic susceptibility to addiction comes from different aspects. (1). Inherited temperament, such as impulsivity, sensation-seeking, and novelty-seeking, may interact with environmental factors, thus increase the likelihood of substance exposure. (2). Deficiency in stress response contribute to the progress from misuse to dependence. These features may be shared by the most kinds of addiction, including non-substance addiction. Genes affect the common reward pathway may include OPRM1, DRD1, DRD2, COMT, and HTR2A *et cetera* (see Table 9.4 for details). (3). Genetic alterations may also render susceptibility of dependence to a specific drug, when they are related to the pharmacokinetics and pharmacodynamics. For example, deficiency of ALDH2 increasesblood concentration of acetaldehyde after alcohol exposure, which evoke intoxicated reactions and low tolerance of alcohol. Approximately 50 % East Asians carry the mutated ALDH2 genes, and they are less likely to develop into alcohol dependence.

Many studies have been done investigating possible genetic markers for certain addictive disorders, signal nucleotide polymorphism (SNP) being the most studied ones. There are mainly two strategies for candidate SNP screening: target gene(s) of interest, and genome wide assessment. In both cases, type I errors are common, and cross verification is needed. Confounding factors, such as ethnics, degree of environmental exposures, and protective environmental features decrease the consistency and increase the possibility of false positive of genetic findings. A study with convincing result would therefore include large cohort of patients with the controls sharing similar environment. It should have robust effects that can stand the strictest multiple testing correction, and can be verified in some other cohorts.

Table 9.4 Summary of SNP studies concerning addictive disorders

Gene category	Gene name	Encoded protein	SNP	Targeted addiction	Diagnostic relevance	First author and year of publication
Opioid	OPRM1	Opium receptor, μ 1	Rs1799971 (A118G)	Opioid (heroin)	118A, 118G related to opioid addiction, depending on studies	Bond 1998; Bart 2004; Kapur 2007; Nagaya 2012; Haerian 2013 [95–98]
			Rs1799971 (A118G)	Not specified	118G increase the risk of drug overdose that needs emergency care	Manini 2013 [64]
			Rs1799972(C17T)	Opioid		Bond 1998 [95]
			Rs9479757 (intronic, pre-mRNA splicing)	Opioid (heroin)	Higher daily consumption and injection	Xu2014 [65]
			rS2075572 (G691C)	Methamphetamine (psychotic)		Ide 2006 [66]
	OPRD1	Opioid receptor, δ 1	Rs1042114 (G80T, exon 1)	Opioid (heroin)		Zhang 2008 [99]
			Rs678849	Opioid	MMT outcome	Crist 2013 [100]
			Rs 581111	opioid	Buprenorphine treatmentoutcome in women	Clarke 2014 [101]
			Rs529520			
			Rs2234918(C921T, exon 3), Rs2236861 (intron 1), rS3766951 (intron 1), rs2236857 (intron 1)	Opioid (heroin)		Levran 2008; Beer 2013; Nelson 2014 [102–104]
	OPRK1	Opioid receptor, κ 1	Rs1051660 (G36T)	Opioid (heroin)	Vulnerability	Yuferov 2010 [105–107]
			Rs6473797 (non-coding)	Opioid (heroin)	Resistance to addiction	Levran 2008 [103, 105–107]
	POMC	Proopiomelanocortin, the precursor of ACTH, β-endorphin, etc	Rs934778 (intron 1), rs10009388 (intron 1)	Opioid	Opioid dependence, but not alcohol or other illicit drug dependence	Xuei 2007 [108]
	PENK	Proenkephalin	rs2609997	Opioid	Opioid dependence, but not alcohol or other illicit drug dependence	Xuei 2007 [108]
			rs1975285 (5′ flanking region), rs1437277 (intron)			
	PDYN	Prodynorphin	Rs1022563	Opioid (heroin)		Clarke 2012 [109]

Catecholamine	DRD1	Dopamine receptor D1	Rs5326 (intron)	Opioid (heroin)		Levran 2009 [110]
			Rs686 (exon 4), rs265975, rs265973	Opioid (heroin)		Jacobs 2013 [111]
			Rs265981 (T800C)	Pathological gambling		Lobo 2007 [112]
			Rs4532 (DRD1 DdeI polymorphism)	Pathological gambling	Homozygous higher in patients	Comings 1997 [113]
	DRD2	Dopamine receptor D2	rS1800497	Opioid	MMT outcome	Lawford2000 [67]
			Rs1800497	Opioid (heroin)	Higher daily consumption	Lawford 2000 [67]
						Vereczkei 2013 [68]
			Rs1800497	Pathological gambling		Comings 1996; Lobo, 2010; Fagundo 2014 [69–71]
			Rs1079597	Opioid (heroin)		Hou 2009 [114]
	DRD3	Dopamine receptor D3	Rs167771	Pathological gambling		Lobo 2015 [115]
	DRD4	Dopamine receptor D4	Rs1800955	Heroin addiction		Vereczkei 2013 [68]
	ADRA1A	Adrenegic receptor α 1A	Rs26779851, 5′ promoter	Opioid (heroin)		Levran 2009 [110]
	COMT	Catechol-O-methyl transferase	Rs4680 (G472A, exon 4)	Opioid		Horowitz 2000[72]
						Oosterhuis 2008 [74]
			Rs4680	Methamphetamine abuse		Li 2004 [73]
			Rs4680	Gambling and drinking at some risk (non clinical)		Guillot 2015 [75]
	DBH	Dopamine β-Hydroxylase	Rs1611115	Opioid (heroin)	Daily consumption and injection	Xie 2013 [116]

(continued)

Table 9.4 (continued)

Gene category	Gene name	Encoded protein	SNP	Targeted addiction	Diagnostic relevance	First author and year of publication
Serotonergic	HTR1B	5-hydroxytryptamine receptor 1B	Rs6296	Opioid (heroin)		Gerra, 2004; Gao, 2011; Cao, 2013 [117–119]
			Rs6297			
			Rs130056			
			Rs130058			
	HTR2A	5-hydroxytryptamine receptor 2A	Rs6311 (A1438G)	Opioid (heroin)		Saiz 2008 [120]
			rs6313 (T102C)	Pathological gambling		Wilson 2013 [121]
	HTR3A	5-hydroxytryptamine receptor 3A	Rs1176724	Opioid (heroin)		Levran 2008 [103]
			Rs897687			
	HTR3B	5-hydroxytryptamine receptor 3B	Rs3758987	Opioid (heroin)		Levran 2009 [110]
Glutamate	GRIN2A	Glutamate Ionotropic Receptor NMDA Type Subunit 2A	rs1070487, rs6497730, rs4587976, rs1650420, rs3219790, rs1102972, rs3104703, rs1071502	Opioid (heroin)		Levran, 2009; Zhao 2013 [110, 122]
	GRIN2A		Rs1650420	Cocaine		Jensen 2016 [78]
	GRIN3A	Glutamate Ionotropic Receptor NMDA Type Subunit 3A	Rs17189632	Opioid (heroin)		Xie 2016 [123]
	GRIN3B	Glutamate Ionotropic Receptor NMDA Type Subunit 3B	Rs2240158	Opioid (heroin)		Xie 2016 [123]
	CNIH3	Cornichon family AMPA receptor auxiliary protein 3	Rs10799590	Opioid		Nelson 2015 [91]
	GABRA1	GABA A receptor protein subunit α1	Rs2279020	Methamphetamineabuse		Lin 2003 [124]

GABAergic	GABRA2	GABA A receptor protein subunit α2	Rs11503014	Opioid (heroin)		Levran 2009 [110]
			Rs279871, rs279826, rs279836, etc.	Alcohol		Edenberg 2004; Covault 2004; Agrawal 2006; Fehr 2006; Soyka 2008 [93, 125–128]
	GABRB3	GABA A receptor protein subunit β3	Rs7165224	Opioid (heroin)		Enoch2010 [129]
			Rs7165224	Cocaine		Hartz 2010 [77]
	GABRG2	GABA A receptor protein subunit γ2	Rs211014	Opioid (heroin)		Levran 2013 [130]
			Rs211013, Rs4480617	Methampheta mineabuse		Lin 2003 [124]
	GAD1	Glutamate decarboxylase 1	rs2058725, rs1978340, rs3791878, rs11542313	Opioid (heroin)		Levran 2009 [130]
	GAD1		Rs2058725	Cocaine		Levran 2016 [130]
			rS1978340			
			rs2241164			
	GAD2	Glutamate decarboxylase 2	Rs8190646	Opioid (heroin)		Wu 2012 [131]
Cholinergic	CHRNA5	cholinergic receptor nicotinic receptor subunit α5	Rs16969968	Cocaine		Gucza 2008 [132]
			rS16969968	Nicotine		Bierut 2007; Saccone 2007; Hartz 2010; Jensen 2016 [76, 79, 85]
						Spitz 2008; Thorgeirsson 2008; Weiss 2008; Berrettini 2008; Saccone 2009 [80–84]

(continued)

Table 9.4 (continued)

Gene category	Gene name	Encoded protein	SNP	Targeted addiction	Diagnostic relevance	First author and year of publication
			rS16969968	Nicotine	Daily consumption	Saccone 2007; Thorgeirsson 2010; Saccone 2010; Liu 2010; Jensen 2015 [79, 81, 85, 88, 133]
			rS16969968	Nicotine	Treatment outcome	Baker 2009; Freathy 2009 [89, 90]
			Rs588765	Nicotine		Spitz 2008; Thorgeirsson 2008; Weiss2008; Berrettini 2008; Saccone 2009, Hartz 2010; Jensen 2016 [80–85]
Neurotrophic factors	BDNF	Brain derived neurotrophic factor	Rs6265 (G196A)	Opioid (heroin)	Early onset age	Cheng 2005; Hou 2010; Jia 2011 [134–136]
			Rs6265	Methamphetamine		Cheng 2005 [134]
			Rs13306221	Opioid (heroin)		Jia 2011 [136]
	GAL	Galanin	Rs694066 Rs3136541 Rs948854	Opioid (heroin)		Levran, 2009; Beer, 2013; Levran, 2014; Levran, 2014 [102, 110, 137, 138]
	NGFB		Rs2239622 (intron)	Opioid (heroin)	Low methadone doses	Levran 2012 [139]
Others	ABCB1	ATP binding cassette transporter 1	Rs1045642	Opioid		Beer 2013 [102]
			Rs1128503 (C1236T) Rs1045642 (C3435T)	Opioid	Methadone doses	Crettol 2006, Levran 2008 [140, 141]
	ADH1B	Alcohol dehydrogenase 1B (class I), β polypeptide	Rs1229984	Alcohol		Chen 1999; Whitfield 2002; Luczak 2006; Edenberg 2007; Bierut 2012; Frank 2012; Park 2013; Gelernter 2014 [142–149]

	ALDH2	Aldehyde dehydrogenase 2 family (mitochondrial)	Rs671	Opioid (heroin)		Wang 2012, Wang 2013 [150, 151]
			Rs671	Alcohol		Thomasson 1991; Chen 1999; Luczak, 2006; Edenberg 2007; Hartz 2010 [77, 143, 144, 147, 152]
	ARRB2	Arrestin β2	Rs1045280	Methamphetamine		Ikeda 2007 [153]
			Rs2036657			
			Rs4790694			
	AKT1	AKT serine/threonine kinase 1	rS3730358	Methamphetamine		Ikeda 2006 [154]
	AVPR1A	arginine Vasopressin receptor 1A	rs3021529, rs3803107, rs11174811, rs10784339, rs1587097	Opioid (heroin)		Levran, 2014; Levran, 2014 [137, 138]
	CDK1	Cyclin-dependent kinase 1	Rs2456778	Cocaine		Gelernter 2014 [155]
	CSNK1E	Casein kinase 1 ε	Rs1534891	Opioid (heroin)		Levran 2008, Wang 2014 [103, 156]
			Rs135745			
	CYP2B6	Cytochrome P450 family 2 subfamily B member 6	Rs2279343(A785G)	Opioid	Lower methadone	Crettol 2005 [157]
			Rs3745274(G516 T)			Levran 2013 [158]
	CAMK2D	Calcium/calmodulin dependent protein kinase II δ	Rs3815072	Pathological gambling	Vulnerability	Lobo 2015 [115]
	DBI	Diazepam binding inhibitor, acyl-CoA binding protein	rS12613135	Cocaine		Kreek 2012 [159]

(continued)

Table 9.4 (continued)

Gene category	Gene name	Encoded protein	SNP	Targeted addiction	Diagnostic relevance	First author and year of publication
	DTNBP1	Dystrobrevin binding protein 1	Rs2619538, Rs3213207	Methamphetamine	Psychosis	Kishimoto2008 [160]
	FAM53B	Family with sequence similarity 53 member B	Rs2629540	Cocaine		Gelernter 2005, 2014 [161, 162]
	FKBP5	FK506 binding protein 5	Rs1360780 Rs380037	Opioid (heroin)		Levran 2014a; Levran 2014b [137, 138]
	GLYT1	Glycine transporter 1	Rs2248829 Rs2486001	Methamphetamine		Morita 2008 [163]
	GSTP1	Glutathione S-transferase P1	Rs1695	Methamphetamine		Hashimoto 2005 [164]
	KCNG2	Potassium voltage-gated channel modifier subfamily G member 2	Rs62103177	Opioid dependence		Gelernter 2014 [155]
	KPNA3	Karyopherin subunit alpha 3	Rs2273816	Opioid (opiate)		Morris 2012 [165]
	MYOCD	Myocardin [155]	Rs1714984	Opioid (heroin)		Nielsen 2008 [166]
	NCAM1	Neural cell adhesion molecule 1	Rs877138 Rs4492854	Opioid (heroin)		Nelson 2013 [167]
	NCOR2	Nuclear receptor corepressor 2	Rs150954431	Cocaine		Gelernter 2014 [161]
	NFAT5	Nuclear factor of activated T-cells 5	Rs1437134	Cocaine		Fernandez-castillo 2015 [168]
	NR3C2	Nuclear receptor Subfamily 3 group C member 2	Rs1040288	Opioid (heroin)		Levran 2014; Levran 2014 [137, 138]

	PICK1	Protein interacting with PRKCA 1	Rs713729	Methamphetamine		Matsuzawa2007 [169]
	SLC22A3	Solute carrier family 22 member 3	Rs3106164	Methamphetamine		Aoyama 2006 [170]
			Rs4709426			
	SNCA	Synuclein α	Rs1372520	Methamphetamine	Psychosis	Kobayashi 2004 [171]
			Rs3756059			
			Rs3756063			
	Near VLDLR		rS1106076	Pathological gambling		Lind 2013 [172]
			rs12305135			
	Nea FZD10		Rs10812227	Pathological gambling		Lind 2013 [172]

One of the classic markers for addiction is A118G in the OPRM1 gene, which encode opioid receptor μ. The μ receptor takes an important role in stress responsive systems. It is expressed in the Corticotropin-releasing factor (CRF) neurons in the hypothalamus. Endogenous opioidlike β-endorphin and enkephalins tonically inhibit CRF neurons via the receptor. The 118G variant has increased affinity of the protein to β-endorphin, but not other agonists or antagonists, and increased binding-induced activation. People with 118G had higher basal cortisol level, and higher cortisol level after challenged by a selective μ opioid agonist, naloxone. Individuals carrying the 118G allele were also more resistant to the analgesia and expiratory depressant effects of alfentanil. Allele 118A has been associated with alcohol dependence, and nicotine or nicotine plus substance abuse. And allele 118G was associated with opiate addiction in a study of male Han Chinese, and a study of Swedish individuals (reviewed by [63]). Allele 118G was also related with risky overdose behaviours [64]. Other mutation in OPRM1, such as intronic rs9479757, was associated with higher consumption and injection of heroin in male Han Chinses cohort [65]. And intronic rs2075572 was associated with methamphetamine dependence in Japanese patients [66].The last two studies, however, failed to find a relationship between A118G and addiction to the drugs of interest.

Dopamine receptors genes DRD1 and DRD2 were suggested in various addictive disorders, with studies of DRD2 yielding more consistent results. A common finding of DRD2 SNP is in the rs1800497, which is also known as the TaqIA and associated with reduced DRD2 expression. Rs1800497 SNP was associated with both daily heroin consumption and required methadone dose in the treatment [67, 68], and was implicated in the development of pathological gambling [69–71].

Rs4680 (G472A) in the exon 4 of COMT is another well studied SNP. The COMT gene encode an enzyme named cathchol-O-methltransferase, which is responsible for degrade catecholamines such as dopamine, epinephrine, and norepinephrine. The 472G allele results in the substitution of methionine to valine in the amino sequence, and therefore a lower enzymatic activity and higher catecholamine levels. A family based study by Horowitz et al 72 found the 472G (val) at higher risk of heroin dependence, which is consistent with a study investigating Chinese Han population with methamphetamine abuse [73]. There was some weaker evidence for the 472A associations with addiction. For example, Oosterhuis and colleagues [74] studied 266 opioid dependent subjects of Caucasian, Hispanics, and African American descendant, and found only Hispanic opiate dependent women associated with G/A and A/A genotypes. Guillot et al. [75] also found a relationship of the A/A genotype with gambling and drinking. However, the participants they identified as positive samples were non-clinical, and included those with minimal risk of developing into pathological gambling and drinking.

The most robust and consistent findings of nicotine associated SNP was rs16969968, a nonsynonymous SNP in the α5 nicotinic receptor subunit gene *CHRNA5* [76–79]. This association has been replicated with either rs16969968 or correlated SNPs in many other independent studies [80–84]. And rs16969968 [79, 85–88] have implicated in the regulation of smoking intensity (eg. cigarettes smoked per day), and was associated with a reduced ability to quit [89, 90].The SNP rs588765 was found to be associated with nicotine dependence and to modify expression of CHRNA5 [77, 80–85].

Credible results may also be seen in those genome-wild analyses of a large number of patients with close-related controls. Nelson et al. [91] conducted a genome-wild association (GWAS) study comparing opioid-dependent daily injectors (N = 1167) with opioid misusers who never progressed to daily injection (N = 161). The participants were from Sydney (the Comorbidity and Trauma Study, CATS). The authors found several intronic CNIH3 single-nucleotide polymorphisms (rs10899590 allele A, rs12130499 allele T, etc.) with protective effect against opioid dependence. These results were then cross-verified with two other genetic studies conducted in the United States of America. CNIH3 encodes the Cornichon Family AMPA Receptor Auxiliary Protein 3, which is part of the AMPA receptor core tetramers. CNIH3, as well as CNIH2, can slow AMPAR deactivation and desensitization in heterologous systems.Epigenetic functionality was suggested for rs10799590, with motif analyses predicting A allele lower binding affinity to the transcription factor Tal1.And fMRI showed the rs10799590 A allele carrier with higher right amygdala habituationto threat-related facial expressions. Thus, the allele A of CNIH3 SNP rs1089590 may protect risky opioid misusers from progression to opioid dependence.

Nishizawa et al. [92] also conducted a GWAS study in 355 Japanese patients with different requirements for post-operative opioid analgesics after cosmetic surgery. They found SNP rs2952768 C allele was associated with more analgesic requirements. Their results were verified with 112 patients after abdominal surgery. The effect of this SNP was examined in psychiatric patients with aberrant reward system. In methamphetamine dependent patients and alcoholic patients, the carriers of C allele were inversely associated with polydrug misuse. In patients with eating disorders, the genotype was associated with comorbid substance dependence. In healthy people, the C allele was associated with reward dependence, but not novelty seeking, harm avoidance, persistence, self-directedness, cooperativeness, and self-transcendence.Post-mortem subject specimens were used to determine the relationship of the genotypes and the mRNA expression of nearby gene CREB1 and METTL21A. They found C/C genotype was related with higher CREB1 expression. This study showed the role of rs2952768 in opioid analgesia and its potential role in reward dependence, possibly via modulation of CREB1 expression.

Several findings have been made utilizing 262 US family genetic data which contained at least three alcoholic members. Edenberg et al. [93] targeted a couple of genes in Chromosome 4p, which encode GABAA receptor subunits: GABRG1, GABRA2, GABRA4, and GABRB1. They found a bunch of three-SNP haplotypes on GABRA2 were significantly associated with alcohol dependence, including 1 of the 5 haplotypes at the 5′ end of the gene and all 43 of the haplotypes starting within exon 3 and extending to the 3′ end of the gene. None of these SNP affect the amino acid sequence. These haplotypes were also correlated with beta oscillations measured by EEG, which was associated with higher susceptibility of alcoholism. Using the same samples, Wang et al. [94] targeted CHRM2 gene, which encode the muscarinic acetylcholine receptor subtype 2. They found a 3-SNP haplotype, rs1824024(G)-rs2061174(C)-rs324650(A) was associated with alcoholism, while haplotype T-T-T may yield some protection against addiction to alcohol and major depression.

In summary, biochemical and genetic markers can be used to detect substance use, monitor abstinence, measure system impairment induced by certain substance or behaviour, and check for susceptibility to addiction. Several biomarkers for alcoholism and nicotine dependence have been well accepted, and ready for clinical use. Biomarker studies for illicit substance and non-substance addictions are relatively underpowered.

Altered stress response may be a common feature of addiction. People with addictive disorders are likely to show tense response in neutral situation, but relatively lower cortisol responsiveness after exposure of stress. However, the procedure of the test is usually complicated, and need a lot of cooperation from the patients.

We observed a spectrum of biochemical changes in patients of addiction, which may reflect the inflammatory impairment that endanger the cardiovascular system. These markers include TNF, IL-6, homocysteine, alpha-synuclein, adipocyte-derived hormones, and several epigenetic markers. Some of these changes may be unspecific, but their existence calls more attention to study the inflammatory react and systematic impairment of cardiovascular and brain diseases caused by addiction.

Acknowledgements This work was supported by the Nature Science Foundation of China (81471350; 81671321) and National Basic Research Program of China (2015CB553504).

References

1. Moeller KE, Lee KC, Kissack JC (2008) Urine drug screening: practical guide for clinicians. Mayo Clin Proc 83:66–76
2. Junghanns K et al (2003) Impaired serum cortisol stress response is a predictor of early relapse. Alcohol Alcohol Oxf Oxfs 38:189–193
3. Meng D et al (2011) Serum NPY and BNDF response to a behavioral stressor in alcohol-dependent and healthy control participants. Psychopharmacology (Berl.) 218:59–67
4. Glahn A et al (2013) Atrial natriuretic peptide, arginine vasopressin peptide and cortisol serum levels in opiate-dependent patients. Neuropsychobiology 67:111–115
5. Geisel O, Panneck P, Hellweg R, Wiedemann K, Müller CA (2015) Hypothalamic-pituitary-adrenal axis activity in patients with pathological gambling and internet use disorder. Psychiatry Res 226:97–102
6. Kiefer F, Jahn H, Schick M, Wiedemann K (2002) Alcohol intake, tumour necrosis factor-α, l eptin and craving: factors of a possibly vicious circle? Alcohol Alcohol 37:401–404
7. Sinha R et al (2011) Effects of adrenal sensitivity, stress- and cue-induced craving, and anxiety on subsequent alcohol relapse and treatment outcomes. Arch Gen Psychiatry 68:942–952
8. Sinha R, Garcia M, Paliwal P, Kreek MJ, Rounsaville BJ (2006) Stress-induced cocaine craving and hypothalamic-pituitary-adrenal responses are predictive of cocaine relapse outcomes. Arch Gen Psychiatry 63:324–331

9. Nakahashi T et al (2000) Vascular endothelial cells synthesize and secrete brain-derived neurotrophic factor. FEBS Lett 470:113–117
10. Choi S-W, Bhang S, Ahn J-H (2011) Diurnal variation and gender differences of plasma brain-derived neurotrophic factor in healthy human subjects. Psychiatry Res 186:427–430
11. Heberlein A et al (2010a) BDNF and GDNF serum levels in alcohol-dependent patients during withdrawal. Prog Neuropsychopharmacol Biol Psychiatry 34:1060–1064
12. Huang M-C et al (2011) Differential patterns of serum brain-derived neurotrophic factor levels in alcoholic patients with and without delirium tremens during acute withdrawal. Alcohol Clin Exp Res 35:126–131
13. Heberlein A et al (2011a) Serum levels of BDNF are associated with craving in opiate-dependent patients. J Psychopharmacol Oxf Engl 25:1480–1484
14. Zhang J et al (2014) Increased serum brain-derived neurotrophic factor levels during opiate withdrawal. Neurosci Lett 571:61–65
15. Angelucci F et al (2007) Chronic heroin and cocaine abuse is associated with decreased serum concentrations of the nerve growth factor and brain-derived neurotrophic factor. J Psychopharmacol Oxf Engl 21:820–825
16. Chen P-H, Huang M-C, Lai Y-C, Chen P-Y, Liu H-C (2014) Serum brain-derived neurotrophic factor levels were reduced during methamphetamine early withdrawal. Addict Biol 19:482–485
17. Corominas-Roso M et al (2013) Brain-derived neurotrophic factor serum levels in cocaine-dependent patients during early abstinence. Eur Neuropsychopharmacol J Eur Coll Neuropsychopharmacol 23:1078–1084
18. D'Sa C, Fox HC, Hong AK, Dileone RJ, Sinha R (2011) Increased serum brain-derived neurotrophic factor (BDNF) is predictive of cocaine relapse outcomes: A prospective study. Biol Psychiatry 70:706–711
19. Ren W et al (2016) Time-dependent serum brain-derived neurotrophic factor decline during methamphetamine withdrawal. Medicine (Baltimore) 95:e2604
20. Geisel O, Banas R, Hellweg R, Müller CA (2012) Altered serum levels of brain-derived neurotrophic factor in patients with pathological gambling. Eur Addict Res 18:297–301
21. Geisel O, Banas R, Schneider M, Hellweg R, Müller CA (2013) Serum levels of brain-derived neurotrophic factor in patients with internet use disorder. Psychiatry Res 209:525–528
22. Hilburn C et al (2011) Is serum brain-derived neurotrophic factor related to craving for or use of alcohol, cocaine, or methamphetamine? Neuropsychiatr Dis Treat 7:357–364
23. Straten G, Eschweiler GW, Maetzler W, Laske C, Leyhe T (2009) Glial cell-line derived neurotrophic factor (GDNF) concentrations in cerebrospinal fluid and serum of patients with early Alzheimer's disease and normal controls. J Alzheimers Dis JAD 18:331–337
24. Wedekind D et al (2011) S100B and homocysteine in the acute alcohol withdrawal syndrome. Eur Arch Psychiatry Clin Neurosci 261:133–138
25. Heberlein A et al (2011b) Decreased galanin serum levels are associated with alcohol-craving during withdrawal. Prog Neuro-Psychopharmacol Biol Psychiatry 35:568–572
26. Reece AS (2007) Evidence of accelerated ageing in clinical drug addiction from immune, hepatic and metabolic biomarkers. Immun Ageing A 4:6
27. Islam SKN, Hossain KJ, Kamal M, Ahsan M (2004) Serum immunoglobulins and white blood cells status of drug addicts: influence of illicit drugs and sex habit. Addict Biol 9:27–33
28. Azarang A et al (2007) T-helper 1 and 2 serum cytokine assay in chronic opioid addicts. Eur Cytokine Netw 18:210–214
29. Shi W-L et al (2012) Serum proteomics of methamphetamine addicts and up-regulation of complement factor H related to methamphetamine addiction. Neurosci Lett 525:23–28
30. Achur RN, Freeman WM, Vrana KE (2010) Circulating cytokines as biomarkers of alcohol abuse and alcoholism. J Neuroimmune Pharmacol Off J Soc NeuroImmune Pharmacol 5:83–91
31. Heberlein A et al (2014) TNF-α and IL-6 serum levels: neurobiological markers of alcohol consumption in alcohol-dependent patients? Alcohol Fayettev N 48:671–676

32. Laso FJ, Vaquero JM, Almeida J, Marcos M, Orfao A (2007) Production of inflammatory cytokines by peripheral blood monocytes in chronic alcoholism: relationship with ethanol intake and liver disease. Cytometry B Clin Cytom 72:408–415
33. Nicolaou C et al (2004) Serum cytokine concentrations in alcohol-dependent individuals without liver disease. Alcohol Fayettev N 32:243–247
34. Bleich S et al (2000) Elevated homocysteine levels in alcohol withdrawal. Alcohol Alcohol Oxf Oxfs 35:351–354
35. Heese P et al (2012) Alterations of homocysteine serum levels during alcohol withdrawal are influenced by folate and riboflavin: results from the German Investigation on Neurobiology in Alcoholism (GINA). Alcohol Alcohol Oxf Oxfs 47:497–500
36. Frieling H et al (2012) Reduced plasma levels of asymmetric di-methylarginine (ADMA) in patients with alcohol dependence normalize during withdrawal. Eur Neuropsychopharmacol J Eur Coll Neuropsychopharmacol 22:836–840
37. Heberlein A et al (2010b) Serum levels of vascular endothelial growth factor A increase during alcohol withdrawal. Addict Biol 15:362–364
38. Heberlein A et al (2012) Association of nerve growth factor and vascular endothelial growth factor A with psychometric measurements of opiate dependence: results of a pilot study in patients participating in a structured diamorphine maintenance program. Eur Addict Res 18:213–219
39. Buydens-Branchey L, Branchey M (2003) Association between low plasma levels of cholesterol and relapse in cocaine addicts. Psychosom Med 65:86–91
40. Lin S-H et al (2012) Association between cholesterol plasma levels and craving among heroin users. J Addict Med 6:287–291
41. Housová J et al (2005) Adipocyte-derived hormones in heroin addicts: the influence of methadone maintenance treatment. Physiol Res Acad Sci Bohemoslov 54:73–78
42. Shahouzehi B, Shokoohi M, Najafipour H (2013) The effect of opium addiction on serum adiponectin and leptin levels in male subjects: a case control study from Kerman coronary artery disease risk factors study (KERCADRS). EXCLI J 12:916–923
43. Bönsch D et al (2005) Alpha-synuclein protein levels are increased in alcoholic patients and are linked to craving. Alcohol Clin Exp Res 29:763–765
44. Feng Y-M et al (2013) Decreased mitochondrial DNA copy number in the hippocampus and peripheral blood during opiate addiction is mediated by autophagy and can be salvaged by melatonin. Autophagy 9:1395–1406
45. Reece A, Elevated S (2013) IGF1 in clinical opiate dependence. Neuro Endocrinol Lett 34:18–26
46. Patkar AA et al (2004) Relationship of serum prolactin with severity of drug use and treatment outcome in cocaine dependence. Psychopharmacology (Berl.) 176:74–81
47. Andersen AM, Dogan MV, Beach SRH, Philibert RA (2015) Current and future prospects for epigenetic biomarkers of substance use disorders. Genes 6:991–1022
48. Nielsen DA et al (2009) Increased OPRM1 DNA methylation in lymphocytes of methadone-maintained former heroin addicts. Neuropsychopharmacol Off Publ Am Coll Neuropsychopharmacol 34:867–873
49. Nielsen DA et al (2010) Ethnic diversity of DNA methylation in the OPRM1 promoter region in lymphocytes of heroin addicts. Hum Genet 127:639–649
50. Chorbov VM, Todorov AA, Lynskey MT, Cicero TJ (2011) Elevated levels of DNA methylation at the OPRM1 promoter in blood and sperm from male opioid addicts. J Opioid Manag 7:258–264
51. Rotter A et al (2013) CB1 and CB2 receptor expression and promoter methylation in patients with cannabis dependence. Eur Addict Res 19:13–20
52. Xu X et al (2016) A significant association between BDNF promoter methylation and the risk of drug addiction. Gene 584:54–59
53. Toledo-Rodriguez M et al (2010) Maternal smoking during pregnancy is associated with epigenetic modifications of the brain-derived neurotrophic factor-6 exon in adolescent offspring. Am J Med Genet Part B Neuropsychiatr Genet Off Publ Int Soc Psychiatr Genet 153B:1350–1354

54. Ge Q et al (2014) miRNA in plasma exosome is stable under different storage conditions. Mol Basel Switz 19:1568–1575
55. Huang S et al (2016) Serum miR-132 is a risk marker of post-stroke cognitive impairment. Neurosci Lett 615:102–106
56. Salta E, De Strooper B (2012) Non-coding RNAs with essential roles in neurodegenerative disorders. Lancet Neurol 11:189–200
57. Lee H-J, Palkovits M, Young WS 3rd (2006) miR-7b, a microRNA up-regulated in the hypothalamus after chronic hyperosmolar stimulation, inhibits Fos translation. Proc Natl Acad Sci U S A 103:15669–15674
58. Eipper-Mains JE et al (2011) microRNA-Seq reveals cocaine-regulated expression of striatal microRNAs. RNA N Y N 17:1529–1543
59. Rodriguez RE (2012) Morphine and microRNA activity: is there a relation with addiction? Front Genet 3:223
60. Wang X et al (2015) Heroin inhibits HIV-restriction miRNAs and enhances HIV infection of macrophages. Front Microbiol 6:1230
61. Zhou Y et al (2015) Heroin use promotes HCV infection and dysregulates HCV-related circulating microRNAs. J Neuroimmune Pharmacol Off J Soc NeuroImmune Pharmacol 10:102–110
62. Long X et al (2016) MiR-582-5p/miR-590-5p targeted CREB1/CREB5-NF-kappaB signaling and caused opioid-induced immunosuppression in human monocytes. Transl Psychiatry 6:e757
63. Kreek MJ, Nielsen DA, Butelman ER, LaForge KS (2005) Genetic influences on impulsivity, risk taking, stress responsivity and vulnerability to drug abuse and addiction. Nat Neurosci 8:1450–1457
64. Manini AF, Jacobs MM, Vlahov D, Hurd YL (2013) Opioid receptor polymorphism A118G associated with clinical severity in a drug overdose population. J Med Toxicol Off J Am Coll Med Toxicol 9:148–154
65. Xu J et al (2014) A heroin addiction severity-associated intronic single nucleotide polymorphism modulates alternative pre-mRNA splicing of the mu opioid receptor gene OPRM1 via hnRNPH interactions. J Neurosci 34:11048–11066
66. Ide S et al (2006) Linkage disequilibrium and association with methamphetamine dependence/psychosis of mu-opioid receptor gene polymorphisms. Pharm J 6:179–188
67. Lawford BR et al (2000) The D(2) dopamine receptor A(1) allele and opioid dependence: association with heroin use and response to methadone treatment. Am J Med Genet 96:592–598
68. Vereczkei A et al (2013) Multivariate analysis of dopaminergic gene variants as risk factors of heroin dependence. PLoS One 8:e66592
69. Comings DE et al (1996) A study of the dopamine D2 receptor gene in pathological gambling. Pharmacogenetics 6:223–234
70. Fagundo AB et al (2014) Dopamine DRD2/ANKK1 Taq1A and DAT1 VNTR polymorphisms are associated with a cognitive flexibility profile in pathological gamblers. J Psychopharmacol Oxf Engl 28:1170–1177
71. Lobo DSS et al (2010) Association of functional variants in the dopamine D2-like receptors with risk for gambling behaviour in healthy Caucasian subjects. Biol Psychol 85:33–37
72. Horowitz R et al (2000) Confirmation of an excess of the high enzyme activity COMT val allele in heroin addicts in a family-based haplotype relative risk study. Am J Med Genet 96:599–603
73. Li T et al (2004) Association analysis of the DRD4 and COMT genes in methamphetamine abuse. Am J Med Genet Part B Neuropsychiatr Genet Off Publ Int Soc Psychiatr Genet 129B:120–124
74. Oosterhuis BE et al (2008) Catechol-O-methyltransferase (COMT) gene variants: possible association of the Val158Met variant with opiate addiction in Hispanic women. Am J Med Genet Part B Neuropsychiatr Genet Off Publ Int Soc Psychiatr Genet 147B:793–798

75. Guillot CR, Fanning JR, Liang T, Berman ME (2015) COMT associations with disordered gambling and drinking measures. J Gambl Stud 31:513–524
76. Bierut LJ et al (2007) Novel genes identified in a high-density genome wide association study for nicotine dependence. Hum Mol Genet 16:24–35
77. Hartz SM, Bierut LJ (2010) Genetics of addictions. Clin Lab Med 30:847–864
78. Jensen KP (2016) A review of genome-wide association studies of stimulant and opioid use disorders. Mol Neuropsychiatry 2:37–45
79. Saccone SF et al (2007) Cholinergic nicotinic receptor genes implicated in a nicotine dependence association study targeting 348 candidate genes with 3713 SNPs. Hum Mol Genet 16:36–49
80. Berrettini W et al (2008) Alpha-5/alpha-3 nicotinic receptor subunit alleles increase risk for heavy smoking. Mol Psychiatry 13:368–373
81. Saccone NL et al (2009) The CHRNA5-CHRNA3-CHRNB4 nicotinic receptor subunit gene cluster affects risk for nicotine dependence in African-Americans and in European-Americans. Cancer Res 69:6848–6856
82. Spitz MR, Amos CI, Dong Q, Lin J, Wu X (2008) The CHRNA5-A3 region on chromosome 15q24-25.1 is a risk factor both for nicotine dependence and for lung cancer. J Natl Cancer Inst 100:1552–1556
83. Thorgeirsson TE et al (2008) A variant associated with nicotine dependence, lung cancer and peripheral arterial disease. Nature 452:638–642
84. Weiss RB et al (2008) A candidate gene approach identifies the CHRNA5-A3-B4 region as a risk factor for age-dependent nicotine addiction. PLoS Genet 4:e1000125
85. Jensen KP et al (2015) A CHRNA5 smoking risk variant decreases the aversive effects of nicotine in humans. Neuropsychopharmacol Off Publ Am Coll Neuropsychopharmacol 40:2813–2821
86. Liu Y et al (2010a) Nucleus accumbens dopamine mediates amphetamine-induced impairment of social bonding in a monogamous rodent species. Proc Natl Acad Sci U S A 107:1217–1222
87. Saccone NL et al (2010) Multiple independent loci at chromosome 15q25.1 affect smoking quantity: a meta-analysis and comparison with lung cancer and COPD. PLoS Genet 6
88. Thorgeirsson TE et al (2010) Sequence variants at CHRNB3-CHRNA6 and CYP2A6 affect smoking behavior. Nat Genet 42:448–453
89. Baker TB et al (2009) Human neuronal acetylcholine receptor A5-A3-B4 haplotypes are associated with multiple nicotine dependence phenotypes. Nicotine Tob Res Off J Soc Res Nicotine Tob 11:785–796
90. Freathy RM et al (2009) A common genetic variant in the 15q24 nicotinic acetylcholine receptor gene cluster (CHRNA5-CHRNA3-CHRNB4) is associated with a reduced ability of women to quit smoking in pregnancy. Hum Mol Genet 18:2922–2927
91. Nelson EC et al (2016) Evidence of CNIH3 involvement in opioid dependence. Mol Psychiatry 21:608–614
92. Nishizawa D et al (2014) Genome-wide association study identifies a potent locus associated with human opioid sensitivity. Mol Psychiatry 19:55–62
93. Edenberg HJ et al (2004) Variations in GABRA2, encoding the alpha 2 subunit of the GABA(A) receptor, are associated with alcohol dependence and with brain oscillations. Am J Hum Genet 74:705–714
94. Wang JC et al (2004) Evidence of common and specific genetic effects: association of the muscarinic acetylcholine receptor M2 (CHRM2) gene with alcohol dependence and major depressive syndrome. Hum Mol Genet 13:1903–1911
95. Bond C et al (1998) Single-nucleotide polymorphism in the human mu opioid receptor gene alters beta-endorphin binding and activity: possible implications for opiate addiction. Proc Natl Acad Sci U S A 95:9608–9613
96. Haerian BS, Haerian MS (2013) OPRM1 rs1799971 polymorphism and opioid dependence: evidence from a meta-analysis. Pharmacogenomics 14:813–824

97. Kapur S, Sharad S, Singh RA, Gupta AK (2007) A118g polymorphism in mu opioid receptor gene (oprm1): association with opiate addiction in subjects of Indian origin. J Integr Neurosci 6:511–522
98. Nagaya D, Ramanathan S, Ravichandran M, Navaratnam V (2012) A118G mu opioid receptor polymorphism among drug addicts in Malaysia. J Integr Neurosci 11:117–122
99. Zhang H, Kranzler HR, Yang B-Z, Luo X, Gelernter J (2008) The OPRD1 and OPRK1 loci in alcohol or drug dependence: OPRD1 variation modulates substance dependence risk. Mol Psychiatry 13:531–543
100. Crist RC et al (2013) An intronic variant in OPRD1 predicts treatment outcome for opioid dependence in African-Americans. Neuropsychopharmacol Off Publ Am Coll Neuropsychopharmacol 38:2003–2010
101. Clarke T-K et al (2014) Genetic variation in OPRD1 and the response to treatment for opioid dependence with buprenorphine in European-American females. Pharm J 14:303–308
102. Beer B et al (2013) Association of polymorphisms in pharmacogenetic candidate genes (OPRD1, GAL, ABCB1, OPRM1) with opioid dependence in European population: a case-control study. PLoS One 8:e75359
103. Levran O et al (2008a) Genetic susceptibility to heroin addiction: a candidate gene association study. Genes Brain Behav 7:720–729
104. Nelson EC et al (2014) Association of OPRD1 polymorphisms with heroin dependence in a large case-control series. Addict Biol 19:111–121
105. Gerra G et al (2007) Human kappa opioid receptor gene (OPRK1) polymorphism is associated with opiate addiction. Am J Med Genet Part B Neuropsychiatr Genet Off Publ Int Soc Psychiatr Genet 144B:771–775
106. Yuferov V et al (2004) Redefinition of the human kappa opioid receptor gene (OPRK1) structure and association of haplotypes with opiate addiction. Pharmacogenetics 14:793–804
107. Yuferov V, Levran O, Proudnikov D, Nielsen DA, Kreek MJ (2010) Search for genetic markers and functional variants involved in the development of opiate and cocaine addiction and treatment. Ann N Y Acad Sci 1187:184–207
108. Xuei X et al (2007) The opioid system in alcohol and drug dependence: family-based association study. Am J Med Genet Part B Neuropsychiatr Genet Off Publ Int Soc Psychiatr Genet 144B:877–884
109. Clarke T-K et al (2012) Genetic association analyses of PDYN polymorphisms with heroin and cocaine addiction. Genes Brain Behav 11:415–423
110. Levran O et al (2009) Heroin addiction in African Americans: a hypothesis-driven association study. Genes Brain Behav 8:531–540
111. Jacobs MM et al (2013) Dopamine receptor D1 and postsynaptic density gene variants associate with opiate abuse and striatal expression levels. Mol Psychiatry 18:1205–1210
112. da Silva Lobo DS et al (2007) Dopamine genes and pathological gambling in discordant sib-pairs. J Gambl Stud 23:421–433
113. Comings DE et al (1997) Studies of the potential role of the dopamine D1 receptor gene in addictive behaviors. Mol Psychiatry 2:44–56
114. Hou Q-F, Li S-B (2009) Potential association of DRD2 and DAT1 genetic variation with heroin dependence. Neurosci Lett 464:127–130
115. Lobo DSS et al (2015) Addiction-related genes in gambling disorders: new insights from parallel human and pre-clinical models. Mol Psychiatry 20:1002–1010
116. Xie X et al (2013) Positive association between--1021TT genotype of dopamine beta hydroxylase gene and progressive behavior of injection heroin users. Neurosci Lett 541:258–262
117. Cao J, LaRocque E, Li D (2013) Associations of the 5-hydroxytryptamine (serotonin) receptor 1B gene (HTR1B) with alcohol, cocaine, and heroin abuse. Am J Med Genet Part B Neuropsychiatr Genet Off Publ Int Soc Psychiatr Genet 162B:169–176
118. Gao F, Zhu YS, Wei SG, Li SB, Lai JH (2011) Polymorphism G861C of 5-HT receptor subtype 1B is associated with heroin dependence in Han Chinese. Biochem Biophys Res Commun 412:450–453

119. Gerra G et al (2004) Association between low-activity serotonin transporter genotype and heroin dependence: behavioral and personality correlates. Am J Med Genet Part B Neuropsychiatr Genet Off Publ Int Soc Psychiatr Genet 126B:37–42
120. Saiz PA et al (2008) Association study between obsessive-compulsive disorder and serotonergic candidate genes. Prog Neuro-Psychopharmacol Biol Psychiatry 32:765–770
121. Wilson D, da Silva Lobo DS, Tavares H, Gentil V, Vallada H (2013) Family-based association analysis of serotonin genes in pathological gambling disorder: evidence of vulnerability risk in the 5HT-2A receptor gene. J Mol Neurosci MN 49:550–553
122. Zhao B et al (2013) Analysis of variations in the glutamate receptor, N-methyl D-aspartate 2A (GRIN2A) gene reveals their relative importance as genetic susceptibility factors for heroin addiction. PLoS One 8:e70817
123. Xie X et al (2016) Association between genetic variations of NMDA receptor NR3 subfamily genes and heroin addiction in male Han Chinese. Neurosci Lett 631:122–125
124. Lin S-K, Chen C-K, Ball D, Liu H-C, Loh E-W (2003) Gender-specific contribution of the GABA(A) subunit genes on 5q33 in methamphetamine use disorder. Pharm J 3:349–355
125. Agrawal A et al (2006) Association of GABRA2 with drug dependence in the collaborative study of the genetics of alcoholism sample. Behav Genet 36:640–650
126. Covault J, Gelernter J, Hesselbrock V, Nellissery M, Kranzler HR (2004) Allelic and haplotypic association of GABRA2 with alcohol dependence. Am J Med Genet Part B Neuropsychiatr Genet Off Publ Int Soc Psychiatr Genet 129B:104–109
127. Fehr C et al (2006) Confirmation of association of the GABRA2 gene with alcohol dependence by subtype-specific analysis. Psychiatr Genet 16:9–17
128. Soyka M et al (2008) GABA-A2 receptor subunit gene (GABRA2) polymorphisms and risk for alcohol dependence. J Psychiatr Res 42:184–191
129. Enoch M-A et al (2010) The influence of GABRA2, childhood trauma, and their interaction on alcohol, heroin, and cocaine dependence. Biol Psychiatry 67:20–27
130. Levran O et al (2016) Glutamatergic and GABAergic susceptibility loci for heroin and cocaine addiction in subjects of African and European ancestry. Prog Neuro-Psychopharmacol Biol Psychiatry 64:118–123
131. Wu W, Zhu YS, Li SB (2012) Polymorphisms in the glutamate decarboxylase 1 gene associated with heroin dependence. Biochem Biophys Res Commun 422:91–96
132. Grucza RA et al (2008) A risk allele for nicotine dependence in CHRNA5 is a protective allele for cocaine dependence. Biol Psychiatry 64:922–929
133. Liu JZ et al (2010b) Meta-analysis and imputation refines the association of 15q25 with smoking quantity. Nat Genet 42:436–440
134. Cheng C-Y et al (2005) Brain-derived neurotrophic factor (Val66Met) genetic polymorphism is associated with substance abuse in males. Brain Res Mol Brain Res 140:86–90
135. Hou H et al (2010) Influence of brain-derived neurotrophic factor (val66met) genetic polymorphism on the ages of onset for heroin abuse in males. Brain Res 1353:245–248
136. Jia W et al (2011) Polymorphisms of brain-derived neurotrophic factor associated with heroin dependence. Neurosci Lett 495:221–224
137. Levran O et al (2014a) Stress-related genes and heroin addiction: a role for a functional FKBP5 haplotype. Psychoneuroendocrinology 45:67–76
138. Levran O et al (2014b) Drug addiction and stress-response genetic variability: association study in African Americans. Ann Hum Genet 78:290–298
139. Levran O et al (2012) Nerve growth factor beta polypeptide (NGFB) genetic variability: association with the methadone dose required for effective maintenance treatment. Pharm J 12:319–327
140. Crettol S et al (2006) ABCB1 and cytochrome P450 genotypes and phenotypes: influence on methadone plasma levels and response to treatment. Clin Pharmacol Ther 80:668–681
141. Levran O et al (2008b) ABCB1 (MDR1) genetic variants are associated with methadone doses required for effective treatment of heroin dependence. Hum Mol Genet 17:2219–2227
142. Bierut LJ et al (2012) ADH1B is associated with alcohol dependence and alcohol consumption in populations of European and African ancestry. Mol Psychiatry 17:445–450

143. Chen CC et al (1999) Interaction between the functional polymorphisms of the alcohol-metabolism genes in protection against alcoholism. Am J Hum Genet 65:795–807
144. Edenberg HJ (2007) The genetics of alcohol metabolism: role of alcohol dehydrogenase and aldehyde dehydrogenase variants. Alcohol Res Health J Natl Inst Alcohol Abuse Alcohol 30:5–13
145. Frank J et al (2012) Genome-wide significant association between alcohol dependence and a variant in the ADH gene cluster. Addict Biol 17:171–180
146. Gelernter J et al (2014a) Genome-wide association study of alcohol dependence: significant findings in African- and European-Americans including novel risk loci. Mol Psychiatry 19:41–49
147. Luczak SE, Glatt SJ, Wall TL (2006) Meta-analyses of ALDH2 and ADH1B with alcohol dependence in Asians. Psychol Bull 132:607–621
148. Park BL et al (2013) Extended genetic effects of ADH cluster genes on the risk of alcohol dependence: from GWAS to replication. Hum Genet 132:657–668
149. Whitfield JB (2002) Alcohol dehydrogenase and alcohol dependence: variation in genotype-associated risk between populations. Am J Hum Genet 71:1247–1250.; author reply 1250–1251
150. Wang T-Y et al (2012) The aldehyde dehydrogenase 2 gene is associated with heroin dependence. Drug Alcohol Depend 120:220–224
151. Wang T-Y et al (2013) The ADH1B and DRD2 gene polymorphism may modify the protective effect of the ALDH2 gene against heroin dependence. Prog Neuro-Psychopharmacol Biol Psychiatry 43:134–139
152. Thomasson HR et al (1991) Alcohol and aldehyde dehydrogenase genotypes and alcoholism in Chinese men. Am J Hum Genet 48:677–681
153. Ikeda M et al (2007) Possible association of beta-arrestin 2 gene with methamphetamine use disorder, but not schizophrenia. Genes Brain Behav 6:107–112
154. Ikeda M et al (2006) Positive association of AKT1 haplotype to Japanese methamphetamine use disorder. Int J Neuropsychopharmacol 9:77–81
155. Gelernter J et al (2014b) Genome-wide association study of opioid dependence: multiple associations mapped to calcium and potassium pathways. Biol Psychiatry 76:66–74
156. Wang Y et al (2014) A population-based association study of casein kinase 1 epsilon loci with heroin dependence in Han Chinese. J Mol Neurosci MN 53:143–149
157. Crettol S et al (2005) Methadone enantiomer plasma levels, CYP2B6, CYP2C19, and CYP2C9 genotypes, and response to treatment. Clin Pharmacol Ther 78:593–604
158. Levran O et al (2013) CYP2B6 SNPs are associated with methadone dose required for effective treatment of opioid addiction. Addict Biol 18:709–716
159. Kreek MJ et al (2012) Opiate addiction and cocaine addiction: underlying molecular neurobiology and genetics. J Clin Invest 122:3387–3393
160. Kishimoto M et al (2008) The dysbindin gene (DTNBP1) is associated with methamphetamine psychosis. Biol Psychiatry 63:191–196
161. Gelernter J et al (2014c) Genome-wide association study of cocaine dependence and related traits: FAM53B identified as a risk gene. Mol Psychiatry 19:717–723
162. Gelernter J et al (2005) Genomewide linkage scan for cocaine dependence and related traits: significant linkages for a cocaine-related trait and cocaine-induced paranoia. Am J Med Genet Part B Neuropsychiatr Genet Off Publ Int Soc Psychiatr Genet 136B:45–52
163. Morita Y et al (2008) The glycine transporter 1 gene (GLYT1) is associated with methamphetamine-use disorder. Am J Med Genet Part B Neuropsychiatr Genet Off Publ Int Soc Psychiatr Genet 147B:54–58
164. Hashimoto T et al (2005) A functional glutathione S-transferase P1 gene polymorphism is associated with methamphetamine-induced psychosis in Japanese population. Am J Med Genet Part B Neuropsychiatr Genet Off Publ Int Soc Psychiatr Genet 135B:5–9
165. Morris CP et al (2012) KPNA3 variation is associated with schizophrenia, major depression, opiate dependence and alcohol dependence. Dis Markers 33:163–170

166. Nielsen DA et al (2008) Genotype patterns that contribute to increased risk for or protection from developing heroin addiction. Mol Psychiatry 13:417–428
167. Nelson EC et al (2013) ANKK1, TTC12, and NCAM1 polymorphisms and heroin dependence: importance of considering drug exposure. JAMA Psychiat 70:325–333
168. Fernandez-Castillo N et al (2015) Transcriptomic and genetic studies identify NFAT5 as a candidate gene for cocaine dependence. Transl Psychiatry 5:e667
169. Matsuzawa D et al (2007) Identification of functional polymorphisms in the promoter region of the human PICK1 gene and their association with methamphetamine psychosis. Am J Psychiatry 164:1105–1114
170. Aoyama N et al (2006) Association between gene polymorphisms of SLC22A3 and methamphetamine use disorder. Alcohol Clin Exp Res 30:1644–1649
171. Kobayashi H et al (2004) Study of association between alpha-synuclein gene polymorphism and methamphetamine psychosis/dependence. Ann N Y Acad Sci 1025:325–334
172. Lind PA et al (2013) Genome-wide association study of a quantitative disordered gambling trait. Addict Biol 18:511–522

Chapter 10
Development of New Diagnostic Techniques – Machine Learning

Delin Sun

Abstract Traditional diagnoses on addiction reply on the patients' self-reports, which are easy to be dampened by false memory or malingering. Machine learning (ML) is a data-driven procedure that learns algorithms from training data and makes predictions. It is quickly developed and is more and more utilized into clinical applications including diagnoses of addiction. This chapter reviewed the basic concepts and processes of ML. Some studies utilizing ML to classify addicts and non-addicts, separate different types of addiction, and evaluate the effects of treatment are also reviewed. Both advantages and shortcomings of ML in diagnoses of addiction are discussed.

Keywords Addiction • Machines learning • Neuroimaging • Prediction • Training

10.1 Previous Diagnoses of Addiction and Their Shortcomings

The traditional diagnoses on addiction heavily reply on the patient's self-report through using diagnose tools such as Diagnostic and Statistical Manual of Mental Disorders (DSM) and International Classification of Diseases (ICD). These criteria helped to guide clinical interventions on addiction. Unfortunately, subjective reports are highly influenced by patient's memory and cooperation. Moreover, inner changes (e.g. alterations in brain structures or functions) may have occurred before the aforementioned clinical diagnoses. The disorders may be much easier and better to be controlled or treated if the diagnose could be earlier. Therefore, preclinical discrimination between people who will and will not ultimately develop addiction is critical for treatment of the disease in its earliest stages.

D. Sun (✉)
Duke-UNC Brain Imaging and Analysis Center, Duke University Medical Center, Durham, NC, USA
e-mail: sundelinustc@gmail.com

X. Zhang et al. (eds.), *Substance and Non-substance Addiction*, Advances in Experimental Medicine and Biology 1010, DOI 10.1007/978-981-10-5562-1_10

Neuroimaging techniques are developing quickly during these years and are hoped to identify preclinical neural changes that predict subsequent addiction. Magnetic Resonance Imaging (MRI), functional MRI (fMRI), and positron emission tomography (PET) are commonly employed neuroimaging methodologies and have been extensively utilized to understand the structural and functional information of human brain. The neuroimaging methods traditionally allow investigating brain signals that reflect characteristics such as grey matter volume, cortical thickness, blood oxygenation level dependent (BOLD) activations, and metabolic changes. Studies more than 10 years before have begun to use neuroimaging methods to diagnose patients with severe mental disorders such as HIV positive patients [20] and Alzheimer's disease [6]. Previous studies have also detected a series of structural and functional differences in some brain areas between addicted people and healthy controls [12, 14]. A huge amount of clinical and preclinical studies have focused on the roles of midbrain dopamine areas (including the ventral tegmental area and substantia nigra) and the basal ganglia structures (including both ventral and dorsal striatum), which are associated with reward, conditioning and habit formation [12, 32]. Studies in recent years have paid more attention to the role of the prefrontal cortex, which is related with regulation of limbic reward regions and are involved in higher-order executive function such as self-control, salience attribution and awareness [14].

In recent years, due to the significant development of neuroscience studies on addiction, besides the neuroimaging methodologies, several molecular (e.g. genes that modulate novelty seeking and behavioral inhibition), neurobehavioral (e.g. impulsivity and stress reactivity) and neurological (e.g. dopamine D2 receptors level in the striatum, anatomical/functional characteristics of anterior cingulate cortex and orbitofrontal cortex) biomarkers have also been found to potentially classify addicted persons (for a review see Volkow et al. [31]).

However, there are a plenty of limitations when solely using any of these methods to diagnose addiction. Firstly, all the aforementioned findings are based on group-level statistical comparisons. It is still hard to replicate the biomarkers on individual-level. Secondly, these measures often provide high dimensional information that are almost impossible to be classified just through visual inspection. Thirdly, different from those in the severe disorders, the cognitive deficits in some types of addiction are not markedly pronounced. Fourthly, addiction is often accompanied with other mental disorders such as depression, schizophrenia, anxiety, and mania [23]. People suffering from physical or emotional traumas are at higher risk of drug abuse. Nearly 20% veterans from Iraq and Afghanistan have reported symptoms of post-traumatic stress disorder (PTSD) or major depression [22]. About half of the veterans diagnosed with PTSD are also diagnosed as substance use disorder. It has been a challenge to separate the overlapping symptoms of addiction and other mental disorders. The difficulty of accurately diagnosing addiction further dampens the effectiveness of treatment. There is thus an urgent need to find new methods to accurately classify individuals of addiction.

10.2 A Brief Introduction of Machine Learning

During the last 10 years, machines learning (ML) methodology has been developed for the diagnoses of psychiatric disorders based on data from multiple domains. ML is a data-driven procedure that learns construction of algorithms from training data and makes predictions on test data. There are three main categories of ML, i.e. supervised learning, unsupervised learning and reinforcement learning. Supervised learning algorithm targets at learning the general rule through receiving both the example inputs and their output labels from the training dataset. For example, a computer is asked to predict the price of a given house size after being shown with the last 10 weeks' local house prices and the corresponding house sizes. Unsupervised learning algorithm tries to discover hidden patterns in data without knowing the labels of the training dataset. For example, a computer is required to separate the frequent customers of a shopping mall into several distinct clusters based on the customers' shopping habits. Reinforcement learning algorithm aims at interacting with a dynamic environment to achieve a goal. For example, a computer is playing chess against a human or another computer opponent. Diagnosing addicts is a classification problem that belongs to supervised learning. A computer has to learn the relationship between several subjects' data and their clinical group labels (i.e. addicts or non-addicts), and then utilize this rule to predict the classification of dataset from newly presented subjects.

Recent studies have begun to employ ML in clinical usage. High accuracy (78.3%) has been reported to distinguish patients from healthy people, e.g. remitted major depressive disorder (MDD) versus healthy controls [29]. Researchers have also reported differentiations between different types of disorders using ML. Duda et al. [10] separated autism (ASD) and attention deficit hyperactivity disorder (ADHD) through six ML models on 65-item scale score sheets from 2775 individuals with ASD and 150 subjects with ADHD. The findings showed that five of the 65 behaviors measures were sufficient to distinguish ASD from ADHD with high accuracy. A newly published work predicted the persistence and severity of MDD based on subjects' baseline self-reports [18]. This study investigated 1056 respondents with lifetime MDD at baseline and predicted outcome scores that were compared with observed scores assessed 10–12 years after baseline. The ML model outperformed the conventional logistic regression models. Further, the top 20% of the ML-predicted distribution included 34.6–38.1% of respondents with subsequent high persistence chronicity and 40.8–55.8% with the severity indicators, while the lowest 20% of the ML-predicted distribution included only 0.9% of respondents with subsequent hospitalizations and 1.5% with suicide attempts. These examples supported the ideas that ML can be used to classify clinical disorders, separate patients with different disorders and predict the development of a disorder.

ML has so far been used in some addiction classification studies [1, 2, 24, 27, 33, 35] and has been anticipated to be the most promising approach to classify vulnerable persons, addicted individuals, remitted patients, and chronic relapsers [31].

Besides, it may also contribute to advance our understanding of addiction related brain and behavioral changes, and thus helps to develop better biomarkers for clinical diagnose.

10.3 The Steps of Using ML for Diagnose

Generally, there are four steps of ML processing for addiction diagnose: pre-processing, model training, cross-validation and clinical prediction.

10.3.1 Pre-processing

Before formal analysis, the data should be cleaned, reduced and transformed during pre-processing [16]. Data cleaning refers to the treatment of missing data, which is very common in clinical data collection. A few data may be lost due to equipment shifts or errors, participants' noncooperation and experimenters' carelessness. Analyzing data merely on the cases without missing data may bias the results especially when the missing values largely differ from the complete cases or when the rate of missing data is high. The preferable treatment of missing values is to estimate or impute them through statistical methods [9].

Data reduction refers to the approach to reduce the representation of the raw dataset so that the new dataset shares the same or similar results as the raw dataset. Recent behavioral and neuroscience measures of addiction often provide data with dozens or even thousands of features. Some of these features could be removed without much loss of information due to the fact that they are highly correlated with each other (redundant features) or irrelevant to diagnose. Keeping the redundant or irrelevance data in the training dataset can negatively affect the modeling power and undermine the predictive accuracy. Feature selection is the process of selecting a subset of relevant features or proposing new feature subsets for the model to increase prediction accuracy, simplify the model for interpretation, accelerate processing speed and prevent overfitting of the training data [17].

Principal Components Analysis (PCA) is a standard method for reducing data dimensions. It selects the best combinations of variables that linearly fit the raw data and keeps the majority of variance in its first a few components (vectors). A previous study has reported using PCA and Fisher's Linear Discriminant (FLD) [15] methods to classify controls from patients of Alzheimer's disease, schizophrenia and mild traumatic brain injury [11]. Zhang, Samaras et al. [35] utilized PCA method to reduce data dimensions for addiction classification. For the neuroimaging dataset, voxel-based feature selection has also been utilized to select the input features. Voxel is the unit of 3D brain images. Each voxel represents a tidy cube (e.g. $1 \times 1 \times 1$ mm^3) of brain tissue that contains millions of neurons. The high-resolution neural images may have nearly one million voxels. It is difficult and unnecessary to recruit all the voxels for analyses. Values of the most active voxels in some brain areas of interest or over the whole brain may be selected.

Data transformation depends on the specific algorithm, and commonly refers to scaling, decompositions or aggregations [13]. Some ML methods require all predications to share the same scale, e.g. 0 ~ 1. Decomposition is used to split features into parts that are more meaningful. On the contrary, aggregation is applied when it is better to combine several features into a single one for analysis.

10.3.2 Model Training

The data post preprocessing are used to train one or several learning algorithms. In supervised ML, some data (called training data) including both features and category labels are entered to train the program that can be later used to classify the new input data. The best model could be obtained after tuning parameters to get optimal values.

There are a plenty of ML algorithms. For instances, Support Vector Machine (SVM) can model non-linear relationships through constructing hyperplane in a multidimensional space to divide the data points into separate categories [7]. New data are entered into this space and assigned to a category based on which side of the hyperplane they fall into. Adaboost (AB) method [13] combines a number of weak classifiers to form a strong classifier [8, 30]. Logistic Regression (LR) is a particular case of general linear regression models (GLM) that specifies a linear combination of features to predict the labels of new data. Elastic net (EN) [36] is one of penalized regression (supervised learning) methods, encourages a grouping effect in which highly correlated predictors tend to be in or out of the model together. EN is particularly useful when the number of predictors is much bigger than the number of observations. Naive Bayes (NB) is a family of simple classifiers based on the Bayes' Theorem [28], so that it considers each of the features to contribute independently to the probability of an item's label, regardless of any potential correlations between features. Artificial neural networks (ANN) simulate the real neural network processing through adding the features of neurons (nodes) and their interconnections [26]. Each node in the network connects with many others, and the connections can be increased or decreased. Each individual node sums the values of all its inputs together and propagates to other nodes after the summation is above some threshold.

The often-used ML algorithms include but not limit to the aforementioned examples. In their milestone study, Zhang, Samaras et al. [35] applied a number of classifier training methods including NB, SVM and k Nearest Neighbor (kNN) [21], and found that performance of these methods dropped dramatically when there was variability in the sequence of the stimuli. They therefore developed a new boosting algorithm with side information on subject identity to remove the intrasubject variability in order to improve classification. They found that this new algorithm allows for less restrictive data collection conditions with significantly reduced performance penalty, and it can work on combined datasets of different tasks effectively.

10.3.3 *Cross-Validation*

Some data (called test data) including both features and category labels are entered to calculate the accuracy of the program that accurately assigns the labels. The test data should be independent from the aforementioned training data. Researchers in practice often split the dataset into two in which one is used as the training set and the other is the test set.

K-fold cross-validation method is commonly used to train classifiers and select the best one according to the ability to predict outcomes. During this procedure, the training data are randomly divided into K subgroups in which K-1 subgroups are employed as a new training set and the rest one subgroup serves as a test set in K iterations. On the other hand, the "leave-one-out" method is a particular case of "leave-p-out" cross-validation method that employs p observations as the test set and the remaining observations as the training set. Different from K-fold cross-validation, leave-one-out method learns and tests on all possible ways to divide the original sample into a training set and a test set. In practice, each of the K human subjects is used as a test subject while training on the data of the remaining K-1 subjects.

A valid classifier is able to give a high accuracy rate (much larger than chance classification) when estimating correct labels in a new set. Some measures are employed to estimate the prediction accuracy, including sensitivity (also known as true positive rate, e.g. the proportion of addicts who are accurately labeled as addicts) and specificity (also known as true negative rate, e.g. the proportion of non-addicts who are accurately labeled as non-addicts) in the binary classification test. It is suggested to estimate the model based on our research question given that there is usually a trade-off between sensitivity and specificity.

10.4 Previous Studies of Addiction-Related ML

A few studies have utilized ML to separate addicts from persons without addiction [24, 35], to differentiate people addicted to different types of drugs [2], and to evaluate the effects of treatment on addiction [27]. A notable trend is that more and more studies are trying to identify a generalizable risk profile containing information from different domains to increase the accuracy of diagnosing. For example, a recent study on adolescent alcohol misuse utilized ML to a wide range of data incorporating brain structure and function, individual personality and cognitive differences, environmental factors (including gestational cigarette and alcohol exposure), life experiences, and candidate genes [34]. The models could be successfully generalized to new data, and indicated that history (individual area under the curve [AUC] = 0.68; AUC exhibit the model's ability to correctly assign a participant to the corresponding group.), personality (AUC = 0.67) and brain (AUC = 0.63) are important risk factors of binge drinking.

10.4.1 Separating Addicts and Non-addicts

In their milestone study, Zhang, Samaras et al. [35] utilized ML, for the first time, to differentiate the drug addicted subjects from control normal based on the fMRI methodology. They explored a number of classification approaches, and introduced a novel algorithm that integrates side information into the use of boosting. Their algorithm clearly outperformed well-established classifiers as documented in extensive experimental results. They investigated brain activation in the same regions in response to the same task manipulation in all participants. They recruited 16 cocaine dependent individuals and 13 non-drug-using controls matched in sex, race, education and general intellectual functioning. They utilized a 4T Varian MRI scanner to get brain images during a task paradigm that was designed to investigate the neural underpinnings of sensitivity to the salience of monetary reward. The participants were asked to press a button or not based on a picture shown to them, and received a monetary reward for a correct response. There were totally three monetary conditions (high money, low money, no money) and a baseline condition (fixation presented only). The monetary reward related task was used based on previous observations that drug addiction is strongly associated with deficits in reward processing. The authors found that it was difficult to classify the two groups by simply using the general task related brain activation due to the similar fMRI BOLD activation patterns for both subject groups. They then performed classifications on the activation differences between monetary conditions pairs. The authors utilized PCA to reduce data dimensions. They also performed voxel-based feature selection through using the contrast maps as input feature vectors. They divided the whole brain into eight region of interest (ROI), and selected the N most active voxels in each ROI. After that, they selected the N most active voxels over the whole brain. They used several methods including AB, KNN, NB and SVM for learning and classification.

To classify subjects into two groups, they integrated side information into the boosting algorithm by adjusting the weak classifier selection and weight updating steps. The side information contributes to the feature selection process through selecting only those features that enhance the relevant dimensions in the main dataset while inhibiting the irrelevant dimensions in the auxiliary dataset. This method helps to remove the intra-subject variations.

They found that AB outperformed the PCA- and voxel-based methods. Further, boosting with side information outperformed standard AB on the data set containing the contrast maps from different runs. They also found that the classification on the mixture dataset was not as good as on each run separately. It may be due to the intra-subject variability in the mixture dataset. Interestingly, they found that the images from latter runs were easier to be classified than the images from earlier runs. It is possible that drug addicts have different habituation speeds from the healthy controls in reward processing.

Later studies employed not only task-related BOLD contrasts but also several other types of features including resting-state BOLD signals, behavioral performance and demographic information in the models. For example, Pariyadath, Stein et al. [24] applied SVM on resting-state functional connectivity data from nicotine-dependent smokers and healthy controls to identify brain-predictors of nicotine dependence. The authors first identified 16 resting-state brain networks in 21 smokers and 21 non-smoking controls. They then calculated the representativeness of each node with respect to its parent resting-state network, between-network connectivity, and within-network connectivity. After that, they used SVM for training and testing, and employed leave-one-out cross-validation to gain the classification performance.

Through the combination of ML and network analyses, they found that the within-network outperformed both between-network and representativeness of the network in providing information for predicting smoking status. Further, their findings suggested that connectivity within the executive control and frontoparietal networks are particularly informative in predicting smoking status.

10.4.2 Separate Different Subtypes of Addiction

Several lines of evidence suggested that different types of addiction might be underlay by different mechanisms [5]. Therefore, to accurately separate different addiction types may contribute to facilitating the development of individualized prevention and intervention programs for addiction.

Ahn and Vassileva [2] utilized ML method to differentiate heroin dependence and amphetamine dependence. They recruited 39 amphetamine mono-dependent, 44 heroin mono-dependent, 58 polysubstance dependent, and 81 non-substance dependent participants. They employed 54 predictors including demographic, personality (trait impulsivity, trait psychopathy, aggression, sensation seeking), psychiatric (attention deficit hyperactivity disorder, conduct disorder, antisocial personality disorder, psychopathy, anxiety, depression), and neurocognitive impulsivity measures (Delay Discounting, Go/No-Go, Stop Signal, Immediate Memory, Balloon Analogue Risk, Cambridge Gambling, and Iowa Gambling tasks).

They employed the EN method [36] to all of the data aiming at finding multivariate profiles that can be used to classify heroin- and amphetamine-dependence. They split the data into a training set (67% of the data) and a test (validation) set (33% of the data). To check the generalizability of the findings, they further randomly divided the data into training and test sets and checked the model performance 1000 times.

They found substance-specific multivariate profiles that differentiated heroin dependence from amphetamine dependence in new samples with high accuracy. They showed that psychopathy was the best classifier to both types of addiction. Interestingly, the factors classifying heroin dependence from amphetamine dependence often showed opposite patterns.

10.4.3 Evaluating the Effects of Treatment

The ML method can also be used to evaluate the effects of particular treatment on addicts. Rish, Bashivan et al. [27] investigated the neural responses to methylphenidate in individuals addicted to cocaine. Methylphenidate is an indirect dopamine that has been found to show positive effects on cocaine addicts [19]. It contributed to improving stop signal reaction times that reflect the abilities of inhibitory control. It has also been found to normalize task-related brain activity and resting-state functional connectivity.

Rish, Bashivan et al. [27] generated maps of Pearson correlation coefficients for each voxel that its resting-state BOLD signal time-series were correlated (above a pre-defined threshold) with the other voxels' signal time-series. They then calculated the degree of each voxel as the total number of suprathreshold correlations with that voxel. They further selected a subset of the top-ranked features based on their relevance to the class label, and applied leave-one-out method in cross-validation. Several classifiers were employed in this study, including kNN, SVM, LR, NB, decision tree, random forest and linear discriminant analysis.

They performed two classification analyses. The first analysis aimed at investigating whether the differences between addicts and controls are due to drug administered (i.e. methylphenidate vs. placebo). Reduced classification accuracy under methylphenidate is speculated to reflect the normalized functional connectivity in addicts (i.e. addicts are more similar to controls after methylphenidate administration). The second analysis aimed at clarifying whether the differences between methylphenidate and placebo are due to group identity (i.e. cocaine addicts vs. controls). The findings suggested that methylphenidate tends to normalize network properties in cocaine addicts, providing evidence for potential benefits of methylphenidate in treating cocaine addiction.

10.4.4 Employing Multiple Measures to Classify Subjects

Previous studies focusing on just one type of risk factor have yielded modest predictions of alcohol misuse and have often been dampened due to overfitting. More and more studies have begun to use multiple measures as predictors. Whelan, Watts et al. [34] employed several measures to reflect different risk factors, including fMRI task-related activations (from the stop signal task and the monetary incentive delay task), personality, cognitive tasks (including Wechsler intelligence scale for children and CANTAB tests), history (such as stressful life-events and family history valence), demographics (pubertal status and social economic status for instances), genetics (i.e. several unique single nucleotide polymorphisms) and substance misuse.

To avoid overfitting the imaging data, they calculated a single summary statistic of each imaging contrast for each individual from training data, and nested 10-fold cross-validation for tuning model parameters and final validation. They implemented three levels of nested cross-validation, i.e. the inner, middle and outer levels. An external validation was also used to test the generalizability to different data. Model performance on the test data was computed through the AUC of the ROC curve.

In the inner level, to get the optimized imaging thresholds to differ between groups, they combined all functional and anatomical neuroimaging data within a single, voxel-wise, logistic regression model. They then generated binary masks of classification accuracy over a range of AUC thresholds and a range of cluster extent thresholds. Finally, they generated for each imaging contrast in each subject a summary statistic, which is the average of the AUC-based weight and beta values within the binary mask.

In the middle level, in order to optimize the parameters of the regularization, they employed from 100 models the median best AUC and median best cluster extent that resulted in the highest classification accuracy. They then entered the values of neuroimaging data and those from psychometric and other data into a logistic regression with elastic net variable selection and regularization in the classification procedure. They utilized maximum likelihood estimation to calculate the optimal model in which the median of parameters of the elastic net resulted in the highest AUC.

They quantified the generalizability during the outer level in which the optimized brain and elastic net parameters were used. Finally, to apply to a new group of subjects with slightly different phenotypes, they employed each of the maps from the outer level data to the subjects in the external validation.

The authors identified a generalizable risk profile for alcohol misuse initiation. Classification of current binge drinkers is primarily a function of the History domain. By contrast, the prediction of future binge drinking rely more on a combination of History, Personality and Brain domains. Their findings highlighted the multi-domain analysis for predicting adolescent alcohol misuse.

10.5 Shortcomings

There are several limitations when using ML to diagnose addiction, especially for the neuroimaging data. Firstly, the sample size is often too small for ML. To learn the inner mechanisms of the dataset requires large training data set. However, most of the recent neuroimaging studies collect fewer than 100 subjects per group. To develop algorithms better diagnose addicts requires larger clinical data [25] that may be archieved through cooperation among research teams and/or public datasets.

Secondly, the inter-subject and intra-subject variability are large. Due to the complexity of human brain and human thoughts, the brain activations are different from time to time even for the same person under the same task condition. Further, different individuals are associated with even larger differences in brain response patterns. Moreover, the inclusion/exclusion criteria, task paradigm, data acquisition and analysis methods vary dramatically across studies. These largely limit the usage of ML in single subject, single study or combined dataset from different research projects. New methods are required to increase the accuracy of diagnosing a single subject. On the other hand, resting-state fMRI does not involve any specific stimulus or task, and is a more replicable approach for studying functional connectivity and its disruptions in addiction [24]. Moreover, data collection following the same protocols shared by different sites over periods might contribute to reducing the variability across studies.

Thirdly, the features of neuroimaging data are often too large for ML. There are millions of voxels in the brain. It is a big problem for ML to train and to make predictions through using the intensity of so many voxels. To reduce the data dimensionality, it is important to select a few representing brain regions where the neural responses are tightly associated with the research questions [35]. This selection largely depends on a priori knowledge of the topic specific brain responses.

Fourthly, the application of ML in diagnosing behavioral addiction is still scarce [3, 4]. Future studies should be conducted to investigate the predictors that are either common to all behavioral/substance addictions or specific to a particular disorder.

Fifthly, the current ML is limited to classifying whether a subject belongs to a pre-defined disorder or a state of the disorder, given that the models have been trained in that way. That is to say, it can't predict whether an individual is or will be accompanied with some disorder/addiction without designation. Future studies should incorporate the models from multiple training datasets associated with different types of addiction and states of disorders, so that ML makes more comprehensive predictions. For example, an individual should be diagnosed simultaneously whether he/she is cocaine addicted, whether he/she is also smoking addicted and will be developed into major depression, and whether the treatment to his/her alcohol abuse is effective.

10.6 Conclusion

ML is a hopeful tool in diagnosing addiction. It has been found to contribute to classifying addicts and non-addicts, separating different types of addiction, and evaluating the effects of treatment. Future ML studies based on larger clinical dataset, prior academic or clinical knowledge, and information from multiple domains may help to improve the accuracy of diagnose. Moreover, the validity of ML-based diagnose need to be tested in more sub-types of substance addiction and behavioral addiction.

References

1. Ahn WY, Ramesh D, Moeller FG, Vassileva J (2016) Utility of machine-learning approaches to identify behavioral markers for substance use disorders: impulsivity dimensions as predictors of current cocaine dependence. Front Psychol 7:34
2. Ahn WY, Vassileva J (2016) Machine-learning identifies substance-specific behavioral markers for opiate and stimulant dependence. Drug Alcohol Depend 161:247–257
3. Alavi SS, Ferdosi M, Jannatifard F, Eslami M, Alaghemandan H, Setare M (2012) Behavioral addiction versus substance addiction: correspondence of psychiatric and psychological views. Int J Prev Med 3(4):290–294
4. Albrecht U, Kirschner NE, Grusser SM (2007) Diagnostic instruments for behavioural addiction: an overview. Psychosoc Med 4:Doc11
5. Badiani A, Belin D, Epstein D, Calu D, Shaham Y (2011) Opiate versus psychostimulant addiction: the differences do matter. Nat Rev Neurosci 12(11):685–700
6. Bookheimer SY, Strojwas MH, Cohen MS, Saunders AM, Pericak-Vance MA, Mazziotta JC, Small GW (2000) Patterns of brain activation in people at risk for Alzheimer's disease. N Engl J Med 343(7):450–456
7. Burges CJC (1998) A tutorial on support vector machines for pattern recognition. Data Min Knowl Disc 2(2):121–167
8. Collins M (2002) Ranking algorithms for named-entity extraction: boosting and the voted perceptron. 40th Annual Meeting of the Association for Computational Linguistics, Proceedings of the Conference: 489–496
9. Ding YF, Simonoff JS (2010) An investigation of missing data methods for classification trees applied to binary response data. J Mach Learn Res 11:131–170
10. Duda M, Ma R, Haber N, Wall DP (2016) Use of machine learning for behavioral distinction of autism and ADHD. Transl Psychiatry 6:e732
11. Ford J, Farid H, Makedon F, Flashman LA, McAllister TW, Megalooikonomou V, Saykin AJ (2003) Patient classification of fMRI activation maps. Med Image Comput Comput-Assist Interv Miccai 2879(Pt 2):58–65
12. Fowler JS, Volkow ND, Kassed CA, Chang L (2007) Imaging the addicted human brain. Sci Pract Perspect 3(2):4–16
13. Freund Y, Schapire RE (1997) A decision-theoretic generalization of on-line learning and an application to boosting. J Comput Syst Sci 55(1):119–139
14. Goldstein RZ, Volkow ND (2011) Dysfunction of the prefrontal cortex in addiction: neuroimaging findings and clinical implications. Nat Rev Neurosci 12(11):652–669
15. Hastie T, Tibshirani R, Friedman J (2009) The elements of statistical learning: data mining, inference, and predictions, Springer, New York
16. Iniesta R, Stahl D, McGuffin P (2016) Machine learning, statistical learning and the future of biological research in psychiatry. Psychol Med 46(12):2455–2465
17. James G, Witten D, Hastie T, Tibshirani R (2014) An introduction to statistical learning: with applications in R. Springer, New York
18. Kessler RC, van Loo HM, Wardenaar KJ, Bossarte RM, Brenner LA, Cai T, Ebert DD, Hwang I, Li J, de Jonge P, Nierenberg AA, Petukhova MV, Rosellini AJ, Sampson NA, Schoevers RA, Wilcox MA, Zaslavsky AM (2016) Testing a machine-learning algorithm to predict the persistence and severity of major depressive disorder from baseline self-reports. Mol Psychiatry 21:1366
19. Li CSR, Morgan PT, Matuskey D, Abdelghany O, Luo X, Chang JLK, Rounsaville BJ, Ding YS, Malison RT (2010) Biological markers of the effects of intravenous methylphenidate on improving inhibitory control in cocaine-dependent patients. Proc Natl Acad Sci U S A 107(32):14455–14459
20. Liow JS, Rehm K, Strother SC, Anderson JR, Morch N, Hansen LK, Schaper KA, Rottenberg DA (2000) Comparison of voxel- and volume-of-interest-based analyses in FDG PET scans of HIV positive and healthy individuals. J Nucl Med 41(4):612–621

21. Mitchell TM (1977) Machine learning, McGraw-Hill Science/Engineering/Math, Boston
22. Morey RA, Gold AL, LaBar KS, Beall SK, Brown VM, Haswell CC, Nasser JD, Wagner HR, McCarthy G, Workgrp MM (2012) Amygdala volume changes in posttraumatic stress disorder in a large case-controlled veterans group. Arch Gen Psychiatry 69(11):1169–1178
23. NIDA (2010) Comorbidity: addiction and other mental illnesses
24. Pariyadath V, Stein EA, Ross TJ (2014) Machine learning classification of resting state functional connectivity predicts smoking status. Front Hum Neurosci 8:425
25. Passos IC, Mwangi B, Kapczinski F (2016) Big data analytics and machine learning: 2015 and beyond. Lancet Psychiatry 3(1):13–15
26. Ripley BD (1996) Pattern recognition and neural networks. Cambridge University Press, Cambridge
27. Rish I, Bashivan P, Cecchi GA, Goldstei RZ (2016) Evaluating effects of methylphenidate on brain activity in cocaine addiction: a machine-learning approach. Med Imaging 2016-Biomed Appl Mol Struct Funct Imaging 9788
28. Russell SJ, Norvig P (2010) Artificial Intelligence: a modern approach. Pearson, Boston
29. Sato JR, Moll J, Green S, Deakin JFW, Thomaz CE, Zahn R (2015) Machine learning algorithm accurately detects fMRI signature of vulnerability to major depression. Psychiatry Rese-Neuroimaging 233(2):289–291
30. Viola P, Jones MJ (2004) Robust real-time face detection. Int J Comput Vis 57(2):137–154
31. Volkow ND, Koob G, Baler R (2015) Biomarkers in substance use disorders. ACS Chem Neurosci 6(4):522–525
32. Volkow ND, Li TK (2004) Science and society – drug addiction: the neurobiology of behaviour gone awry. Nat Rev Neurosci 5(12):963–970
33. Weygandt M, Schaefer A, Schienle A, Haynes JD (2012) Diagnosing different binge-eating disorders based on reward-related brain activation patterns. Hum Brain Mapp 33(9):2135–2146
34. Whelan R, Watts R, Orr CA, Althoff RR, Artiges E, Banaschewski T, Barker GJ, Bokde ALW, Buchel C, Carvalho FM, Conrod PJ, Flor H, Fauth-Buhler M, Frouin V, Gallinat J, Gan G, Gowland P, Heinz A, Ittermann B, Lawrence C, Mann K, Martinot JL, Nees F, Ortiz N, Paillere-Martinot ML, Paus T, Pausova Z, Rietschel M, Robbins TW, Smolka MN, Strohle A, Schumann G, Garavan H, Consortium I (2014) Neuropsychosocial profiles of current and future adolescent alcohol misusers. Nature 512(7513):185. –+
35. Zhang L, Samaras D, Tomasi D, Volkow N, Goldstein R (2005) Machine learning for clinical diagnosis from functional magnetic resonance Imaging. IEEE Comput Soc Conf Comput Vision Pattern Recog Proc 1:1211–1217
36. Zou H, Hastie T (2005) Regularization and variable selection via the elastic net. J R Stat Soc Ser B-Stat Method 67:301–320

Part IV
Comparison Between Substance and Non-substance Addictions in Treatment

Chapter 11
Drug Therapy

Ri-Hui He and Ran Tao

Abstract This chapter first summarizes the therapy of addiction disorder, and elaborates on the progress of medication. First, the difference between dependency and addiction are introduced. The basic principles of the therapy of substance and non-substance addiction are then put forward. It is also pointed out in this chapter that with the progress of the study, the goal of addiction disorder therapy is expected to transfer from reducing the relapse and harm of the addiction to completely eliminating and recovering from it. This chapter also introduces the progress of psychological addiction elimination technology, especially the "Unconditioned Stimulus Retrieval Extinction Paradigm and Conditioned Stimulus Retrieval Extinction Paradigm" and PITDH technology. Finally it is pointed out that in addiction disorder therapy, comprehensive intervention has become a trend. With regard to the medication for addiction disorders, this chapter also includes the progress and deficiencies of substance and non-substance addiction. In terms of addiction disorder rehabilitation, the foundation of substance addiction is medication which is, however, limited for non-substance addiction. The key to the rehabilitation of addiction disorder is psycho-behavioral therapy, which is especially effective in eliminating craving.

Keywords Dependence • Addiction • Addiction disorder • Psychological addiction (craving) • Psychological addiction elimination • Medication

---

R.-H. He
RiHuiAddiction and Mental Disorders Medical Center, Guangzhou 510000, China

R. Tao (✉)
Department of Psychological Medicine, PLA Army General Hospital,
Beijing 100700, China
e-mail: bjptaoran@126.com

X. Zhang et al. (eds.), *Substance and Non-substance Addiction*, Advances in Experimental Medicine and Biology 1010, DOI 10.1007/978-981-10-5562-1_11

11.1 Introduction

Addiction is a widely used term in our daily life and academic field. It refers to the positive emotional experience such as relaxation, pleasure, euphoria, excitement etc. that the patient feels after repeated use of psychoactive substances or being engaged in addictive behaviors and the patient is unable to control himself even if he is aware of the harm. Dependence is generally divided into physical dependence and mental (or psychological) dependence. However, the long-term use of a substance can produce physiological dependence (such as beta-blockers for the therapy of hypertension), and may include tolerance and withdrawal syndrome etc., but it is associated neither with the harmful effects of addiction, such as those that affect the vital functions of daily life, nor with the craving symptom.

Therefore, dependence has more to do with physical dependence, such as tolerance and withdrawal symptoms, whereas addiction more often refers to behavioral control and psychological craving. Of course, the process of addiction will also have adverse effects on individuals afflicted with the problem, their friends, their family and society.

In 1964, the World Health Organization (WHO) Committee of Experts proposed to replace addiction with dependence. Since then, the term "addiction" has no longer been used in classification of mental disorders and diagnostic criteria in both ICD and DSM systems. However, DSM-5 has changed the decades-long practice back to addiction, combining the terms of dependence and abuse in DSM-IV, and this combination is "substance use disorder" in DSM-5, in which "non-substance related disorder" was first proposed. Originally listed in impulse control disorder listed, gambling disorder was listed in the new category. While online game disorder related to the internet is classified into "other conditions potential to be the focus of clinical attention", which illustrates that internet-related disorder is a mental health problem, but its nature and characteristics still need further study [1]. Non-substance related disorder or non-substance addiction, also known as "behavioral addictions (BAs)" refers to repeated and uncontrolled impulsive behavior that can produce undesirable consequences, and feelings of pleasure, relaxation or even excitement or other positive emotional experience generally appear after implementing the behaviors, such as gambling addiction, Internet addiction, food addiction, sexual addiction, shopping addiction, etc.

Clinically, behavioral addiction and substance addiction share many important features, and some studies have suggested that BAs has similar natural progress of disease, impulsivity of personality and compulsive characteristics with substance addiction. Some studies have investigated the comorbidity rate of substance addiction and gambling disorders, compulsive shopping, sexual addiction and internet addiction, and indicate the prevalence of comorbidities between these two kinds of diseases, which supports the view on common pathophysiological basis between BAs and substance addiction [2].

In the diagnostic criteria of ICD-10 substance dependence, dependent elements have three the key features: out-of-control of behavior; psychological craving;

tolerance and withdrawal state. If the above characteristics, namely the clinical manifestations of basic pathophysiological mechanisms, can be regarded as the key elements of addiction, especially the "repeated psychological craving" which is a part of the diagnostic criteria of addiction disorder in DSM-5, it seems to be reasonable to consider some behavioral problems (such as Internet addiction, pathological gambling, shopping addiction, etc.) as non-substance addiction. It is because these behavioral problems are involved in the key features of pathopsychological process.

According to modern mainstream medical science, substance addiction is a chronic recurrent encephalopathy, and brain science studies have found that the changes of brain function areas in non-substance addicts, such as gambling addicts, are similar to those of substance addiction, such as drugs. From this perspective, non-substance addiction can also be seen as a chronic recurrent encephalopathy. Therefore, addiction therapy can help patients stop the use of addictive substances or addictive behavior through medical, psychological, social and other multi-dimensional interventions, in order to restore normal psychosocial function, which is actually a systematic work [3]. With the author's (Ri-Hui He, similar hereinafter in this chapter) years of clinical research and practice, it also believes that the rehabilitation of addictive diseases requires at least the accomplishment of ten objectives, namely physical rehabilitation, comorbidity rehabilitation, elimination of craving, trauma repair, personality remodeling, family reconstruction, reconstruction of the overlooks on the world, life and values, goal setting, efficient learning (or work) and reintegration into society.

However, if we recognize the differences between "addiction" and "dependence" and the key elements of diagnostic criteria for addiction disorders, there should be a clear distinction between the concepts, theories and methods of the therapy of substance and non-substance addictions. In short, the therapy of substance and non-substance addictions will be expected to transfer to elimination of psychological craving (i.e, addiction) based on physical rehabilitation, and the behavior is controllable after elimination of psychological addiction. Elimination of psychological addiction makes complete rehabilitation of substance and non-substance addictions possible, and neither addictions will be seen as chronic recurrent encephalopathy, which will change our understanding of this disease and the whole therapy system.

11.2 Basic Principles of Addiction Therapy

According to modern mainstream medicine, the goals and principles of substance and non-substance addiction therapy are similar with other chronic diseases. Three-level prevention and cure strategies should be adopted. Stress should be put on the prevention of diseases, early detection and early therapy. When the condition develops to addiction, comprehensive intervention should be carried out from the medical, psychological and social and other dimensions, and the main therapeutic target is to reduce the degree of addiction, prevent relapse, restore physical and psychosocial function, and reduce the harm. As for addicts who have strong desire for withdrawal, allowable economic conditions, and good social support system such as work and family, if the addiction rehabilitation therapy institutions have mature systematic therapy technology, especially mature technology to eliminate psychological addiction, the therapeutic target can be complete rehabilitation.

Based on the main goal of addiction therapy, the National Institute on Drug Abuse organized a panel of experts, who established 13 basic principles of substance addiction therapy [4]. Yang Yongxin, a Chinese psychiatrist, took electric shock and electroconvulsive shock for the therapy of teenagers with Internet addiction, which gave a false impression on elimination of Internet addiction behavior, but these teenagers suffered from PTSD after the therapy. In view of this, the author puts forward 12 basic principles of substance and non-substance addiction therapy for reference.

(1) The prerequisite to therapy is that it should be harmless and shouldn't reduce patients' quality of life. It should be strictly prohibited to use electric shock therapy, electroconvulsive shock therapy and ablative procedure of cerebral nuclei when the patients are conscious; (2) The therapy should be personalized, humanized, efficient, integrated and multi-dimensional; (3) The therapy should be convenient and accessible; (4) The therapy should be flexibility; (5) The therapy time should be adequate and follow-up visits should be on a long-term basis; (6) Great importance should be attached to psychological therapy, especially the elimination of psychological addiction; (7) Medication can be encouraged in substance addiction therapy but used with discretion in non-substance addiction therapy; (8) Aggressive treatment should be administered for comorbid psychiatric disorders; (9) Therapy is divided into three stages: acute withdrawal therapy, psychosomatic rehabilitation therapy, prevention of relapse and return to society; (10) Effective therapy does not require voluntary conduct; (11) Assessment of effects not only relies on monitoring explicit behavior, such as regular monitoring of the use of addictive substances or addictive behavior, but also on the patient's inner feelings and emotions; (12) Assessment and consultation should be undertaken on disease complications, such as HIV/AIDS, and therapy complications, such as PTSD.

11.3 Core Content of Addiction Therapy

Addiction therapy is a long-term process, and can be divided into three stages: acute withdrawal therapy, psychosomatic rehabilitation therapy, prevention of relapse and return to society. The core content of addiction therapy includes diagnostic assessment, therapy plan and various comprehensive therapy measures. Addiction therapy is based on the medical model of "biology - psychology - society", and takes comprehensive prevention and therapy measures as a general rule. Therapy mainly includes medical therapy, psychological and behavioral therapy and social support therapy.

As in DSM-5 the term "dependence" has been replaced by "addiction", and gambling addiction categorized into non-substance addiction, the author believes that psychological craving, commonly known as psychological addiction, is considered the key characteristic of addiction, and the key content of substance and non-substance addiction therapy is expected to be transferred to the complete elimination of psychological addiction.

The team led by Professor Lu Lin, a Chinese researcher has found that when addicts once again encounter with the addiction memory-related clues, the original addiction memory will be aroused, and come into an unstable state. The duration can be up to 6 h, which provides an opportunity for eliminating, processing and updating addiction memory. During the unstable state of addictive memory, they repeatedly expose the addiction-related stimulus or give low-dose addictive substances to the addict to eliminate pathological emotional memory of addition through "CS memory retrieval-extinction paradigm" and "UCS memory retrieval-extinction paradigm" [5]. It has laid a scientific foundation for the elimination of craving.

11.4 Trends in Addiction Therapy

Nowadays, comprehensive intervention has gradually become the mainstream of addiction disorder therapy. In fact, the addiction disease therapy has showed systematized, comprehensive, effective and humanized trend in China at present.

For instance, Prof. Tao Ran proposed the Tao diagnostic criteria for Internet addiction, which was adopted by DSM-5. He put forward a "five-in-one" intervention model for Internet addiction, including medical therapy, psychological therapy, health education, military training and social experience activities. The basic characteristics of this model are personalized, progressive and all-round system therapy. The total effective rate was 71–94% for patient with Internet addiction, and the satisfactory rate was up to 60-71% for the parents and patients after 1 month of follow-up [6].

Details of medication and psycho-behavior therapy will be elaborated in the following chapters. Interventions models of military training, education and social experience in the therapy are significant, in which patients are taught how to live and love their life and realize the value of life.

The multidimensional addiction rapid therapy (MDART), which was created by the author targeting prescription drug addiction, has made a significant breakthrough in the efficacy [7]. This therapy integrates (non-methadone replacement therapy), general anesthesia (UROD), craving elimination, personality remodeling, family relationship reconstruction and efficient learning (or work), return to society. It is referred to as "5+2" mode for short. To put it in simple terms, firstly, it applies non-addictive drugs to control withdrawal reaction; and then conduct UROD therapy when withdrawal reaction becomes slight. Conduct deep hypnosis during UROD period and repeatedly implant new programs. While rapidly expelling the residual opioids, patents rapidly recover in body and brain, and at the same time, the negative reinforcement effect induced by withdrawal reaction shall be quickly eliminated and cravings will be reduced rapidly. Secondly, intensify the effect of eliminating cravings through program implantation technology under deep hypnosis (PITDH) and create new and healthy conditioned reflex; carry out trauma repair and personality remodeling through deep hypnosis; harmonious family relationship can be reconstructed through dealing with family conflicts by family therapy; for school-age teenagers, PITDH can be used to reconstruct conditioned reflex for optimum learning, whereas for adults, conditioned reflex for efficient working state shall be reconstructed. Finally, guide the patient to set a goal for life and help them return to society faster. Social work agency's help may be elicited when necessary. Naltrexone can be taken during the rehabilitation period after discharge, to eliminate the positive reinforcement effect caused by addictive drugs abused by patients occasionally. The treatment system is characterized by fast recovery, complete elimination of cravings, no trauma, non-addiction and high abstinence rate. Clinical practice shows that prescription drug addicts who have strong withdrawal desire, certain economic strength and good family support system can achieve more than 80% abstinence rate in a year, which can be regarded as an established method for prescription drug addiction. The improved method has been used in the therapy of various addiction disorders. The author personally believes that psychological addiction i.e. cravings is in nature the pathologic conditioned reflex to addictive substance or behaviors on subconscious level. The pathologic conditioned reflex can be eliminated rapidly on subconscious level through PITDH, so that new and healthy conditioned reflex, even positive behavior reaction can be established to achieve the effect of rapid elimination of psychological addiction. PITDH is promise to be the fourth generation of CBT therapy after DBT. The technology system is upgraded constantly and may become an efficient rehabilitation method to addictive disorders.

11.5 Drug Therapy of Substance Addiction

11.5.1 Drugs for Alcohol Addiction Rehabilitation

There are two ways to treat alcohol addiction: one is to reduce or stop alcohol consumption directly through the adverse effects of drinking alcohol, or by weakening the neurotransmitter system enhanced by alcohol. The second is the therapy of persistent psychotic symptoms associated with alcohol addiction, with the aim of preventing or reducing the enhancement of "self-drinking" behaviors of individuals as a result of these symptoms. Commonly used drugs to reduce or stop drinking are:

11.5.1.1 Alcohol Sensitizer

Alcohol sensitizers can alter body's response to alcohol, thereby rendering uncomfortable sensation or toxic reaction to the individual after drinking. Disulfiram is the only alcohol sensitizer approved for the therapy of alcohol addiction in the United States, and is widely used clinically. The mechanism of action is to inhibit aldehyde dehydrogenase, which catalyzes the oxidation of acetaldehyde to acetic acid. The effect is to raise the concentration of acetaldehyde in the blood, and trigger Disulfiram – alcohol reaction (DER). The intensity of DER varies with the dose of Disulfiram and alcohol consumption. Most DERs are self-limiting and can last about 30 min. Studies have shown that Metronidazole can also play the similar role as Disulfiram.

More double-blind, placebo-control studies are needed to investigate the effectiveness of alcohol sensitizers in preventing the relapse of alcohol addiction. In the completed control study, the difference in outcome between subjects who received Disulfiram therapy and those given placebo was generally insignificant. Disulfiram may be of clinical value for special groups who need special efforts to ensure compliance to alcohol addiction.

11.5.1.2 Opioid Receptor Antagonist

Naltrexone and Nalmefene have been used to treat alcohol addiction. Naltrexone is approved with effective prevention of relapse of severe alcohol abuse based on two independent studies. The study found that Naltrexone was well tolerated and significantly reduced alcoholic craving and number of days of drinking than placebo. It is effective with Naltrexone therapy for more than 12 weeks, but oral Naltrexone is with poor compliance that will reduce the potential benefits of drugs. This increases the interest in developing and evaluating long-acting injected preparations.

Now long-acting Naltrexone has been launched into the market. Compared with daily administration, monthly administration will improve the adherence of medication, and parenteral administration can avoid the first pass metabolism that will increase the bioavailability of drug. Long-acting Naltrexone approved by FDA can be taken 380 mg monthly: The built-in package says it is limited to alcohol addicts at initial stage of alcoholism therapy (at least 7 days after withdrawal), because the study indicates that long-acting Naltrexone has the best effect at this time. The drug is approved for patients already abstaining from alcohol and receiving psychosocial therapy.

Nalmefene is an opioid receptor antagonist without characteristics of agonist. In a recent multicentre, placebo-control, randomized trial, the target dose of Nalmefene (encourage subjects to take 10–40 mg drug without hesitation before drinking) combined with the most basic psychosocial intervention have a similar effect with Naltrexone.

11.5.1.3 Acamprosate

Acamprosate is an amino acid derivative. The study on more than 4000 patients carried out in Europe provided good evidence of the efficacy of acamprosate in preventing re-drinking and reducing alcohol consumption in relapsed patients. There are two multicentre trials in the United States. In the first trial, patients take two therapeutic doses of Acamprosate, and in the second trial, patients take combination therapy (combining medication and behavioral intervention for the therapy of alcoholism), which is currently the largest trial of alcohol therapy, but intent-to-treat analysis fails to show that Acamprosate is superior to placebo. A comparative trial among Acamprosate, Naltrexone and combination therapy showed Naltrexone was superior to placebo or combined therapy of Acamprosate and Naltrexone, and can significantly increase the percentage of days of alcohol withdrawal.

11.5.1.4 Anticonvulsive Drugs

Anticonvulsive drugs include Carbamazepine, Divalproex and Topiramate. Although these drugs have different mechanisms of action, they may play a therapeutic effect on alcohol addiction through antagonism on glutamate neurotransmitter and agonist on GABA, thereby prompting the normalization of abnormal neurotransmitter activities induced by long-term alcohol abuse.

11.5.1.5 Baclofen

In a small sample research, individuals who had initial success in alcohol abstinence receive at ramdeom 30 mg Baclofen or placebo, three times per day. The results showed that Baclofen was well tolerated and patients who took it were more likely to keep abstinence than the placebo group during the course of more than a month (and also showed more days of alcohol withdrawal). In one of the largest observational studies to date (100 subjects) 92% of patients reported craving suppression and long-term relapse rates were low. Baclofen, presents with great promise, particularly in patients with more severe forms of AUD. The side-effects of oral baclofen (i.e., somnolence, insomnia, dizziness, paresthesia, etc.) though, pose a principle limitation to its administration in alcohol addiction [8].

11.5.2 Drug Therapy of Benzodiazepine Sedative and Hypnotic Drug Addiction

Benzodiazepine sedative and hypnotic drug addiction is commonly found in individuals with other substance use disorders. From this point of view, sedative hypnotics are often used to enhance other drug effects and relieve adverse reactions during the use of other substances or withdrawal. Most patients addictive to benzodiazepine and sedative hypnotics start to get therapy due to sleep disorder, anxiety disorders or other problems [9].

The most direct way for sedative addiction withdrawal is to take gradual reduction that gradually reduces the dose in 6–12 weeks. This method is suitable for patients that have used sedative hypnotics for a long time and developed physiological dependence. It is not urgent to stop the current medication. For the effectiveness of this regimen, patients must be able to follow complex dosing regimens, adhere to routine follow-up appointments, and have no other active substance use disorders. When a lower dose is reached, the reduction in dose at each stage should be moderate, especially for short half-life drugs. At late stage, more frequent dosing intervals help prevent any withdrawal symptoms.

Another strategy for withdrawal is to replace therapeutic drug with long-acting agents of the same dose, and then gradually reduce the dose of the latter based on the aforementioned medication principle. Because the drug with long half-life has slow onset, the possibility of abuse is relatively limited, especially for Clonazepam, which is a good choice.

Another option for withdrawal therapy is the use of Carbamazepine. Two open studies have confirmed the efficacy of the drug in the therapy of complex Benzodiazepines withdrawal. There is evidence to support the use of Gabapentin and Divalproex in the therapy of alcohol withdrawal symptoms, and both two drugs has better efficacy than Carbamazepine, which indicates that the two drugs may be equally effective in the therapy of sedative hypnotic withdrawal symptoms.

For patients with severe sedative-hypnotic addiction, antipsychotics such as Olanzapine and Quetiapine can be used to improve sleep during withdrawal period, and general anesthesia can be taken as auxiliary therapy if necessary. The author has treated an addict who took daily dose of 240 tablets of Alprazolam. When he suffered from severe insomnia during withdrawal, the patient was treated with intravenous anesthesia Propofol for sedation under ECG monitoring in three consecutive nights. His sleep was then improved and he recovered rapidly.

11.5.3 Drug Therapy of Opioid Drug Addiction

Opioid addiction withdrawal syndrome has a two-phase character: (1) a relatively short initial phase, i.e., acute withdrawal phase experienced by opioid addicts; (2) Followed by long-term chronic protracted withdrawal syndrome. Acute withdrawal syndrome usually lasts 5–14 days, including: (1) Gastrointestinal symptoms (such as diarrhea and vomiting). (2) Temperature regulation disorder. (3) Insomnia. (4) Muscle pain. (5) Joint pain. (6) Significant anxiety. (7) Bad mood. Protracted withdrawal syndrome is mainly manifested by bone pain, intractable insomnia and irritability. Withdrawal symptoms, whether acute or chronic, although not life-threatening, will cause significant physical discomfort, is an important reason for relapse. Therefore, the body's highly effective rehabilitation is not only necessary, but will reduce the relapse rate.

11.5.3.1 Drug Therapy of Opioid Drug Addiction at Acute Withdrawal Stage

Opioid Receptor Agonist and Some Receptor Agonist

Drug withdrawal therapy of opioid addiction under medical surveillance is based on the principle of "cross-resistance", that is, an opioid is replaced by another opioid, and gradually reduces the dosage until withdrawal.

Methadone is the most commonly used alternative, and commonly used in the therapy of heroin addiction, because of its long half-life and is administered only once a day. The initial dose is 15–30 mg/day, and generally should not exceed 40 mg/day. In emergencies, the initial dose of Methadone should be maintained on the second or third day after the drug peak, and then is gradually reduced by 10–15% per day. For outpatients, Methadone withdrawal therapy must follow national approved opioid addiction therapy program.

Buprenorphine is slowly dissociated from the μ opioid receptor, with long duration of drug efficacy. It has minor withdrawal symptoms and signs after stopping administration compared with complete agonist. This feature gives Buprenorphine advantages for patient without apparent pain for rapid withdrawal during drug withdrawal therapy. Several studies support the efficacy of Buprenorphine and Buprenorphine/Naloxone in opioid withdrawal therapy. The best dose of Buprenorphine has not been established for inpatients with acute heroin withdrawal.

Non-opioid Drug Therapy

Non-opioid drugs used in opioid abstinence therapy mainly include two α2 adrenergic agonist, Clonidine and Lofexidine. They act on presynaptic receptors in locus coeruleus to relieve the symptoms induced by increase of norepinephrine activity during opioid withdrawal.

Early clinical studies have confirmed that Clonidine can reduce withdrawal symptoms in patients with opioid addiction. It is generally recommended the initial dose of 1.2 mg/day every 24 h administered by split doses. However, due to the obvious hypotensive side effect of Clonidine, the second generation of α2 adrenergic agonist, Rofecoxib, is widely studied and used, and becomes popular in patients and clinicians for its obviously lessened hypotensive side effect.

Ultra-rapid Opioid Detoxification Under General Anesthesia

Ultra-rapid opioid detoxification under general anesthesia (UROD) was firstly proposed by Presslich and Lominer in 1988, and began to be used in clinical practice. It was recognized in the 1990s, began to be promoted, and is considered to create a new era of physical withdrawal for opioid drug addicts. UROD is to make opioid addicts taking intolerant high-dose opioid receptor antagonist at awake state, in order to elute morphine and other opioid receptor agonist from opioid receptor, induce acute opioid withdrawal reaction, thereby greatly reducing the physical withdrawal duration. The general anesthesia technology can help patients get through the acute withdrawal period without pain. Opioid receptor antagonist can be given for maintenance therapy after UROD [10]. At present, this method is mature, with high security, and physical withdrawal success rate of 100%, and has good application prospects.

The method is mainly used for the therapy of heroin, morphine and other traditional opioid addiction. The author, as a psychiatrist, once received a master's degree in anesthesiology, Fudan University in China, is the first one in China to conduct more in-depth study of this technology. In combination with the proprietary traditional Chinese medicine and anesthesia-specific technology for drug addiction therapy, UROD was greatly improved. The improved UROD has minor side effects and its unique advantages of quick physical withdrawal, less pain, mild addiction, short recovery period and no trauma are fully reflected. The technology was applied to promote the use of methadone, buprenorphine, tramadol, compound tramadol, oxycodone hydrochloride, compound oxycodone and cough syrup containing codeine prescription drug addiction therapy; after modification, it was also applied to methamphetamine, ketamine and other new synthetic drug addiction therapy, and achieved very good clinical results. Now the author will integrate deep hypnosis with UROD. The initial clinical results show more obvious reduction in addiction, the effect is more lasting.

Psychiatrists often consider UROD contains high risk, is difficult to operate, produce high cost and is hard to popularize. They also believe that the problem of physical dependence on opioid addiction has been addressed. The technique is only suitable for patients with severe opioid addiction, and it is suspected of killing the chicken with a chopper for taking this technology to treat mild to moderate opioid addicts. The author has treated more than 3000 cases with various substance addictions by this technology, including mild to moderate opioid addiction patients. Clinical practice has proved that the technology has low risk if operated by a professional anesthesiologist. More importantly, UROD can help patient recover quickly, and greatly reduce the recovery period, eliminate negative reinforcement induced by acute and chronic withdrawal syndrome of patients with substance addiction at withdrawal phase, obviously reduce addiction, and provide a solid foundation for further psychological therapy. Further psychological intervention and family intervention after UROD, especially applying PITDH technology can significantly shorten the rehabilitation period, increase the proportion of rehabilitation, and even completely cure the patient. For patients with opioid addiction and alcohol addiction, UROD combined with Naltrexone maintenance therapy during the rehabilitation period can eliminate the positive reinforcement caused by substance abuse, and improve the chances of rehabilitation. The average cost of UROD is relatively low for successful withdrawal compared with cost for relapse by other therapy methods. Of course, the method is more suitable for substance addicts with better economic conditions and strong withdrawal desire.

11.5.3.2 Long-Term Therapy of Opioid Drug Addiction

Protracted Abstinence Syndrome

For patients with opioid addiction, acute withdrawal is only the beginning of therapy. Protracted abstinence syndrome at rehabilitation stage, especially refractory insomnia, bone pain and dysthesia, is an important factor for relapse. Methadone maintenance therapy (MMT) has been recommended for patients with PA, but the methadone withdrawal syndrome is more serious. Chinese medicine, acupuncture and other therapies may help alleviate PA. In fact, the above-mentioned UROD can perfectly solve this problem. Studies have confirmed that the sensitivity of opioid receptors returns to normal within 24 h after UROD therapy. Clinical research and experience confirmed that subacute stage was ended within 72 h after UROD therapy, and followed by chronic stage, during which bone pain, dysphoria and other symptoms of protracted abstinence syndrome basically disappeared, symptoms of intractable insomnia improved significantly. The residual symptoms can last for 1–2 months, and be treated with sedative sleeping pills and exercise therapy.

Naltrexone Maintenance Therapy

Naltrexone is an oral long-acting opioid antagonist. It can completely block opioid receptors with administration at least three times per week or daily, and the total dose of about 350 mg a week. However, because of the persistence of opioid craving during naltrexone maintenance therapy, patients are less likely to take medication, which may influence the rate of abstinence. One solution is to use a long-acting Naltrexone injection.

Methadone Maintenance Therapy

For heroin addicts, the best option is to receive Methadone maintenance therapy (MMT) if they are with poor abstinence repeatedly, have poor family economic conditions, lack family and social support, are unemployed, and have a serious epidemic such as HIV/AIDS. The patient's physical health can get a certain recovery, and provide opportunity for social and psychological stability that can significantly reduce criminal activity and social harm.

In addition to male sexual dysfunction and QT interval prolongation, it is generally not accompanied by serious adverse reactions during MMT. Large-scale studies had shown that 60 mg/day or higher dose of Methadone was more effective than that less than 60 mg/day. In general, the time required for successful rehabilitation is several years for Methadone therapy.

Although MMT has the above benefits, it needs daily medication, is controlled by the relevant government departments, as well as with side effects that can cause inconvenience to patients. At a certain time after successful withdrawal of heroin,

Methadone addiction withdrawal is also a problem. UROD may be a better choice for patients to recover if they have strong withdrawal desire and good economic conditions.

Buprenorphine Maintenance Therapy

Buprenorphine is an opioid receptor partial agonist. Buprenorphine sublingual tablets include two dosage forms. One only contains buprenorphine, and the other is called Suboxone. The latter is formulated with buprenorphine and opioid receptor antagonist naloxone at a ratio of 4:1, with the aim to resist illegal smoking and intravenous injection. Single dosage form is mainly used under clinical monitoring, and Suboxone is more suitable for home use. Many years of clinical studies have demonstrated that Buprenorphine, the mixture of buprenorphine and naloxone, is a safe and effective agent to substitute Methadone during opioid agonist maintenance therapy. It is worth noting that the ceiling effect of Buprenorphine reduces the risk of overuse and limits the tendency for drug abuse. A new formulation of buprenorphine, a transdermal and sustained drug, has been developed to provide continuous relief of palliative opioid withdrawal, reduce the number of medical visits, and improve compliance, and with less possibility of smoking and abuse.

11.5.3.3 Therapy by Chinese Herbs

In China, opioid addiction (called "detoxification" in China) can be divided into three categories: opioids, non-opioids and Chinese herbs. Chinese herbs have multi-target effect, no addiction and other characteristics.

On July 7, 1995, the Chinese Ministry of Health issued the approval certificate of the first state-level traditional Chinese detoxification medicine – Fukang Tablets, followed by the development of Chinese detoxification medicine, which has aroused widespread concern at home and abroad. As of January 21, 2015, there were nine detoxification drugs approved by China's State Food and Drug Administration (SFDA). Compared with the most commonly used Methadone replacement therapy, Chinese herbal medicine has the advantages of good compliance, no addiction, accurate curative effect on protracted symptoms and high conduct rate. With the establishment of China's drug addiction therapy, as the supplement and extension of "compulsory detoxification, reeducation through labor and voluntary detoxification" for drug addicts, Chinese herbal detoxification drugs can be purchased directly in pharmacies because of no addiction, so they play an increasingly important role for drug addicts to get rid of drug addiction and return to the society.

For the frequency of single herbal medicine used in these nine kinds of Chinese patent medicines, it showed that Corydalis, Datura Flower, Ginseng, Radix Aconiti, Radix Salviae Miltiorrhizae, Radix Paeoniae Alba and Rhizoma Coptidis were more frequent. Studies have shown that Corydalis and Aconite can significantly reduce body shaking, head and limbs tremble in withdrawal rats; Datura Flower can significantly reduce limb stretching, facial tremors, diarrhea, polyuria and abnormal posture; Ginseng can prevent morphine tolerance and addiction, and can adapt to the original status, but also regulate the state of the body and relieve withdrawal symptoms. However, there are still limited studies on the active ingredients of traditional Chinese medicine (TCM) for the therapy of addiction and withdrawal symptoms.

During the therapy of acute opioid withdrawal symptoms, traditional Chinese medicine can significantly reduce the doses of opioid or non-opioid drugs and have a unique advantage in the therapy of protracted symptoms. Studies have shown that Banxia Houpu Decoction can improve protracted withdrawal symptoms after the detoxification of heroin addiction, and early medication seems to reduce the relapse rate. Studies have shown that Anjun Ning, Ji Tai Tablets, YiAn Huisheng Oral Liquid can effectively control the protracted withdrawal symptoms.

Mental dependence is the primary cause of relapse. Professor Yang Zheng, a professor from Chinese Academy of Military Medical Sciences, carried out a preliminary study and found that Corydalis extract L-tetrahydropalmatine has an advantage in the late detoxification, in particular, to reduce addicts' desire, and have a natural "cocktail"-like effect. Jitai Tablets can inhibit the drug-seeking behavior induced by psychological dependence, and have a certain role in prevention of opioid re-absorption [11].

In short, traditional Chinese medicine is used for detoxification at a long term, with no addiction, multi-target therapy, and obvious efficacy in therapy of protracted symptoms and anti-relapse. Opioid addiction, withdrawal symptoms, relapse and other problems involve into multiple parts of the brain, the mechanism is complex, and Chinese herbal detoxification compound can give full play to its multi-target, and advantages of comprehensive therapy. High-throughput screening technology and rapid separation of natural products become more mature, and are expected to screen the most active components of Chinese herbal detoxification compound. The active components of Chinese herbal detoxification compound are many and complex, and their mechanisms of drug detoxification are relatively vague. Studies on pharmacological and toxicological characteristics are relatively small, so it needs further studies of comprehensive observation and evaluation on pharmacology and toxicology of active components of Chinese herbal detoxification compound.

11.5.4 Drug Therapy of Cocaine, Amphetamine and Other Stimulant Addiction

In the United States, stimulants (such as cocaine and amphetamine) are the most widely used illegal drugs after cannabis. At present, there is no universally accepted and widely used drug therapy for stimulant addiction.

11.5.4.1 Pharmacological Mechanisms of Drugs Treating Cocaine Addiction

Among the drug therapy of cocaine addiction, four approaches have potential effects: (1) Cross-tolerance substitution therapy with stimulants, similar to Methadone maintenance therapy (MMT) opioid addiction. (2)Antagonists therapy blocking cocaine targets. (3) Drug therapy with pharmacological action to antagonize the effects of cocaine (to reduce the "craving"effect of cocaine). (4) Reduce the relative action targets of cocaine in the brain by changing pharmacokinetics of cocaine. There is currently no drug approved by the US FDA or other national health authorities for the therapy of cocaine addiction, primarily because none of the drugs have a consistent, significant effect in repeated, controlled studies. Most of the current studies focus on the second and third methods of drug therapy described above to reduce and block cocaine or to act directly on cocaine-binding neuronal targets (drug antagonists) or indirectly reduce the effect of cocaine by other means [4].

11.5.4.2 Selection on Drug Therapy

(Antidepressants)

Tricyclic and heterocyclic antidepressants are the most widely used and in-depth clinically studied drugs to treat cocaine addiction. Dexamipramine is the most in-depth studied tricyclic antidepressant, but the meta-analysis showed no statistically significant for its therapeutic effect. Recent clinical studies have shown that Citalopram (20 mg/day) is significantly more effective than placebo. Different from previous studies, this study used the management of contingency in addition to cognitive behavioral therapy, suggesting that psychosocial intervention had important effect on drug therapy, and there is no other evidence on effective therapy of SSRl.

Dopamine Agonist (Anti-Parkinson's Agent)

Dopamine agonist can improve the activity of dopamine caused by discontinuation of cocaine by stimulating synaptic dopamine activity. The clinical manifestations of dopamine agonist are loss of pleasure, inability, depression and cocaine craving.

Clinical trials have shown that the clinical efficacy of all commercially available therapeutic agents for the therapy of Parkinson syndrome, including bromoc-riptine, pergolide and amantadine are inconsistent. Recent studies have found that Disulfiram significantly reduces positive reinforcement of cocaine in individuals, suggesting that, despite the potential adverse drug interactions, it may be a promising drug for the therapy of cocaine addiction.

Stimulants

Several psychomotor stimulants or appetite suppressants for the therapy of attention-deficit hyperactivity disorder (ADHD) and narcolepsy can be used as substitutes for stimulants. Two small clinical studies have found that sustained release of D-amphetamine slow-release (30–60 mg/day) can significantly reduce the use of cocaine, but no significant difference is found between low-dose and placebo.

Antipsychotics

Traditional antipsychotics (mainly antagonist postsynaptic dopamine D2 receptors) do not significantly improve the "craving"and "abuse" of cocaine. The second generation of new antipsychotics has no conclusive evidence of significant reduction in the "carving" and "abuse" of cocaine. Because cocaine can induce the depletion of dopamine neurotransmitters, concurrent use of antipsychotics may cause cocaine "abusers" having neurologic blockers-induced malignant syndrome, so much attention should be paid.

Anti-seizure Drugs

Anti-seizure drugs have been tried for the therapy of cocaine addiction, and most extensive research is carried out in carbamazepine. Carbamazepine had no effect on cocaine "abusers", nor did Gabapentin and Baclofen. Recent clinical trials have shown that Tiagabine, Tobramix and Vigabatrin have some beneficial effects on cocaine "abusers", and Phenytoin (300 mg/day) can significantly reduce the risk of cocaine in clinical trials.

11.5.4.3 Amphetamine Addiction

Because of their common pharmacological principles, many drugs used to treat cocaine addiction are also investigated for the therapy of amphetamine addiction. For cocaine addiction, most control clinical studies do not show that these drugs have a therapeutic effect on amphetamine addicts.

The most promising therapy to date has been the use of agonist replacement therapy (agonist and GABA neuronal activity potentiators). In three clinical control trials, it was found in two trials that D-amphetamine significantly reduced use of amphetamine compared with placebo. It was found in another clinical control study that Methylphenidate (54 mg/day) sustained-release agent significantly reduced use of amphetamine. In two open clinical trials, it was found that anti-spastic drug Vigabatrin can inhibit degradation of GABA and increase GABA neuronal activity by GABA transferase, and ultimately reduce the use of Methamphetamine.

11.5.5 Drug Therapy of Nicotine Addiction

Drug therapy of nicotine addiction is mainly divided into nicotine replacement therapy and non-nicotine drugs.

11.5.5.1 Nicotine Replacement Therapy (NRT)

Nicotine replacement therapy (NRT) remains the major method of tobacco addiction therapy. So far, the US FDA approved five nicotine replacement products: nicotine gum, nicotine patches, nicotine nasal spray, nicotine inhalants and nicotine sugar. Nicotine gum, nicotine patches and nicotine sugar can be purchased directly at the drugstore counter, while nicotine nasal spray and nicotine inhalants cannot be purchased without a doctor's prescription.

11.5.5.2 Non-nicotine Products Include

Bupropion

In the whole process of trying to quit smoking, the emergence of depression is related with relapse. Of all the evaluated antidepressants, Bupropion is the first non-nicotine drug approved for the therapy of nicotine addiction, and a monocyclic antidepressant to inhibit reuptake of norepinephrine and dopamine. This drug is safe. The long-term use of Bupropion for the prevention of relapse is effective in smokers with or without a history of depressive disorder.

Varenicline

Varenicline is a selective agonist antagonist of the $\alpha4\beta2$ nicotinic acetylcholine receptor. In two important clinical trials, Varenicline was found to be more effective in nicotine withdrawal than placebo or Bupropion. Varenicline showed better nicotine withdrawal effect compared with nicotine replacement therapy (NRT), and adverse reactions were comparable to NRT.

Nortriptyline

Nortriptyline is a tricyclic antidepressant, and recommended as a second-line drug for the therapy of nicotine addiction.

Clonidine

Clonidine is a central $\alpha 2$ receptor agonist, and is a second-line drug for therapy of nicotine addiction. It is easy to be absorbed by skin, and the recommended dose is 0.2 mg/day for duration of 3–10 weeks. Clonidine patches began to be used 1 week before the tobacco addiction patient plans to stop smoking, and then the dosage should be changed weekly.

11.5.6 Pharmacological Interventions for Other Drug Addiction and Multi-drug Mixed Addiction

11.5.6.1 Cannabis

For cannabis addicts, there are no proven short-term or long-term therapies. Cannabinoid receptor CB1 mediates the psychoactive effects of cannabis, and agonist or antagonist studies may contribute to the therapy of cannabis addiction. 9-THC is a cannabinoid receptor agonist that relieves cannabis-associated withdrawal symptoms. Studies have shown that CB1 receptor antagonist, rimonabant, can block the physiological and psychological effects of cannabis.

11.5.6.2 (Phencyclidine,PCP)

At present, PCP addiction is still lack of systematic experience in drug therapy. A small sample outpatient study showed that Demipramine and Buspirone significantly improved symptoms of depression and other psychiatric symptoms, but neither of them can decrease the use of PCP in a double-blind, placebo-control study,

11.5.6.3 Hallucinogens

Hallucinogens include Lysergic Acid Diethylamide (LSD), Psilocybin, Dimethyltryptamin (DMT), and so on, and often affect the neurotransmission of serotonergic neurons. Some hallucinogens, such as 3,4-methylenedioxymethamphe tamine (MDMA), exert their effects by affecting catecholaminergic neurotransmission.

At present, there is no effective therapy of hallucinogens abuse. Human experimental studies have shown that serotonin reuptake inhibitor, citalopram, or serotonin 2A/C receptor antagonist, Ketanserin, can reduce the acute psychological effects of MDMA. Persistent psychiatric symptoms often suggest that individuals have psychiatric disorders before using hallucinogens. Low dose of antipsychotics is recommended. LSD can lead to sensory abnormalities, such as illusions, hallucinations, and so on. This persistent or recurrent sensory abnormality lasts for several years after LSD withdrawal. Abnormal perception can be found after a period time with normal perceptual function, and known as illusion rendition or flashbacks. There are case reports suggesting that Sertraline, Naltrexone, Clonidine, or Benzodiazepines can be used for the therapy of abnormal perception and flashback symptoms, whereas antipsychotics (eg, Haloperidol, Risperidone) and SSRI can worsen this situation.

11.5.6.4 Ketamine

The therapy of ketamine addiction and associated disorders follows the principles of individual and integrated therapy. For acute poisoning patients, symptomatic therapy is recommended. Antipsychotics are suggested for symptomatic therapy of psychotic symptoms; SSRI new antidepressants and anti-anxiety drugs are recommended for symptomatic therapy of depression and anxiety symptoms, respectively. Ketamine addicts generally have urinary system damage, causing bladder contracture and aseptic inflammation. Antibiotic therapy can be used for patients with leucopenia; adrenergic receptor blockers, such as Tamsulosin, or cholinergic receptor blockers, such as Tolterodine Tartrate, can be used to relieve symptoms of sterile inflammation.

11.5.6.5 Prescribed Drug Addiction

The abuse of prescription drugs, including controlled substances, is on the rise in the world, especially among young people. Commonly abused prescription drugs mainly include codeine cough syrup, Tramadol Hydrochloride or Compound Tramadol, and compound containing oxycodone ingredients, such as Tylerine (China).

The author has treated more than 3000 cases of prescription drug addiction, and found that prescription drug addiction therapy has its particularity. On the one hand, these addictive drugs are prescription drugs, mainly affect opioid receptors to produce excitatory effect, but have less toxicity than simple opioid receptor agonist, such as methadone or heroin, so drug therapy is recommended, but not alternative therapy, in particular, Methadone Alternative therapy. The author takes take non-addictive drug therapy or UROD-based "multi-dimensional addiction rapid therapy" (MDART); on the other hand, most addicts are teenagers, and even students, so therapy is not only to solve physical dependence or psychological dependence, but also for efficient therapy of their learning disabilities after psychosomatic rehabilitation. While the PITDH shows the superiority that can quickly eliminate the original anxiety conditioned reflex to learning, rebuild the excitable conditioned reflex, establish efficient and happy learning state, and even create a similar genius learning state, so that students quickly return to school. With high rehabilitation speed, this therapy method can quickly help young people return to school and are welcomed by students and their families.

In addition, medical personnel need to strengthen their understanding of prescription drug abuse and addiction. The author found that addicts who abuse Tramadol are more associated with seizures in clinical practice, easy to be misdiagnosed as primary seizures. In fact, the seizures can disappear after cure of tramadol addiction; Most addicts who abuse cough syrup have complication of hypokalemia. It should firstly take antidiarrheal drugs for symptomatic therapy of diarrhea, one opioid withdrawal symptom before taking potassium supplement; the severe cases may be associated with osteoporosis. The author had treated a case with abuse of cough syrup for 10 years, with average daily dose of 1200 ml, and had been misdiagnosed as "ankylosing spondylitis". The patient suffered from severe osteoporosis, whose bone mineral density was only equivalent to 34% of normal value. The patient also had "S"-shape bending spine, barrel chest, pain, limp and other symptoms and signs, and the height decreased 12 cm over the last 2 years. After 5 months of therapy, height recovered 6 cm, the patient got psychosomatic rehabilitation, and completely recovered at 5-year follow-up, and finally had a healthy boy.

For iatrogenic drug addicts, it is best to cure the patient's primary disease before treating his substance addiction.

11.6 Drug Therapy of Non-substance Addiction

Non-substance addiction is a behavioral addiction. Drug therapy of behavioral addiction can learn from the study and clinical experience of substance addiction, even with different addiction behavior theory models and mechanisms. Current pharmacological studies are still at initial stage and lay focus on effectiveness of therapy. In general, a variety of studies have proved that some drugs have a certain effect on behavioral addiction, but are featured by small sample, less control, lack of long-term follow-up and other deficiencies. There are two types of drugs in clinical research; the first one is to take symptomatic therapy according to the theoretical model or the pathogenesis of behavioral addiction or the symptoms of behavioral addicts. The second is to take therapy for the comorbid mental disorder with behavior addiction. In behavioral addicts with comorbid psychiatric disorders, the rational use of psychiatric therapy drugs can improve the compliance of patients with behavioral addiction therapy, and hope to break the vicious circle between behavior addiction and comorbid psychosis, thereby enhancing the overall efficacy. Therefore, the therapy of behavioral addiction is necessary for diagnosis and therapy of comorbidities. The following introduces research advances on drug therapy of different behavioral addictions.

11.6.1 Drug Therapy of Gambling Addiction

Gambling addiction (GA) is also called gambling disorder, problematic gambling, pathological gambling, etc., and classified as impulse control disorder at the earliest.

The new DSM-5 considers GD as a behavioral addiction, sharing neurobiological and clinical similarities with substance use disorders [12].

In addition, network gambling has become a new trend nowadays. "Internet gambling addiction" should belong to "gambling addiction" rather than "Internet addiction", but it relies on internet, and has some new features compared with traditional offline gambling, such as the amount of money won or lost in the gamble is just a number for patients who are lack of real feelings. So it is easier to for implementation and make people more addictive.

Nowadays there are no therapy guidelines approved by the Food and Drug Administration (FDA). Therefore, pharmacologic therapies should be focused on clinical dimensions (i.e., impulsivity, compulsivity) or on the contingent comorbid psychiatric disorders and individualized in relation to the specific characteristics of the patient [12].

In recent years, several controlled clinical trials have been conducted on a variety of pharmaceutical classes, establishing an evidence-based background for the disease [12].

In the first case, pharmacological approach is based on antiobsessive or antidepressant drugs, in order to improve serotoninergic transmission. Drug dose is usually medium-high and the therapy lasts longer than in depression. Controlled trials have shown positive results, in particular for fluvoxamine, paroxetine, escitalopram, and sertraline, but shown conflicting results as well [12].

According to the second perspective, the most used compounds are opioid antagonists. In particular, controlled studies have been conducted for naltrexone and nalmefene on larger samples and with the best results [13]. Several studies have showed a reduction of urges to engage in the addictive behavior and longer periods of abstinence. A genetic predisposition has been hypothesized to regulate response to opioid antagonists across diagnostic groups and a family history of alcoholism was associated with positive therapy response to these drugs [12].

In the third approach, therapy is based on mood stabilizers such as lithium and atypical antipsychotics, as in the therapy of resistant depression and bipolar disorder. Mood stabilizers showed anti-impulsive properties as well as efficacy in reducing craving and preventing relapse in different substance-related disorders [12].

Pharmacological therapy has only been examined using randomized clinical trial methodologies in adults with pathological gambling, and, therefore, there is no direct evidence of either safety or efficacy of these therapies in adolescents with the disorder. Developmental issues are important to consider when prescribing medication for adolescents [14].

Future investigations should be addressed to detect differences in outcome among specific subgroups of GD patients. While empirically validated therapies for GD have varying degrees of support, little is known about their mechanisms of action or how specific therapies might work better for specific individuals. In clinical practice, clinicians are accustomed to using combinations of different drugs, in particular to address the comorbid conditions, such as major depression, bipolar disorder, and substance-related disorders. Combination strategies need to be studied, with the goal of providing validated therapeutic algorithms and more effective therapy strategies [12].

For patients with Parkinson's disease, there are some special features for gambling addiction induced by taking dopamine agonist. You can reduce the dosage, even withdrawal if necessary, and the symptoms of gambling addiction are improved or disappear.

11.6.2 Drug Therapy of Internet Addiction

Internet addiction, also known as the problematic network use, pathological network use, forced network use, etc., is characterized by a strong desire for Internet access, feeling of pleasure, excitement and other emotional experience in Internet, as well as spending more time and difficult to self-control In Internet, can lead to social dysfunction and heart pain. Internet addiction can be divided into computer, mobile phone, flat-panel or digital TV addiction according to the different tools. Internet addiction falls into addictions to online games, gambling, pornography, information search or social addiction, mainly to online games. Internet gambling addiction belongs to the category of gambling addiction rather than internet addiction. Internet addiction may have a variety of sub-type overlaps [15].

Internet addiction has not been classified as a behavioral addiction by DSM-5, so there is no FDA-approved pharmacological therapy. Similar to gambling addiction, so far there have been two main categories on the drug addiction therapy researches, the first category is for the theory or mechanism of Internet addiction, or symptomatic therapy of symptoms of addicts, and the second category is direct therapy of comorbid mental disorder of Internet addicts.

In the diagnosis and therapy of Internet addiction, Chinese Professor Tao Ran goes ahead. He led a team to make systematic diagnosis and therapy on a large number of adolescents with Internet addiction, and accumulated rich experience in drug therapy. Their experience, in combination with other literature reports, suggests that antidepressants, including tricyclic antidepressants and selective 5-HT reuptake inhibitors, are effective in the therapy of depressive disorder secondary or concomitant to patients with Internet addiction. Internet addicts are often accompanied by anxiety, irritability and other emotions, often confuse day and night, are secondary or concomitant with anxiety disorder or social phobia, therefore, the methods to eliminate symptoms of Internet addiction include resetting the body clock and taking anti-anxiety medication. Currently benzodiazepines and non-benzodiazepine drugs are mainly used for treating Internet addiction patients; the latter mainly includes β-receptor antagonists and azalopalone class, on behalf of Propranolol and Buspirone. Among them, Propranolol can alleviate the neuropsychological symptoms associated with social phobia in patients with internet addiction, but has limited effect on generalized anxiety and panic disorder. Emotional stabilizer can stabilize the mood of patients with Internet addiction. Some Internet addicts are with bipolar disorder, and emotional stabilizer has become the preferred first-line drugs. Sodium valproate is commonly used. Some Internet addicts may have severe mental symptoms, such as auditory hallucinations, overvalued ideas, delusion of persecution, exaggerated delusions and relationship delusions, and the severity of symptoms even get to the level at early stage of schizophrenia in some patients. Clinically, atypical antipsychotic drugs can be used appropriately for therapy, and with significant effect. Such drugs mainly include Olanzapine, Risperidone, Quetiapine Fumarate Tablets and so on. This kind of drugs is with broad spectrum, little sedative effect, and less interference on patient's thought and daily activities. They are with good drug tolerance and compliance that are conducive to maintain therapy and prevent relapse. Other sedative drugs are recommended for combined therapy of patients with acute agitation Internet addiction. Internet addiction patients can take administration of Oryzanol and Vitamin B1 to adjust the function of inter-brain and autonomic nervous system for playing weak stability-like, hormone-like and vitamin E-like effects, thus improving physical and mental symptoms caused by long-term Internet-induced autonomic dysfunction. Use of Run Shu, Run Jie, Le Dun and other eye drops can reduce eye pain and improve visual fatigue [6].

It is noteworthy that Internet addicts shall be carefully while using the above drugs for the therapy of Internet addiction, because most of the symptoms of Internet addiction are often minor than the severity of mental disorders, and some addicts are juveniles, so psychotropic drugs should be used with caution, and start with a small dose. For those young Internet addicts with poor self-control ability, serious resistance and refusing therapy, it should firstly take drug to stabilize mood, and then take psychological therapy. Psychological and social intervention should be considered to prevent relapse for long-term prognosis.

11.6.3 Drug Therapy of Sexual Addiction and Sexual Preference Disorders

Sexual addiction is also known as "orgasm addiction", impulsive sexual behavior or impulsive-compulsive sexual behavior, refers to individuals with strong, forced continuous or periodic sexual impulses, and feeling anxiety and pain if these sexual impulses are unmet. Of course, it needs to rule out organic diseases, such as sexual desire hyperthyroidism induced by pituitary tumors, adrenal tumors, hyperthyroidism, etc., but also needs to rule out sexual desire hyperactivity induced by manic episode of bipolar disorders in patients. However, sexual addiction has not been fully recognized by the academic community, due to lack of research evidence and existence of controversy. The latest revision of DSM-5 still follows the previous traditions, does not list it as a mental disease, but gives a certain interpretation in unexplained sexual dysfunctions: some people have multiple sexual intercourses with many people, who are regarded as goods by the former, and they also feel painful for this pattern of behavior.

Since the emergence of Internet pornography, people's network activity has increased significantly, and its means of sexual stimulation is mainly through the electronic tool to get access to the Internet, which also has two addiction elements as "Internet" and "sex", but the key is sexual addiction. Problematic cybersex is usually defined as difficult to self-control, excessive network pornography, and often accompanied by: failure of self-control; characteristic cognitive symptoms (persistent or intrusive pornography related ideas or obsessive intentions); regulate moods with pornography; withdrawal symptoms; increase of tolerance; and negative results.

Sexual preference disorders, also known as psychological disorders, sexual perversion, para-philias, etc., refer to a variety of forms of sexual preference and sexual behavior disorders, including: fetishism, transvestism, exhibitionism, frotteurism, scopophilia, bestiality addiction, pedophilia, sadism and masochism, necrophilia, etc. These sexual preference barriers also have "repeated psychological craving"and are in line with key items of diagnosis of addiction in DSM-5. Therefore, the author carefully puts forward the view to re-examine sexual preference barriers, and even called "sexual preference addiction" (such as change "fetishism" to "fetish addiction"; In Chinese, the word "proclivity" is a derogatory term that implies low morality and low willpower, so change to "addiction" can eliminate ethical evaluation), which contributes to understanding of the disease and highly effective therapy.

Sexual addiction and sexual preference addiction are often associated with mood disorders, anxiety disorders and substance use disorders and other mental disorders, so it is recommended to take symptomatic therapy based on comorbidities and dominant symptoms. Benzodiazepines may promote sexual impulses that should be used with caution, unless there is irritation or aggressive behavior.

There are some special characteristics for the drug therapy of targeted addiction and sexual preference addiction due to the close relationship between sexual impulses and sex hormones. The current research is mainly focused on the following:

1. Antidepressant

   A common strategy for drug therapy of sexual addiction and sexual preference addiction is to treat common irritability and other emotional symptoms at the initial stage of abstinence therapy (also long lasting). 5-HT dysfunction may lead to the development of sexual addiction, which lays the foundation for the use of selective 5-HT reuptake inhibitors (SSRIs) to treat sexual addition. In addition, it is recommended for relieving the symptoms of sexual preference addiction due to its side effects on sexual function. The effect is diverse, and the sample is relatively small for the study on sexual preference disorder, the study is open designed and easy to be affected by other confounding factors, so it is difficult to get conclusion for evaluation.

2. Naltrexone

   Naltrexone is an opioid receptor antagonist, can take effect on opioid receptors for treating sexual addiction, but not produce activity, and ultimately reduce the release of dopamine caused by the activation of opioid receptors, thereby reducing abnormal sexual behavior of excitement. Unfortunately, there is no randomized control trial of sex addiction therapy.

3. Hormone therapy

   Hormone therapy involves the use of anti-androgen therapy to prevent production of testosterone by chemical method. This method is commonly referred to chemical castration, in fact, with injecting estrogen (female hormones), anti-androgen (testicular inhibitor), gonadal stimulating hormone inhibitor or luteinizing hormone releasing hormone (LHRH) agonist into the body of criminal to lose sexual impulses, no longer erection, and ultimately lose unique human body reaction of man. Poland and South Korea enacted laws In June 2010, Russia in February 2012, respectively, for implementing chemical castration by force in male perpetrators of pedophilia.

Androgen antagonists in anti-androgen can combine with androgen receptor for competitive inhibition of DHT, including two kinds, steroids and non-steroids. The former includes chloramphenicol progesterone (CPA) and Megestrol; the latter includes Flutamide, Casodex, Nilutamide and so on. Flutamide is mostly used.

It should be noted that there are many clinical studies on the use of anti-androgen for the treatment of sexual preference addiction in addition to pedophilia addiction, and have some clinical effects, but PITDH technology gradually becomes mature in rehabilitation of sexual preference addiction, and anti-androgen therapy has more side effects. The author believes that so it is not advisable to implement anti-androgen therapy for patients with sexual preference addiction except pedophilia addiction. And even for pedophilia addiction, hormone therapy should not be used as the first, but a punitive treatment to pedophiles who commit sexual abuse in accordance with the law.

11.6.4 Drug Therapy of Food Addiction

Food addiction can be roughly defined as the enjoying eating behavior. The patient often eats delicious food, such as high salt, high sugar and high fat food, and the food intake significantly exceeds their own energy demand limit, but the patient cannot control it. From this definition, both binge eating disorder (BED) and bulimia nervosa (BN) are food addictive manifestation, and consistent with the key feature of "relapse of psychological craving"for the diagnosis of addictive disorder. BED has been listed as an independent disease of eating disorders in DSM-5. Binge eating is also a symptom of BN. The main difference is that BN patients have some inappropriate compensatory actions to prevent obesity, such as self-induced vomiting, misuse of laxatives, excessive exercise, and so on. In current clinical studies, food addiction still needs more researches and hopes to be eventually included in the type of "non-substance-related disorder".

People with BN are often accompanied with mental disorders. Most experience at least one mental disorder, and many of them experience a variety of comorbidities, which are not limited to any particular category, but involve a variety of mental disorders. For example, the lifetime morbidity rate of BN in individuals with substance addiction, especially alcohol or stimulant use disorders, is at least 30%, and the use of stimulants usually begins with an attempt to control appetite and weight. BED is also associated with some mental disorders. Therefore, mental disorder in patients with food addiction should be given special drug therapy, and it is an important aspect.

Studies have shown that antidepressants have significant effect on BN and most antidepressants are effective. Fluoxetine hydrochloride is the only drug approved by FDA for the treatment of BN, and studied with the largest sample. A multicenter study showed that 60 mg/day fluoxetine was the most effective daily dose (no statistical difference was found between 20 mg/day and placebo). An open study of fluoxetine also showed effective results in the therapy of BN adolescents. People largely ignore the persistence of drug therapy, and the current data is frustrating. Once the antidepressant therapy stops, most patients relapse soon, and even those who remain to take drug therapy at the maintenance phase still relapse or fall off. Drug therapy lacks of long-term effect, and is coupled with a very high rate of falling off, and many patients are reluctant to only receive drug therapy, all of which have emphasized the importance of effective psychotherapy for food addiction. In fact, the comparative study between drug therapy and psychological therapy found that drug therapy should be combined with psychological therapy, and CBT is currently recognized as the major therapy of BN. Therapy of BED still needs more studies, and the current studies suggest that a variety of drugs can inhibit the behavior of overeating [16].

Some evidence showed that short-acting opioid receptor antagonists (such as Naloxone) have a certain effect on overeating behavior, but randomized placebo-controlled trials showed no specific efficacy of Naloxone at conventional dosage. Uncontrolled trials found a certain effect of high-dose Naloxone (200–300 mg/day), but which should be used with caution because of potential liver toxicity of high dose Naloxone.

11.6.5 Drug Therapy of Shopping Addiction

Shopping Addiction, also known as forced shopping, compulsive consumption, or shopaholic, etc., has not yet been included in the Diagnostic Classification of Mental Disorders and is naturally not included into the category of "addictive disorder" in DSM-5. However, shopping addiction still has the key feature of "relapse of psychological craving"in the diagnostic criteria for addictive disorder. As a result, shopping addiction is expected to be listed as non-substance addiction. At present, there are few studies on drug therapy addiction for shopping addiction, and some studies have confirmed that opioid receptor antagonists, Naloxone and Nalmefene, are effective for shopping addicts [17].

References

1. APA (2013) Diagnostic and statistical manual, 5th edn. American Psychiatric Association, Arlington
2. Wei H (2016) Addiction medicine: theory and practice. People's Medical Publishing House, Beijing
3. Jianguo S (2002) Addiction medicine. Science Press, Beijing
4. Christopher A (2004) Cavacuiti. In: Wei H, Tieqiao L (eds) Summary of addiction medicine. Main translator. People's Medical Publishing House, Beijing
5. Luo Y-X, Xue Y-X, Liu J-F et al (2015) A novel UCS memory retrieval-extinction procedure to inhibit relapse to drug seeking. Nat Commun 14(6):7675
6. Ran T (2007) Investigation and invention of internet addiction. Shanghai People's Publishing House, Shanghai
7. Jing J Can (2016. June 24) "Drug Addiction" be eliminated from Brain? Sci Technol China. http://www.stdaily.com
8. Azevedo CA, Mammis A(2017) Neuromodulation therapies for alcohol addiction: a literature review. Neuromodulation Jan 5
9. Glen O Gabbard MD (2010) Treatment of mental disorders. Main translator, Jingping Z. Beijing: People's Medical Publishing House
10. Wang X, Shaoyong H, He R (2011) A preliminary study on the effect of rapid withdrawal combined with psychological intervention under general anesthesia for the improvement of symptoms in patients with cough mixture dependence. Acta Acad Med 20(4):284–287
11. Yu X, Xiaolong H, Jianguo F et al (2016) Recent advances on chinese herbal detoxification drug. Chin Tradit Herb Drug., February 47(3):519–525
12. Lupi M1, Martinotti G1, Acciavatti T1 et al (2014.) Epub 2014 May 18) Pharmacological treatments in gambling dsorder: a qualitative review. Biomed Res Int
13. Grant JE, Odlaug BL, Potenza MN et al (2010) Nalmefene in the treatment of pathological gambling: multicentre, double-blind, placebo-controlled study [J]. Br J Psychiatry 197(4):330–331
14. Grant JE1, Potenza MN (2010) Pharmacological treatment of adolescent pathological gambling. Int J Adolesc Med Health. Jan-Mar 22(1):129–138
15. Huang XQ, Li MC, Tao R (2010) Treatment of internet addiction.Oct. Curr Psychiatry Rep 12(5):462–470
16. Robert E, Hales MD, MBA(2010) Textbook of pathergasiology. Main translator, Mingyuan Z. Beijing: People's Medical Publishing House
17. Barlow DH (2004) Clinical manual of psychological disorders. China Light Industry Press, Beijing

Chapter 12
Physical Therapy

Li-Jun Xiao and Ran Tao

Abstract Physical therapy has the evidence-based science knowledge to address a wide range of physical and psychological problems of addiction. Neuromodulation techniques are becoming more and more important in the treatment of addiction. Here, the efficacy of different neuromodulation techniques in addiction, such as transcranial direct current stimulation (tDCS), repetitive transcranial magnetic stimulation (rTMS), deep brain stimulation (DBS), is critically evaluated. Other physical therapy methods including Biofeedback, Physical Activity and Acupuncture are also presented.

Keywords Neuromodulation techniques • Transcranial direct current stimulation • Repetitive transcranial magnetic stimulation • Deep brain stimulation • Biofeedback • Physical • Activity • Acupuncture

12.1 Introduction

Addictions are complex disorders conventionally represented by substance use disorders (SUDs). Other behaviors without any substance use share many clinical similarities, and are therefore categorized as addictions without drug use, −more commonly called behavioral addictions (BAs). BAs also encompass video games addiction, Internet addiction, sexual addiction, compulsive buying, sports addiction, and eating disorders. Some clinical features suggest that BAs and SUDs may share similar neurophysiopathological abnormalities. The central reward pathway involves the dopaminergic system such as the mesolimbic cortical ventral tegmental area and projections to the nucleus accumbens and the prefrontal cortex. Neuroimage studies underlined the important function of the prefrontal cortex, especially the dorsolateral prefrontal cortex (DLPFC), in both SUDs and BAs. The pharmacological and psychotherapeutic treatments of addictions and of the craving in

---

L.-J. Xiao • R. Tao (✉)
Department of Psychological Medicine, PLA Army General Hospital, Beijing 100700, China
e-mail: bjptaoran@126.com

X. Zhang et al. (eds.), *Substance and Non-substance Addiction*, Advances in Experimental Medicine and Biology 1010, DOI 10.1007/978-981-10-5562-1_12

particular, have shown their limits, which indicates the need for new treatment possibilities [1].

More recently, new treatment modalities such as non-invasive brain stimulation (NIBS) have been explored in the field of addiction, such as Transcranial Direct Current Stimulation (tDCS) and repetitive transcranial magnetic stimulation (rTMS). rTMS and tDCS, applied to the DLPFC, may transiently modify decision-making, risk-taking, and impulsivity, processes that are directly linked to behavioral disorders. It has thus been shown that applying tDCS on prefrontal areas modifies the decision process not only in sane subjects, but also in addicted subjects. The decision-making process shares common mechanisms with the impulsive behaviors observed in addictions [1].

Physical therapy can be divided into two categories, one is based on using various physical factors (such as sound, light, cold, heat, electricity, magnetism, etc.)as the main means, including rTMS, tDCS, DBS, Biofeedback and Acupuncture mentioned in this article. The other is based on the function of training and treatment as the main means, also known as the exercise therapy, including exercise, Yoga and mindfulness mentioned in this article.

12.2 Neuromodulation Techniques

12.2.1 Transcranial Direct Current Stimulation (tDCS)

Transcranial direct current stimulation (tDCS) , as a non-invasive brain stimulation technique, induces plasticity via generation of s sub-threshold, stimulation polarity-dependent alteration of membrane potentials modifying spontaneous discharge rates [2]. It is a promising tool in neuroplasticity research as well as a therapeutic instrument in neurological disorders [3].

Previous studies have demonstrated that the application of tDCS is helpful for substance disorders [4] and psychiatric diseases [2]. But the use of tDCS has not been clearly investigated in substance addition except one study about alcohol-dependent patients by Boggio [5], which investigated the effect of tDCS in alcohol dependence. Researchers found anodal stimulation of either left or right DLPFC was found to reduce alcohol craving temporarily. In the future, the parameters of tDCS and the underlying neuobiological mechanisms in the substance addition should be further optimized and clarified.

The application field of tDCS in behavioral addictions (BAs) is for now restricted to the study of food craving, mainly in so-called "sane" participants, i.e., who do not fulfill the diagnostic criteria of characterized eating disorders. These studies show that stimulating the right DLPFC and inhibiting the left DLPFC reduce the induced food craving. Therefore, there is a clinical interest in having a symptomatic treatment of craving, by considering tDCS as a complementary therapy to the standard treatment of eating disorder. The tDCS techniques offer many undeniable advantages in treating BAs: they are non-invasive, well-tolerated, implemented through a portable, and compact device, and relatively cheap compared to other techniques (such as rTMS). Thus, tDCS could be implemented in outpatient structures specialized in addictions. Several research avenues must be explored, in line with the research conducted with rTMS. The effects of tDCS in other BAs could be explored, like pathological gambling, sports addiction, sexual addictions, or video games. It would also be particularly interesting to evaluate the effects of tDCS in the longer term, whether on craving or on other BA symptoms, such as maintained abstinence. In summary, the main goals of tDCS application in Bas are all at once therapeutic, which are achieved by modulating craving, impulsivity, executive functions and physiopathological and by enhancing the knowledge of neurophysiological basis of Bas [1].

12.2.2 Repetitive Transcranial Magnetic Stimulation (rTMS)

Repetitive transcranial magnetic stimulation (rTMS), also as a non-invasive brain stimulation technique, has gained notable attention in neurologic and psychiatric research in recent years. It involves the use of a wire coil through which brief pulses of electrical current are passed, leading to the generation of magnetic fields that pass through the skull [6]. It can change the brain's neuronal activity depending on the target area and stimulation parameters, such as intensity and frequency.

Abuse and addiction to substances, such as alcohol, nicotine, cocaine, or other drugs, are major health issues. These disorders are difficult to treat and the relapse rate is high, even following detoxification, pharmacological and psychological interventions. The rationale to use rTMS as a treatment for substance addiction and craving is that the DLPFC, which plays a major role in top-down inhibitory control mechanisms and reward mechanisms, is dysfunctional in these disorders. The first case that evaluated rTMS in substance addiction patients in natural environment was by Herremans [7]. The HF-rTMS could be a potential treatment option for alcohol-dependent patients and may result in a faster clinical response. But in the evidence-based guidelines on the therapeutic use of rTMS, according to their criteria, only a Level C recommendation can be proposed for the possible efficacy of HF rTMS of the left DLPFC in reducing cigarette consumption. Regarding alcohol and food craving, data from placebo-controlled studies published to date are insufficient to consider any therapeutic recommendation [8].

In an Update about Human Studies of rTMS in the Treatment of Drug Addiction, Papers published prior to December 2012 were found through an NCBI PubMed search. A total of eleven studies were identified that met review criteria. There is nascent evidence that rTMS could be effective in reducing cocaine craving and nicotine and alcohol craving and consumption and might represent a potential therapeutic tool for treating addiction. Further studies are needed to identify the optimal parameters of stimulation for the most executive treatment of drug addiction, to improve our comprehension of the treatment neurophysiological effects, and to conduct rigorous, controlled efficacy studies with adequate power [9] and concentrates on the associated methodological and technical issues. Of the potential 638 articles, 18 met the criteria for inclusion. Most of these (11 of the 18) supported the efficacy of rTMS, especially in the short term. In most cases, the main assessment criterion was the measurement of craving using a Visual Analogue Scale. Clinical and non-clinical predictors of treatment outcomes must be determined to provide more detailed indications of neuromodulation. To our knowledge, only one study reached the conclusion that rTMS had a more powerful effect on heavy smokers than light smokers.

12.2.3 Deep Brain Stimulation (DBS)

Deep brain stimulation (DBS) has mainly been used in medication-refractory movement disorders, such as Parkinson's disease and essential tremor, and in treatment resistant obsessive-compulsive disorder and depression [10]. DBS utilizes 3D imaging technologies to precisely position the Implanted Pulse Generator (IPG) at specific regions in brain, such as Subthalamic Nucleus (STN) and NAe. Upon stimulation, the IPG interferes neural activities by high-intensity pulses to achieve desired therapeutic outcomes [11]. There have been several animal (i.e rats) and human studies (i.e. alcohol-dependent) that evaluated the effect of DBS [12]. Although in human cases, there was almost no direct effect on patient's primary disorder, alcohol consumption of patients decreased after a month [13]. Jiwen and other researchers at Shanghai Renji Hospital reported a 24-year-old male who has been suffering from drug addictions for 5 years experienced five episodes of relapses during the treatment process. However, after being treated with the DBS surgery on both-side NAc (nucleus accumbens), the patient recovered from addiction almost instantly, and relapse was not observed clinically for 3 months after the surgery. In addition, no impact on memory, intelligence, personality, etc., was observed according to different psychological assessments. Evidence from self-rating scales that were given to the patient has shown that the short-term outcome of the treatment has been achieved.

To provide background and rationale for applying this therapeutic option to obesity and addiction, researchers performed a structured literature review of the animal studies of DBS, which revealed attenuation of food intake, increased metabolism, and decreased drug seeking. They also reviewed the available radiologic evidence in humans, implicating the hypothalamus and nucleus in obesity and addiction. The available evidence of the promise of DBS in these conditions combined with significant medical need, and supported pursuing pilot studies and clinical trials of DBS in order to decrease the risk of dietary and drug relapse. They think that Well-designed pilot studies and clinical trials enrolling carefully selected patients with obesity or addiction should be initiated [14]. In conclusion, comparing to traditional Radiofrequency Ablation, DBS has certain advantages because it is reversible, controllable, minimal-invasive, and has rapid recovery rate. Therefore, DBS has the potential to become one of the optimal physiotherapies for substance addiction (i.e.

alcohol addiction). However, only limited researches and clinical trials were conducted on the use of DBS in the treatment of substance. Therefore, Researchers are expected to confirm the optimal stimulation place and parameters of DBS in the future.

12.3 Biofeedback

The neuro-feedback training has been widely used in the treatment of many diseases and disorders [15]. Studies have shown that neuro-feedback training is a good way to quit drug addiction whereas long-term use of drugs has a profound effect on the individual's EEG. Temptation and craving for drugs could be reduced by neurofeedback training in patients addicted to cocaine [16]. This treatment can also be used to treat alcoholism [17]. Basically, there are two classical directions in neurofeedback training. It is either focusing on low frequencies (alpha or theta) to strengthen relaxation and focus [18] or emphasizing on high frequencies (low beta, beta, and theta) for reinforcing activation, organizing, and inhibiting distractibility [19]. Alpha/theta is an indicator between awareness and sleep. Alpha/theta training is used for deep levels of depression, addiction, and anxiety while it increases creativity, relaxation, musical performance, and promotes healing from trauma reactions.

In patients with alcohol dependence leading to autonomic imbalance with neurovascular and cardiac dysfunction, the latter results in reduced heart rate variability (HRV). So, researchers assess the effects of HRV biofeedback training on HRV, vasomotor function, craving, and anxiety with a randomized controlled study. Their data indicate that HRV biofeedback might be useful to decrease anxiety, increase HRV, and improve vasomotor function in patients with alcohol dependence when complementing standard rehabilitative inpatient care [20].

To date the sequence of events that inevitably leads to addiction has not been defined, Naisberg presents a new model for 'biophysical synchronization and desynchronization' in relation to addiction-induced and addiction-free states. For high risk children and adolescents, five predetermined factors from birth contribute to the establishment of an addiction entity from the addiction spectrum, which can be counteracted with five protective factors for addiction-free states: (1) Sleep correction under EEG for night-time transitory homeostatic resynchronizaion(THR); (2) Biofeedback training under EEG for daytime THR; (3) Standardized competent coping skills training;(4) Standardized anti-stress techniques training for stress-free social encounters; (5) Addiction aversion training with socially compatible placebos. Biophysical synchronization ascertains the homeostatic flow of all cognitive bioinformation-processing to be navigated to appropriate channels of communication to secure health and well-being. Biophysical synchronization can be detected by subjective sensations and with the assistance of objective neurophysiological measurements. In this respect, subjects feel positive emotions and tranquility from one site and may display non-invasive homeostatic biophysical parameters at another site. This principle can be applied in prevention programs to induce subjects to function in a manner that produces positive emotions and relaxation. The same principle can be applied as objective biofeedback programs(Naisberg 2002).

Biofeedback has been used in the treatment of internet addiction for adolescents. Treatment utilizing EEG biofeedback typically includes 16 sessions, which are evenly spread into 8 weeks, and each treatment session usually lasts for 30 min. Internet Addiction Scale, Self-Rating Depression Scale, as well as individual interviews are given to adolescents both before and after the treatment to assess the impact that EEG biofeedback may have on these participants. Results show that EEG bio effect is effective in reducing anxiety and depression symptoms, improving physical health conditions, and helping with sleeping difficulties [21].

12.4 Physical Activity

12.4.1 Yoga and Mindfulness

Yoga practices, including postures and meditation, direct attention towards one's health, while acknowledging the spiritual aspects of one's nature. Mindfulness comes from ancient Buddhist philosophy, and mindfulness meditation practices, such as gentle Hatha yoga and mindful breathing, are increasingly integrated into secular health care settings [22].

There is a growing number of clinical experiments and cases about substance addiction such as alcohol dependence and smoking showing the effect of yoga and mindfulness. One recent study [23] on 168 adults with substance use disorders found that MBRP (Mindfulness-Based Relapse Prevention), compared to a treatment-as-usual control group, resulted in significantly lower rates of substance use at 2-month follow-up.

Adolescence is a key developmental period for preventing substance use initiation. Recent research suggests that mind–body practices such as yoga may have beneficial effects on several substance use risk factors, and that these practices may serve as promising interventions for preventing adolescent substance use. Results revealed that participants in the control condition were significantly more willing to try smoking cigarettes immediately post-intervention than participants in the yoga condition. Immediate pre- to post-intervention differences did not emerge for the remaining outcomes. However, long-term follow-up analyses revealed a pattern of delayed effects in which females in the yoga condition, and males in the control condition, demonstrated improvements in emotional self-control. The findings suggest that school based yoga may have beneficial effects with regard to preventing males' and females' willingness to smoke cigarettes, as well as improving emotional self-control in females [24].

Some systematic reviews and theoretical papers propose multiple overlapping mechanisms to explain how mind–body practices may prevent or reduce substance use, including: (1) Reduction of stress (and/or tension) and its overt behavioral and underlying neuroendocrine components; (2)Improvement of impaired mood such as reduction of depression and anxiety and a resulting increase in psychological well-being; (3)Induction of a peak experience or higher state of consciousness, effectively replacing the attraction of a substance-induced high; (4) Improvement in self-awareness and self-regulation of psychological and psychophysiological states allowing for improved self-efficacy through the ability to intervene and prevent destructive or maladaptive behavior before its onset; and (5) The establishment of improved self-esteem and a better philosophical relationship and understanding between the individual and his/her internal and external (social) worlds [24].

12.4.2 Exercise

There is a relationship between dopamine and all behavioral aspects that involve motor activity and it has been demonstrated that exercise leads to an increase in the synthesis and release of dopamine, stimulates neuroplasticity and promotes feelings of well-being. Exercise and drugs of abuse activate overlapping neural systems. Thus, Researchers study the influence of chronic exercise in the mechanism of addiction using an amphetamine-induced conditioned-place-preference in rats. The results conclude that a previous practice of regular physical activity may help preventing amphetamine addiction in the conditions used in this test [25].

Mechanistically, physical activity and exercise activate the same reward pathway as drugs of abuse, through increases in dopamine concentrations and dopamine receptor binding. These effects may be particularly beneficial at preventing drug use and reducing initial vulnerability to drug use (Lynch et al. 2013).

Longitudinal studies demonstrated that high levels of physical activity predict lower levels of cigarette and illicit drug use during both adolescence and early adulthood [26]. And such study also showed that an increase in levels of exercise participation from adolescence to adulthood predicts a decrease in rates of smoking and use of marijuana and other illicit drugs during adulthood [26].

Physiological stimulation in the form of exercise has been shown to stimulate cell prolife ration and adult hippocampal neurogenesis. The plastic nature of the mammalian brain, especially neurogenesis continuing in the hippocampus well into adulthood, has allowed for exercise to exert its effects at the cellular level [41]. Among traditional aerobic exercises (walking, running, swimming and ball sports with modulate intensity), anaerobic exercises (strength training with high exercise intensity), and body-mind exercises (Yoga, Tai Chi and Qigong with very mild physical and mental exercise), aerobic exercises are most commonly used for intervening drug addiction, while body-mind exercises are becoming a very popular practice in improving cognition.

Track and field, swimming, ball games, and body – building exercises are all believed to be effective interventions for cellphone addiction. There has been evidence suggests that exercise can increase body temperature, and thus induce the release of endorphin, which is capable of decreasing anxiety levels for patients with addiction [27]. Exercises can also improve patients' mood state, cultivate confidence in patients, and redirect patients back to their normal studying and working routines.

Exercise rehabilitation has the evidence-based exercise science knowledge to address a wide range of physical and psychological problems. It uses exercise programs for patient rehabilitation based on exercise science. It follows the scientific process. In the clinical subfield, baseline such as physical capacity, health information, medical history, work status, previous exercise experience need to be set. After assessment, supervised rehabilitation sessions are conducted for achieving the stated goals. Exercise rehabilitation aims to recover not only muscular-articular rehabilitation after surgery, chronic pain or fatigue, neurological or metabolic conditions but also even psychological conditions such as depression and anxiety. Smartphone addiction is a psychological disorder having both physical and psychological signs and symptoms. People who are addicted to the internet or smartphone and do not do much physical activites (they generally disregard their health) and also negative physical signs like carpal tunnel syndrome, poor posture, backaches, migraine headaches, poor personal hygiene, irregular eating, sleep deprivation, eye strain, dry eyes, lack of sleep (which can affect immune functioning and hormone secretion patterns, cardiovascular and digestive patterns). Exercise rehabilitation can employ the first goal of recuperating their physical health on the surface. Moreover, if they indulge in specific exercise program such as horseback riding or exercise gymnastics, treatment can be going on to the second stage. Exercise rehabilitation could seek mental changes through feeling of confidence, satisfaction, and new feeling of happiness [28].

12.5 Acupuncture

Starting from the 1970s, acupuncture has been utilized in the treatment of opioid addiction, and its efficacy has been proved by clinical trials [42] Acupuncture, a key component of Traditional Chinese Medicine, involves the penetration of the skin with thin metal needles, and is controlled by an appropriately trained practitioner or further stimulated by electrical stimulation (Electroacupuncutre). Hsiang-Lai Wen, a neurosurgeon in Hong Kong, discovered, serendipitously in 1972, that needles inserted in the ear–intended as a preoperative anesthetic–abated physical withdrawal symptoms from opium. Jisheng Han et al. at Chinese Academy of Sciences also observed that acupuncture can trigger the synthesis and release of Endogenous Opioid Peptides from the Central Nervous System, relieving physical withdrawal symptoms and decreasing relapse rate among opioid-addicted patients. The researchers further developed a transcutaneous electrical nerve stimulation device – "Han's Acupoint Nerve Stimulator" (HANS). They randomly assigned 28 volunteer addicts into either the Buprenorphine group or Hans with low-dose BPN group. Both groups were treated continuously for 14 days with the same standard of not observing physical symptoms of withdrawal. The results showed that 2/100-Hz electrical stimulation of HANS can significantly reduce the dosage of BPN during detoxification periods by 92.3 %, accelerate the elimination of physical scars of injecting heroin, and decrease the relapse rate at some level [29, 30].

Acupuncture has been used to treat addiction for three decades [31]. In 1997, the National Institutes of Health accepted acupuncture therapy as an acceptable procedure complementary to Western medicine. In 1985, Dr. M. Smith finalized the National Acupuncture Detoxification Association (NADA) protocol that is currently practiced in over 250 hospitals in the United Kingdom and United States. In 1996, the World Health Organization accepted acupuncture as a treatment for drug abuse. The latest modification to this treatment protocol was developed in 2005 by Dr. Ji Sheng from Peking University, Beijing, China. Currently, more than 700 addiction treatment centers use acupuncture as an adjunctive procedure. Prominent effects of acupuncture are increases in the levels of enkephalin, epinephrine, endorphin, serotonin, norepinephrine, and dopamine in the central nervous system and plasma that might mediate substance abuse.

In terms of the parameters used by electro acupuncture, the most common ones are 2 Hz, 100 Hz, and 2/100 Hz. The minimal galvanic current for the electric stimulation to work is 0.5 mA, with the incremental change being 1–3 mA. Different frequencies of electro acupunctures also have different degrees of specificity. Related researches have indicated that while the 2 Hz electric stimulation is capable of inducing the release of encephalin and endorphin, the 100 Hz electric stimulation can increase the release of dynorphins. In terms of treating Morphine relapses, 100 Hz electric stimulation has proved to be the most effective, with the 2/100 Hz being the second, and 2 Hz stimulation being the least effective [31].

Auricular acupuncture has been used extensively in substance abuse treatment programs, hospitals, and prisons throughout the U.S. and the world for the past 30 years [32]. The National Acupuncture Detoxifcation Association (NADA)-standardized 3-to 5-point ear acupuncture protocol has evolved into the most widely implemented acupuncture-assisted protocol. In 2000, Avants [33]published results of a well-designed study in which 82 cocaine-dependent, methadone-maintained patients were randomly assigned to receive the NADA protocol, a sham acupuncture control condition, or a non-needle relaxation control condition five times weekly for 8 weeks with three times weekly urine drug screen monitoring. Analysis of longitudinal urine toxicology data indicated that the NADA protocol was significantly more effective in reducing cocaine use than either a relaxation control ($p = 0.01$)or a needle-insertion control ($p < 0.05$).This landmark study generated a significant amount of interest in NADA as a potential treatment and was followed by an attempt to replicate the findings in a larger, multisite trial of 620 patients in six sites throughout the U.S.

Tianmin Zhu and other researchers [34] simulated patients' Baihui, Sishencong, Guhe, Taichong, Neiguan and Sanyinjiao acupoints every other day for a total of 20 times, and concluded from patients' decreased ISS scores that the combination of psychological therapies and electro-acupuncture may be more effective than only psychological therapies. Chen and other researchers [35] claimed that acupuncture has positive impact on Internet addiction.

Addiction Disorder, based on their observations of 17 adolescents who suffered from the disorder. Also, some researchers applied 2/100 Hz transcutaneous electrical stimulation on 18 adolescents with Internet addiction, and treated 9 adolescents in the control group by mock-HANS. After a three-day therapy, the average time spent on Internet of experiment group was significantly shorter than that of the control group, and their scoring for the Internet Addiction Test was also significantly lower than that of the control group. The researches implied that 2/100 Hz transcutaneous electrical nerve stimulation can efficiently reduce Internet time of adolescents with Internet addiction, and also significantly inhibit Internet Addiction Disorder (IAD) [36]. Moreover, researchers observed that acupuncture could alter the grey matter density in IAD patients' brain [37].

Although Internet Addiction Disorder mostly manifests itself in mental symptoms, according to Traditional Chinese Medicine, it also involves multiple organs inside the body. Addicted Internet users originally experience pleasure and a sense of accomplishment from the Internet through vision, audition, and somatosensation; however, in order to gain more satisfaction, they tend to increase the usage of Internet, a behavior that eventually leads to addiction. Such addiction can further pose harm to the heart, which is expected to have influence on other organs in the system of traditional Chinese medicine. Patients whose heart have believed to be harmed are suspect to symptoms such as dizziness, depression, and though retardation. Multiple successful attempts have been conducted by researchers to treat Internet addiction, based on the principle from traditional Chinese medicine. Studies indicate that Acupuncture is fairly effective in treating Internet addiction, and it enjoys high degree of feasibility. Unfortunately, other researches found that Acupuncture fails to eliminate addiction, despite the fact that it does alleviate symptoms to certain extent. Possible reasons for the failure are inadequate treatment time or intensity, the complexity of the disorder, and the persistent nature of the symptoms. In addition, since physical, psychological, socioeconomic, and cultural factors all contribute to the development of Internet Addiction Disorder, future treatment may require the integration of different treatment methods including Acupuncture.

12.6 Conclusion

rTMS generates a magnetic field in a coil that is placed on the scalp. The magnetic field induces an electrical current in the brain tissue beneath the coil, resulting in alterations of neural excitability. tDCS is another NIBS method capable of modulating cortical excitability. By modulating it, we could decrease impulsivity in addicted patients, and, indirectly, act on the craving. Anodal tDCS over the DLPFC may enhance executive function and provide improved cognitive control, and thus reduce the probability of relapse to drug use. rTMS and tDCS applied to the DLPFC may therefore indirectly modulate dopaminergic pathways and may consequently have an impact on the symptoms of addiction. Cognitive control could be improved and/or cravings could be reduced. So far, tDCS have proven its efficacy to decrease craving in SUDs and Bas [1].

DBS is an adjustable, reversible, and non-destructive neurosurgical intervention using implanted electrodes to deliver electrical pulses to areas in the brain. Recent researches in both animals and humans have indicated that DBS may be an effective intervention for patients with treatment-refractory addiction. Our analysis of the literature suggests that the NAc is currently the most promising DBS target area for patients with treatment-refractory addiction. The medial prefrontal cortex (mPFC) is another promising target, but needs further exploration to establish its suitability for clinical purposes [38].

Two current medical paradigms continue to debate addiction causality: addiction is seen as a pure psychological misbehavior responding to cravings; addiction is genetically or biologically predetermined contributing to the emergence of a special vulnerability leading to addictive craving for substances. Two pronounced biophysical integrity-attached wiring webs regulate human homeostatic survival. Autonomous neuronal networks (ANN) control the inner organs and systems and the mental neuronal networks (MNN) engage in bioinformation-processing to produce all mental (emotional, cognitive and behavioral) operations. 'Biophysical synchronization' provides a macroscopic biophysical interpretation for the homeostatic state of all underlying operations mediated by biophysical pulses that are ultimately integrated into forms of positive emotions and tranquility. Levels of biofeedback recapturing transitory homeostatic resynchronization, this new conceptualized preventive model will counteract addiction spectrum development [39]. Similar to other treatments, neuro-feedback has its own pros and cons. Although it is a safe and non-invasive procedure that showed improvement in the treatment of addiction, its validity has been questioned in terms of conclusive scientific evidence of its effectiveness. It is also time-consuming and its benefits are not long-lasting. Finally, it might take several months to see the desired improvements [15].

Mindfulness-based interventions, such as yoga, have sound conceptual underpinnings and growing empirical support for enhancing addiction treatment, prevention, and recovery. Individual differences in treatment and underlying psychological, biological, and behavioral mechanisms across different addictions should be further tested. Research suggests that mind–body practices may have beneficial effects on several psychological risk factors for substance use including stress, mood impairment, and emotional dysregulation, and that an amelioration of these risk factors may prevent or reduce substance use itself. In other words, it is possible that by addressing risk factors for substance use, yoga may prevent actual substance use even in the absence of explicit instruction or behavioral skills training regarding illicit substances.

More and more studies have revealed the direct efficacy of exercise as a prevention for addiction. Physical activity, and specifically exercise, is a potential non-pharmacological treatment for addiction that targets systems implicated in both early and late stages of the addiction process and has secondary health benefits.

Future work is needed to determine the conditions that produce the most beneficial effects, and to characterize the neurobiological mechanisms by which exercise, alone or in combination with other treatments, exerts its efficacy as a function of stage of the addiction process.

The current prevalent hypothesis for the use of acupuncture in the treatment of substance is the relationship between acupuncture and the Cascade Theory of Rewarding. By increasing the amount of 5-HT in the hypothalamus, acupuncture can recover the complicated functions of the brain reward cascade. Patients after acupuncture have not only reported alleviated symptoms of withdrawals, but also shown less demand of addicted drugs. When the brain reward cascade system goes back to normal functioning, patients may feel peaceful and pleasure after acupuncture therapies. Thus, besides being used to relieve withdrawal symptoms, acupuncture can also be used as a treatment for the relapse of substance addictions [40]. Studies indicate that Acupuncture is fairly effective in treating Internet addiction, and it enjoys high degree of feasibility. In addition, since physical, psychological, socioeconomic, and cultural factors all contribute to the development of Internet Addiction Disorder, future treatment may require the integration of different treatment methods including Acupuncture.

References

1. Sauvaget Anne, Trojak Benoît, Bulteau Samuel, Jiménez-Murcia Susana, Fernández-Aranda Fernando, Wolz Ines, Menchón José M, Achab Sophia, Vanelle Jean-Marie, Grall-Bronnec Marie (2015)
2. Kuo MF, Paulus W, Nitsche MA (2014) Therapeutic effects of non-invasive brain stimulation with direct currents (tdcs) in neuropsychiatric diseases. NeuroImage 85(Pt 3(2)):948
3. Liebetanz D, Nitsche MA, Tergau F, Paulus W (2002) Pharmacological approach to the mechanisms of transcranial dc-stimulation-induced after-effects of human motor cortex excitability. Brain 125(10):2238–2247
4. George MS, Aston-Jones G (2010) Noninvasive techniques for probing neurocircuitry and treating illness: vagus nerve stimulation (VNS), transcranial magnetic stimulation (TMS) and transcranial direct current stimulation (tDCS). Neuropsychopharmacology 35:301–316
5. Boggio PS, Sultani N, Fecteau S, Merabet L, Mecca T, Pascual-Leone A, Basaglia A, Fregni F (2008) Prefrontal cortex modulation using transcranial DC stimulation reduces alcohol craving: a double blind-sham controlled study. Drug Alcohol Depend:55–60
6. Barker AT, Freeston IL, Jalinous R, Jarratt JA (1987) Magnetic stimulation of the human brain and peripheral nervous system: an introduction and the results of an initial clinical evaluation. Neurosurgery 20:100–109
7. Herremans SC, Baeken C (2012) The current perspective of neuromodulation techniques in the treatment of alcohol addiction: a systematic review. Psychiatr Danub 24(supplement 1):S14–S20
8. Lefaucheur J-P, André-Obadia N, Antal A, Ayache SS, Baeken C, Benninger DH et al (2014) Evidence-based guidelines on the therapeutic use of repetitive transcranial magnetic stimulation (rTMS). Clin Neurophysiol 125:2150–2206
9. Bellamoli Elisa, Manganotti Paolo, Schwartz Robert P, Rimondo Claudia,Gomma Maurizio, Serpelloni Giovanni (2014) rTMS in the Treatment of Drug Addiction:An Update about Human Studies.Behavioural NeurologyVolume, Critical review aims to summarise the recent

developments with respectto the efficacy of rTMS for all types of addiction and related disorders (including eating disorders)
10. Knapp CM, Tozier L, Pak A, Ciraulo DA, Kornetsky C (2009) Deep brain stimulation of the nucleus accumbens reduces ethanol consumption in rats. Pharmacol Biochem Behav 92:474–479
11. Wang R, Hong-Yu Z (2010) Application of deep electrical stimulation in the treatment of drug addiction. Chin J Clin Neurosurg 15(3):189–192
12. Henderson MB, Green AI, Bradford PS, Chau DT, Roberts DW, Leiter JC (2010) Deep brain stimulation of the nucleus accumbens reduces alcohol intake in alcohol-preferring rats. Neurosurg Focus, E12
13. Kuhn J, Lenartz D, Huff W, Lee SH, Koulousakis A, Klosterkoetter J, Sturm V (2007) Remission of alcohol dependency following deep brain stimulation of the nucleus accumbens: valuable therapeutic implications? Neurol Neurosurg Psychiatry 78:1152–1153
14. Halpern Casey H, Torres Napoleon, HurtigHoward I, Wolf John A, Stephen James, Oh Michael Y, Williams Noel N, Dichter Marc A, Jaggi Jurg L, Caplan Arthur L, Kampman Kyle M, Wadden Thomas A, Whiting Donald M, Baltuch Gordon H Expanding applications of deep brain stimulation: a potential therapeutic role in obesity and addictionmanagement.Acta Neurochir (2011) 153:2293–2306
15. Marzbani H, Marateb HR, Mansourian M (2016) Neurofeedback: a comprehensive review on system design, methodology and clinical applications. Basic Clin Neurosci 7(2):143–158
16. Horrell T, El-Baz A, Baruth J, Tasman A, Sokhadze G, Stewart C et al (2010) Neurofeedback effects on evoked and induced EEG gamma band reactivity to drug-related cues in cocaine addiction. J Neurother 14(3):195–216
17. Moradi A, Pouladi F, Pishva N, Rezaei B, Torshabi M, Mehrjerdi ZA (2011) Treatment of anxiety disorder with neurofeedback: case study. Procedia-Soc Behav Sci 30:103–107
18. Gruzelier J (2009) A theory of alpha/theta neurofeedback, creative performance enhancement, long distance functional connectivity and psychological integration. Cogn Process 10(1):101–109
19. Ros T, Moseley MJ, Bloom PA, Benjamin L, Parkinson LA, Gruzelier JH (2009) Optimizing microsurgical skills with EEG neurofeedback. BMC Neurosci, 10(1), 87
20. Penzlin AI, Siepmann T, Illigens BM-W, Weidner K, Siepmann M (2015) Heart rate variability biofeedback in patients with alcohol dependence: a randomizedcontrolled study. Neuropsychiatr Dis Treat 11:2619–2627
21. Pan S-j, Xiu-ying D (2010) Efficiency of electroencep halographic biofeedback treatment in middle school students with internet addiction disorder. J Ningxia Med Univ 32(1):71–73
22. Khanna S, Greeson JM (2013) A narrative review of yoga and mindfulness as complementary therapies for addiction. Complement Ther Med 21(3):244
23. Witkiewitz K, Bowen S (2010) Depression, craving, and substance use following a randomized trial of mindfulness-based relapse prevention. Consult Clin Psychol 78(3):362–374
24. Butzer B, LoRusso A, Shin SH, Khalsa SBS (2017) Evaluation of yoga for preventing adolescent substance use risk factors in a middle school setting: a preliminary group randomized controlled trial. J Youth Adolesc 46:603–632
25. Fontes-Ribeiro CA, Marques E, Pereira FC, Silva AP, Macedo TRA (2011) May exercise prevent addiction? Curr Neuropharmacol 9:45–48
26. Terry-McElrath YM, O'Malley PM (2011) Substance use and exercise participation among young adults: parallel trajectories in a national cohort-sequential study. Addiction 106(10):1855–1865
27. Murphy, Arnsten, Jentsch et al (1996) Dopamine and spatial working memory in rats and monkeys:pharmacological reversal of stress-induced impairment. J Neurosci 16(23):7768–7775
28. Kim H (2013) Exercise rehabilitation for smartphone addiction. J Exerc Rehabil 9(6):500–505
29. Han JS, Zhang RL (1993) Suppression of morphine abstinence syndrome by body electroacupuncture of different frequencies in rats. Drug Alcohol Depend 31(2):169–175

30. Wu L-Z, Cai-Lian C, Ji-Sheng H (1999) Treatment on heroin addicts by four channel Han' s acupoint nerve stimulator(HANS). J Beijing Med Univ 31(3):239–242
31. Motlagh FE, Ibrahim F, Rashid RA, Seghatoleslam T, Habil H (2016) Acupuncture therapy for drug addiction. Chin Med 11(1)
32. Xu J, Yu L, Lei Z, Wen-guang H, Liang Y (2012) Research progress of acupuncture on the treatment of opioid addiction by dopamine in the midbrain border. Chin J Drug Depend 21(1):1–5
33. Stuyt EB, Voyles CA (2016) The national acupuncture detoxification association protocol, auricular acupuncture to support patients with substance abuse and behavioral health disorders: current perspectives. Subst Abuse Rehabil 7:169–180
34. Avants SK, Margolin A, Holford TR, Kosten TR (2000) A randomized controlled trial of auricular acupuncture for cocaine dependence. Arch Int Med 160:2305–2312
35. Zhu T-m, Rong-jiang J, Xiao-ming Z, Chen J, Hui LI (2009) Clinical effect of electroacupunctum combined with psychologic interference on patient with internet addiction disorder. Chin J Integr Tradit West Med 29(3):212–214
36. Wei C, Jiang-hong L, Jing-mei W (2014) Clinical study of acupuncture on adolescents with internet addiction disorder. J Gannan Med Univ 34(2):247–249
37. Wu L, Junjuan Y, Jisheng H (2007) Treatment on 27 adolescents with internet addiction by 2/100 hz han's acupoint nerve stmulator (hans). Chin J Drug Depend 16(1):32–35
38. Xi-xi Chen (2012) A fmri study on the effect of electro-acupuncture on the brain function of IAD suffere. Doctoral dissertation, Chengdu University of Traditional Chinese Medicine
39. Luigjes J, van den Brink W, Feenstra M, van den Munckhof P, Schuurman PR, Schippers R, Mazaheri A, De Vries TJ, Denys D (2012) Deep brain stimulation in addiction: a review of potentialbrain targets. Mol Psychiatry 17:572–583
40. Naisberg Y (2002) Long-term biophysical synchronization for prevention of addiction spectrum formation in high risk children and adolescents: theory and practice. Med Hypotheses 58(6):433–438
41. Wang H-j, Wang li (1998) Biochemical hypothesis of acupuncture treatment of substance addiction. Int J Tradit Chin Med 4:18–20
42. Jun H, Hussaini SMQ, Rigby MJ, Jang M-H (2012) Functional role of adult hippocampal neurogenesis as a therapeutic strategy for mental disorders. Neural Plast:1–20
43. Xiang-Ling Z, Long-Ming L, Lu Y-H, Wang Z-T (2004) Clinically controlled study on deintoxication of heroin dependence by acupuncture on points of the governor vessel. Chin Acupuncte Moxibustion 24(6):385–387

Chapter 13
Traditional Chinese Medicine (TCM) Therapy

Li-Jun Xiao and Ran Tao

Abstract Traditional Chinese Medicine(TCM) has been utilized in China for more than 2,000 years, and it has been practiced in treatment of substance addiction and non-substance addictions. TCM have efficacy in the rehabilitation of abnormal physical problems induced by chronic drug use, including improving immune function, increasing working memory, and protecting against neurological disorders. Given that TCM is potentially effective in the prevention of relapse, it has been suggested that TCM may be the ideal choice in the future for the treatment of opiate addiction. This review examines the significance of effective Chinese herbs and prescriptions for Drug Addiction, Alcohol addiction and food addiction.

Keywords Traditional Chinese medicine • Ginseng • Yanhusuo • Kudzu • Pediculus melo • Acupuncture • Patterns of TCM

13.1 Introduction

Traditional Chinese medicine (TCM) is one of typical ethno-medicine that derives from the regular experiences of Chinese in the early age. It is also a systematic principle of thousands of years of clinical practice. Compared with modern medicine, experts believe that TCM has less side effects, is safe enough and has ideal effects in treating refractory chronic illnesses. Generally, TCM contains Chinese herbal medicine and acupuncture. Chinese herb medicine consists of natural products including plants, animals and minerals. TCM has been utilized in China for more than 2,000 years, and for the past 200 years it has been practiced in treatment of substance addiction, such as drug addiction. The general therapeutic principle of TCM was based on its unique theory of "reinforcing healthy Qi and resolving and removing effects of toxicity". In addition, TCM can be co-administered with modern medicine or other traditional medicine to reduce toxicity. The advantages of TCM

L.-J. Xiao • R. Tao (✉)
Department of Psychological Medicine, PLA Army General Hospital, Beijing 100700, China
e-mail: bjptaoran@126.com

X. Zhang et al. (eds.), *Substance and Non-substance Addiction*, Advances in Experimental Medicine and Biology 1010, DOI 10.1007/978-981-10-5562-1_13

have generated its widespread use in acute opiate detoxification during the past decade in China. TCM have efficacy in the rehabilitation of abnormal physical problems induced by chronic drug use, including improving immune function, increasing working memory, and protecting against neurological disorders.

13.2 Drug Addiction

Given that TCM is potentially effective in the prevention of relapse, the core characteristic of addiction [1–3], it has been suggested that TCM may be the ideal choice in the future for the treatment of opiate addiction.

13.2.1 Traits of Chinese Medicines

Previous studies suggested that a majority of Chinese medicines have the following traits: (1) sedation, pain relief, local anaesthesia, hypnosis and anti-convulsion; (2) stabilization of blood sugar, improvement of protein metabolism, protection of liver, blood pressure control and anti-hypoxemia; (3) anti-fatigue, anti-stress and anti-shock; (4) cardiovascular system protection and modulation of immune function. For example, Radix aconite (fuzi) is effective in relieving body reeling and head and extremities tremble in opiate withdrawal of rats. Radix ginseng is effective in stopping morphine tolerance, addiction and adaptation in the regulation of bodily functions and in relieving withdrawal symptoms [4]. Rhizoma (yanhusuo) can control nervous vomiting, dilate the coronary artery, antagonize arrhythmia and regulate the function of the GI tract [5]. Scopolamine, the major chemical ingredient of Daturae albae, flos (yangjinhua), is effective in restraining the cerebral cortex, encouraging anesthetic action, activating the respiration center and improving the metabolism of morphine. It also has benefits in controlling the withdrawal symptoms of morphine and other opiate drugs, preventing the establishment of morphine tolerance and restoring the pain relief effect of morphine in mouse models [6].

Clinicians who hope to exert herbal remedies into therapies for drug withdrawal and addiction will be confronted by a surprisingly lack of data from well-constructed clinical trials investigating the efficacy and effectiveness of these substances. Though in general the efficacy of Chinese medicine in controlling opiate withdrawal symptoms can be summarized as follows: (1) less than narcotic detoxification agents; (2) similar or even better than non-narcotic detoxification agents (e.g. clonidine, lofexidine hydrochloride); and (3) moderately effective with limitations in treating patients with severe drug addiction. The therapeutic effects could be subtly different because of their different composition and formulas. For instance, some may be more effective in controlling rhinorrhea, lacrymation and sweating, some for relieving pain, whereas others might be more effective in addressing gastrointestinal (GI) symptoms such as nausea and vomiting or suitable for treating insomnia.

Patients with opiate abuse and addiction usually experience withdrawal symptoms from d1 to d4 following Chinese medicine administration.

13.2.2 Effective Chinese Medicines and Papaveraceae Herbs

Currently, the State Food and Drug Administration (SFDA) of China has issued approval of ten Chinese medicines for use in clinical practice for the treatment of addiction, including the Fukang tablet, Lingyi capsule, Yian Liquid, Jitai tablet, Fuzhengkang granule, Anjunning mini pill, Kangfuxin, Xuanxia detoxification capsule, Shifusheng capsule and Zhengtongning granule for opiate acute detoxification [7–9]. Clinical trials of six Chinese medicines which include the Taikangning capsule, Jiedukang capsule, Yanshen liquid, Fuyuan granule, Jingan Jiedu pill, Jinjiawang granule and Junfukang capsule, are currently underway and pending approvals by SFDA. Several additional Chinese medicines are also undergoing preclinical trials. Chinese medicines act by targeting multiple processes in the human body. There are several papaveraceae herbs being used in traditional medical practice, such as Rhizoma corydalis (yanhusuo), Flos daturae, Semen hyoscyami, Herba chelidonii, snake venom for pain relief [10], Radix ginseng, Radix astragali, Radix panacis quinqueflii, Radix aconite lateralis praeparata, Radix angelicae sinensis and Cordyceps for healthy Qi reinforcement, Rhizoma pinelliae, Semen ziziphi spinosae, Radix polygalae for sedation and tranquilization, Flos lonicerae japonicae, Herba taraxaci, Gossampinus malabarica (mumian), pumpkin, Radix glycyrrhizae, pine leaves, small flower milkwort herbs with roots (Jinniucao) and Hedyotic diffusa (baihua sheshecao) for body toxin-removing [11–13]. The treatment effects of these herbs show synergy when used in reasonable mixture. A couple of TCMs are taken as examples below to illustrate TCM's effects on drug addiction.

yanhusuo is a herbal analgesic with sedative, hypnotic and antihypertensive properties [14]. The active component, the natural product levotetrahydropalmatine (l-THP), inhibits in a murine model the locomotor hyperactivity induced by oxycodone [15]. In addition, treatment with l-THP can attenuate morphine-induced withdrawal syndromes and conditioned place preference in mice [16]. These promising preclinical findings have triggered clinical investigations of l-THP. In a randomized, double-blind, placebo-controlled clinical trial, l-THP was administered to 120 heroin addicts over a 4-week period [17]. Although l-THP effectively reduced opioid craving, withdrawal syndromes and relapse rates in heroin-dependent patients, several other features of this trial deserve mentioning [17]. First, all study participants had completed a 7-day detoxification period before enrolment in the study; virtually none, therefore, were likely to have been in acute withdrawal. Second, of the 59 participants randomized to the l-THP group, only 44 'survived' 2 of the 4 weeks of treatment whereas 59 of 61 participants who received placebo completed 2 of 4 weeks of treatment. Third, the authors never stated the number of participants in either group who completed the 4-week treatment period. Less than half of the l-THP-treated individuals who remained in the study at 2 weeks remained abstinent at 3 months; any beneficial outcomes, therefore, could have been driven by increased motivation to overcome addiction in a subset of the treatment population as much as from the effect of l-THP.

Ginseng is another botanical commonly used in Chinese traditional therapies [18]. Two major types of ginseng are, Panax ginseng (Asian ginseng) and Panax quinquefolium (American ginseng). The main active natural products in Panax ginseng are called ginsenosides, of which more than 20 have been characterized [19]. Ginsenosides have putative effects on the CNS and cardiovascular system, and could improve metabolism and immune function [20]. Panax ginseng attenuates the physiological effects of drugs of abuse including morphine in pre-clinical studies [21]. A multi-center clinical trial found that Radix ginseng (the root of Panax ginseng) was both safe and effective for the treatment of moderate-to-severe acute heroin withdrawal [22]. In this double-blind study, 212 heroin addicts were randomized to treatment with either the ginseng herbal mixture or lofexidine over a 10-day period. The researchers found that the herbal medicine combination was as useful as lofexidine at relieving the symptoms of opioid withdrawal, with patients reporting only gradual improvement in symptomatology over the researching period. Unfortunately, lofexadine is minimally effective at treating acute opioid withdrawal, and the study design did not incorporate a placebo arm. It is therefore possible that what was observed in the patients was untreated opioid withdrawal, the natural history of which is improvement over time. Significant adverse effects included elevations of liver transaminases, but because Chinese herbal mixtures are manufactured with little regulatory oversight, the specific cause of hepatic injury can not be ascribed [23].

The active chemical in Panax quinquefolium, not found in Panax ginseng, is pseudoginsenoside-F11 (PF11), a saponin [19]. Panax quinquefolium, especially PF11, exerts distinct effects following morphine administration. PF11 attenuates memory impairment in the Morris water maze test, analgesia measured by tail pinch, locomotor sensitization and, at higher doses, the expression of conditioned place preference [24]. Neuro-chemically, PF11 antagonizes opioid receptor signaling and decreases the concentrations of dopamine and its metabolites in the brain of test animals treated with morphine ([24, 25]. While these findings suggest that formulations containing PF11-elaborating herbal products can be applied to the management of opioid withdrawal, the supporting clinical evidence is mixed. For example, WeiniCom is a herbal product mixture that contains Corydalis and Panax quinquefolium; some formulations that are sold online appear also to contain kratom (Mitragynia spp.) [26]. A double-blind, clinical trial compared WeiniCom (also called Xian Xu Qudu Jiaonang) treatment with buprenor- phine in 42 heroin addicts entering treatment [27]. WeiniCom not only relieved opioid cravings more rapidly than buprenorphine, but also treated subjective measures of withdrawal symptoms such as abdominal pain, diarrhoea, rhinorrhoea, myalgias and piloerection. Adverse effects from WeiniCom were not reported. The beneficial effects from WeiniCom could result from kratom because (1) kratom seems to treat opioid withdrawal; (2) some WeiniCom formulations contain kratom; and (3) the exact composition of Chinese herbal remedies is often poorly defined.

13.2.3 A Short Conclusion

Chinese medicine may not be as effective as methadone, but most residual symptoms are tolerable. In some cases tranquilizers are required as a supplement. Therefore, TCM therapy should start in advance or as early as possible in order to control withdrawal symptoms if it is used without other medicines. For drug users with opiate dependence and symptoms (e.g. those with a long history, high dose and long term IV drug users), only Chinese medicine is not enough. It is recommended that low-dose, narcotic detoxification drugs, such as methadone or buprenorphine be co-administered coupled with Chinese medicine. To fully incorporate herbal interventions into existing treatment regimens, Jeanine Ward et al. suggested that several guidelines of evidence should be developed. First, the pharmacology, pharmacokinetics and toxicology of herbal materials in humans should be described in greater detail. For example, the pharmacological basis of kratom-associated seizure activity should be investigated, as well as the clinical features that place individuals at risk for this outcome. In addition, the impact of ibogaine administration on the QT interval and other cardiac effects should be investigated rather than discounted. Second, sufficient rigor should be incorporated into clinical research studies to allow meaningful assessments of outcome. For example, a study design that compares a remedy with an intervention does not allow clinicians to determine if that remedy is more effective than placebo. If herbal substance has no benefit, then its use carries only risk; using proper study design would ensure that the potential for benefit can be incorporated into decisions related to medication selection. Third, clinicians should develop the social and cultural contexts in which herbal remedies have utility. For example, methadone-based opioid detoxification in some cultures has been associated with high rates of relapse and treatment failure. Studies that examine the contexts in which culturally relevant interventions may be incorporated into evidence-based treatment regimens may improve outcomes. Ultimately, herbal therapies for opioid addiction and withdrawal can complement existing treatments, and future studies should explore the relationship between evidence-based pharmacotherapies and traditional remedies [28].

13.3 Alcohol Addiction

The manner in which alcohol addiction is treated depends on several factors including age, sex, personality structure, coexisting somatic illnesses, social situation, level of intoxication, intensity of withdrawal, and coexisting psychiatric illnesses. The phase of acute withdrawal can be treated either in or out patient with medications, such as benzodiazepines and clonidin, which diminish the most prominently experienced symptoms. When necessary, low potency, typical antipsychotic medications can be given in addition. After this acute phase, a period of massive craving sets in.

Over the past decades, the number of drinking problems and alcohol-related illness has increased quickly in China, so did the demand for effective treatments. Herbal medicine and acupuncture have been capturing increasing attention in the clinical practice of alcohol-related problems. Comparing western medicine and Chinese herbal medicine, it is easy to observe that huge differences exist in the understanding of disease, the mechanism of treatment, as well as the definition of cure [29]. According to traditional Chinese medicine, alcohol is recognized not only as a therapeutic substance that can invigorate the blood circulation and reduce petechiae, but also as a potential etiological factor that is "extremely hot," causing "depletion of energy in stomach and spleen" (related disorders such as hepatitis, depression, et al.). Based on this understanding of ethanol, TCM treatment of alcohol addiction generally focuses on "clearing away the heat and restoring balance between Yin and Yang" [30].

What is noteworthy is that most studies on TCM are now carried out under the guidance of western experimental philosophy, of which the rationality is still controversial [29]. In the present study, some researchers intend to introduce the Chinese alcohol culture and the basic TCM philosophy concerning alcohol intake. Meanwhile, from a modern scientific perspective, Qing Liu et al. summarized clinical and primary studies that investigate the efficacy and mechanism of major herbs and acupuncture used in the treatment of alcohol use disorder.

Researchers working on alcoholism in China have concluded that several aspects about alcohol drinking in China are quite different from the western countries [31]. First of all, Chinese cultural norms encourage social drinking and discourage solitary drinking. Researchers believe it may prevent most Chinese population from drinking too much, thus explaining the relatively low alcoholism in China before 1970s (Hao et al. 1999). Second, the Chinese population possesses abundant experience in the production of grain-made alcohol, including distilled liquor, yellow rice wine, etc. Thus, a wide range of alcoholic beverage is consumed in China than in most other countries. Compared with most western countries, the Chinese consume more spirits but less beer and wine [31]. Thirdly, TCM is a crucial element in shaping the understanding toward alcohol drinking in China. Except from the recognition of physical harm caused by alcoholism (such as impairment of fertility, birth defects and liver diseases), TCM also ranks alcohol as an important medicine which improves circulation, treats arthritis, increases blood production, and functions as "a leader of medicines" that "can guide other medicines to the place of disease" [32]. Thus, moderate alcohol is considered to be beneficial in Chinese culture. Because of the particularity of the Chinese drinking culture and its culture impact, TCM is considered to be one indispensable solution for alcoholism problems in China.

13.3.1 The Mechanism of TCM Used in the Treatment of Alcohol Use Disorder

Unlike the modern therapy and western treatment approaches, the theory of traditional Chinese medicine is closely linked to ancient Chinese philosophy, which emphasizes the combination of heaven and humankinds. According to the Yellow Emperor's Internal Canon (Pinyin: Huangdi Neijing), everything in the universe is composed of two opposite energies: Yin (阴) and Yang (阳). Yin means all things that are motionless, cold, downward, inner, inhibitory; while Yang represents active, hot, upward, outside, stimulatory. The accumulation of Yin forms heaven, of Yang forms earth (Huang and Zhu 2007). As for the human body, all structures can be divided into two opposite parts, either Yin or Yang. In a broad sense, the upper and external parts of body belong to Yang while the lower and inner parts belong to Yin. Furthermore, the nature of heart and lung is Yang while that of liver, kidney and spleen is Yin. More specially, there is also Yin and Yang within each organ. For example, heart includes heart Yin and heart Yang [32].

The philosophy of TCM indicates that balance between Yin and Yang is vital to sustain optimal body function, which mainly refers to the ability of self-adjustment. In TCM, diseases are common products of both etiological agents and maladjustments in the body. Different from Western medicine, TCM concerns the body's reaction to etiological agents and tries to settle the internal maladjustment to retain homeostasis. Therefore, treatment via TCM principally relys on the visible signs and symptoms of patients, which have been sorted into various "patterns" [29]. For example, nausea, headache and alternate heat or cold feeling, which are responses of the human body to external etiological stimulants, are diagnosed as a "damp-pattern" [33]. As for alcohol, TCM states that the nature of ethanol is extreme heat with toxicity, and chronic binge drinking can deplete the energy and damage the spleen and stomach, causing a deficiency of energy in the spleen and stomach [34, 35]. People with alcoholism mainly have signs that suggest a damp pattern. Based on the development of disease and the theory of "pattern-differentiated therapeutic strategies," alcohol induced damp pattern could be divided into several sub-patterns, including heat-damp and cold-damp patterns. Li Dongyuan, in his book Treatise on Spleen and Stomach (Pinyin: Pi Wei Lun) summarized that the illness of alcoholism is accumulated in the gallbladder, stomach and spleen. Appropriate treatment should clear away heat and promote dieresis so as to resolve the dampness, thus restore a balance of Yin and Yang [30].

13.3.2 Effective Chinese Herbs to Alcohol Addiction

Kudzu Kudzu originated in China and has been valued for its healing power in traditional herbal medicine for thousands of years (Shen Nung Pen Ts'ao Ching, 2800 B.C.) [31]. The use of Kudzu for alcohol treatment was first documented by

Sun SimiaoV in Formulas of a Thousand Gold Worth (Pinyin: Beiji Qian Jin Yao Fang, 581–682 A.D.), and then by Li Dongyuan (1180–1251) as an antidipsotropic agent, widely known in the famous prescription Ge-hua Jie-cheng Decoction (literally: Decoction with Pueraria lobata Flower for Alcohol Detoxication) [36, 37]. Currently, Kudzu is employed as the monarch drug (Pinyin: Jun-yao) in different prescriptions for alcohol abstinence, such as Jie-jiu Oral Liquid [38]. Employing the principles of modern science, Liu et al. carried out two random clinical trials to compare the anti-dipsotropic effect of Jie-jiu Oral Liquid and diazepam [39, 40]. The study revealed that the overall effective rate was significantly higher in Jie-jiu Oral Liquid group than in the diazepam group. What's more, patients in the diazepam group exhibited various complaints, such as lethargy, hypodynamia and dermatorrhea, which were not apparent in the Kudzu treatment group.

Besides the current clinical evidence of Kudzu in ethanol-dependence treatment, significant progress has also been made in primary studies concerning this medicinal substance. The anti-dipsotropic activity of Kudzu and its extracts have been confirmed in golden hamsters, Wistar rats, Fawn-Hooded (FH) rats and alcohol-preferring (P) rats under various experimental practice, including two-lever choice, two-bottle free choice (ethanol/water), limited access, and ethanol-deprived paradigms [37, 41, 42]. With modern experimental techniques, researchers were able to separate several active components from Kudzu, among which Daidzin is discovered to be especially effective in reducing alcohol intake in rodent models [43]. As for the mechanisms of action, several studies on Kudzu and its extracts have proposed that Daidzin may suppress ethanol intake by inhibiting mitochondrial aldehyde dehydrogenase (ALDH-2), which is also involved in serotonin (5-HT) and dopamine (DA) metabolism [37, 44]. Indeed, based on the structure of Daidzin, a new compound has been produced that can reduce heavy alcohol drinking in preclinical screens [45]. Pediculus melo (Pinyin: gua di) or Pedicelli Melonis Powder (Pinyin: gua di feng) Pediculus melo (also called musk melon base) is the fruit stem of muskmelon, which is grown in most regions of China. According to Pi Wei Lun, the nature of Pediculus melo is bitter, cold, and poisonous. It mainly works in the stomach, inducing vomiting as well as reducing jaundice [46]. In the treatment of alcoholism, Pediculus melo is mostly used to produce an emetic effect, thus forming a conditioned taste aversion toward ethanol. In a cohort study [47], 97.3% patients showed a significant drop in alcohol intake after taking Pediculus melo wine (wine mixed with Pediculus melo).

Pediculus melo It is important to notice that **Pediculus melo** holds several advantages when compared to apomorphine, namely small dosage, long residual actions well as convenient administration (oral). Shang's study evaluated the potency of a Pediculus melo capsules (Guadi Capsule, containing 0.2 g Pediculus melo) in clinical practice [48]. His results confirmed the study of Wang and highlighted the usage of Pediculus melo with fewer side effects than apomorphine. Hypericum perforatum L. (St. John's wort, Pinyin: guanyelianqiao) The plant Hypericum perforatum L. (HPE, St. John's wort) has been introduced as a"heat-clearing and detoxifying" drug in several TCM classic books, including the Folk Medicinal Herbs of Nanjing (Pinyin: Nanjing Minjian Yaocao 1956), Guizhou Civil

Bark Collect (Pinyin: Guizhou Minjian Fangyaoji 1978), Chinese Medicine Record of Sichuan (Pinyin: Sichuan Zhongyaozhi 1979), etc. The extracts of HPE have been successfully used for the treatment of depression in both TCM and western medicine [49]. Though it is not declared in the records of ancient Chinese practice that HPE may be helpful in the treatment of alcoholism, modern medicine has made it quite clear that because of the similarity in the pathogenesis of depression and alcoholism, antidepressant drugs may reduce pathological alcohol intake in people [50–53]. As for HPE, its in uence on voluntary alcohol intake has been studied by many laboratories in alcohol-preferring rats with different strains, such as Fawn-hooded (FH) rats, Marchigian Sardinian alcohol-preferring (msP) rats and high-alcohol drinking (HAD) rats [54–56]. Results from those research studies showed that HPE extract (3-day pre-treatment) could attenuate the intake of alcohol, ranging from a 30% reduction compared to the baseline in FH rats to a 72% reduction in HAD rats [55].

The adoption of HEP in clinical treatment of alcoholism is not novel. Back in 1993, Krylov and Ibatov confirmed that St. John's wort is beneficial for alcoholic patients. As for the mechanism of action, most studies have linked HEP with several neurochemical systems in brain, for example the dopaminergic or glutamatergic systems [57–60]. However, none of the pathways has been generally accepted as the functional mechanism of HEP.

13.3.3 Several Kinds of TCM Prescription

In addition to the above herbs, other herbal medicines such as Flos ddaturae (Pinyin: Yangjinhua) and Ginseng (Pinyin: Renshen) are also extensively used in various folk prescriptions to treat alcohol addiction. Many of them, almost most of them are allowed to make into decoction, which might enable these herbs to be absorbed more easily and effectively. Several kinds of decoction are displayed below:

1. Huanglian Wendan Decoction (Literally "Coptis Decoction for Warming the Gallbladder").

Composition: rhizoma coptidis (coptis, 6 g); rhizoma pinelliae (10 g); pericarpium citrireticulatae (10 g); rhizoma zingiberis recens (10 g); caulis bambusae in taenia (15 g); rhizoma gastrodiae (15 g); poria (15 g); rhizoma acori (10 g); tatarinowii (10 g); bombyx batryticatus (10 g); radix curcumae (6 g); radix et rhizoma glycyrrhizae (6 g). According to Li Dongyuan (1180–1251), excessive drinking of wine induces moist heat in several organs (including the pancreas, stomach, liver, spleen), causing jaundice, excessive phlegm, dementia or tremor [34]. Within the prescription of Huanglian Wendan decoction, coptis is the monarch drug with potency in eliminating dampness and heat. Other herbs are ministerial drugs that can dissipate phlegm or invigorate the spleen [35]. From a modern perspective, one case report study revealed that Huanglian Wendan decoction (500 ml, bid) was effective in treating alcohol dependence patients (overall effective rate: 83.3%) [34].

Furthermore, the potent coptis decoction was also proven to be effective in other aspects, such as sedation, pain control, anti-epilepsy, etc. Laboratory studies revealed that the coptis decoction could alter the action of neurotransmitters and increase excitatory amino acids in the cortex, which may be an aspect of the mechanism of the coptis decoction in the treatment of alcohol dependence [61, 62].

2. Jiejiu Jiedu Decoction (Literally "Herbal Decoction for Alcohol Detoxication").

Composition: rhizoma coptidis; cortex phellodendri chinensis; radix angelicae sinensis; radix aconiti lateralis praeparata; rhizoma cimicifugae; radix bupleuri; radix aucklandiae; rhizoma pinelliae; radix ophiopogonis; fructus schisandrae chinesnsis; radix et rhizome glycyrrhizae. Within the present formula, herbs with mild medicinal effects are employed. Unlike the"aversion therapy" used in most clinical practice for alcohol abuse, this herbal recipe focuses on the recovery of inner balance in patients through removing heat and nourishing the liver and kidney. In a clinical case-controlled study, a clinical report stated that the antidipsotropic action of Jiejiu Jiedu decoction was as good as furazolidone [63].

3. Jiejiu Decoction (Literally "TCM Decoction for Alcohol Dependence").

Composition: radix dichroae (10 g); alumen (10 g).

Components of jiejiu decoction, including radix dichroae and white alumen, are common herbal drugs with irritant effects on stomach that induce re ex vomiting [64]. Unlike the Jiejiu Jiedu decoction mentioned above, the prescription of Jiejiu decoction is based on the conditioned taste aversion reflex. To evaluate this ancient recipe, Zou and her colleagues carried out a case control study comparing the effects of Jiejiu decoction and apomorphine. The results demonstrated that both Jiejiu decoction and apomorphine were able to induce aversion in all treated patients. And the onset time of aversion was not significantly different between two groups [65].

13.3.4 Principle and Types of the Prescription

One of the special things in TCM is the management of drugs. Almost fundamentally different from the accurate and formally handled prescription in western medicine, the preparation of a TCM prescription could be complicated as well as time-consuming. One prescription in TCM may involve scores of drugs, including monarch components, ministerial components, assistant components and guiding components. The four parts were first illustrated in Huangdi Neijing and were described as the constructional bases for TCM prescription. Yet, there are also some famous recipes containing only one drug that does not fit the "Four Elements Principle." For example, Qing Jin Sang, a recipe commonly used for alleviating cough, is consisted of Scutellaria baicalensis Georgi alone. Understandings toward the principle have evolved over centuries of clinical practice. Now, one dominant opinion considers a therapeutic method as the bottom-line principle and the "Four

Elements Principle" as a flexible principle in the formation of a recipe. It states that a recipe should meet the therapeutic purpose and be constructed accordingly.

The operational theory of TCM states that the action of a monarch drug in a prescription could be enhanced by the other drugs through appropriate management. There are quite a lot of different ways to prescribe through TCM, such as grilling, baking, boiling, etc. TCM clinicians usually instruct patients with the preparation of medicine, except nowadays people may prefer processed TCM medicine. Decoction is the earliest and most commonly used dosage form in TCM due to its simple preparation, high absorption and quick onset of effect [66]. In the therapy of alcohol addiction, several prescriptions of decoction derived from ancient wisdom or created by modern Chinese physicians are applied in clinical practice. The following decoctions introduced here are commonly used and widely researched in clinical or laboratory circumstances.

13.3.5 Acupuncture

Acupuncture originated from TCM and is now practiced, altered and studied with modern techniques all over the world. The use of acupuncture in current treatment of addiction initially began in 1973 by Wen and Cheung, who reported an effect of electro-acupuncture in the treatment towards opium addiction. In America, a mixture of acupuncture and counseling or Alcoholics Anonymous is employed to tackle and treat addictive diseases. Considering the operative site, acupuncture could be divided into two types, auricular stimulation/acupuncture and body acupuncture. While considering the pattern of operation, except from the needle insertion introduced in ancient theory, seed pressure (usually using semen vaccariae), magnetic force as well as electrical current are also used in recent clinical procedures.

The efficacy of various forms of acupuncture in the clinical treatment of alcohol diseases has been reported in many studies with inconsistent results. In a clinical trial alcoholics who were given needle acupuncture at the point of Zhubin(KI9) claimed significant decrease in alcohol craving than control groups (treated with Park Sham Device). In the meantime, another study on alcohol relapse prevention compared auricular laser stimulation, needle acupuncture as well as sham laser stimulation. It indicated that needle acupuncture was effective in reducing the withdrawal duration of alcohol abstinence, while no notable effect was found attributable to auricular laser stimulation. In the case of auricular acupuncture, a randomized, single-blinded control study has confirmed the action of ear stimulation in reducing the anxiety level in female alcoholics during the withdrawal period. In contrast, Milton et al. argued that auricular acupuncture (needle insertion or electro-stimulation) holds no effects either on reducing alcohol craving and self-scaled arousal induced by alcohol, or on increasing treatment compliance of alcoholics.

As for the mechanism underlying the efficacy of acupuncture, many believed that acupuncture basically relates to the biochemical balance in the central nervous system and the maintenance or recovery of homeostasis. Previous reports revealed the effects of needle acupuncture at bilateral Shenmen (HT7) on reducing the DA release in nucleus accumben that increased in chronic ethanol addictive rats.

Debates about the clinical practice of TCM have never stopped since the dominant philosophy of modern medicine generally disapproves the fundamental rule of Chinese medicine [29]. All of the above studies from the clinic and laboratory have enhanced our understanding on the treatment of alcoholism from a TCM perspective. The action of herbs or decoctions on restoring homeostasis is exhibited as an anti-dipsotropic effect that is competitive enough to classic western therapy and with favorable acceptability. The relatively advanced researches on acupuncture showed that needle or electro-stimulation at corresponding points for alcoholic treatment can improve alcohol abstinence and cut down relapse for some people. A number of primary studies have also combined the use of herbs or acupuncture with the neurotransmitters system implicated in alcohol addiction.

13.3.6 A Short Conclusion

The focus continues to address life's circumstances, dealing with problems, gaining and keeping employment, and continued repair of the damage alcohol has done. The rule is that more sobriety is better and that time will help the process. Alcoholics may be at risk for relapse whenever there is a triggering event, but with attention and care, this can be minimized. It is important to realize that the disease itself still exists, even with years of sobriety. Alcoholics who relapse do not slowly spiral back into old habits—they fall almost immediately back to previous levels of consumption. This, along with a phenomenon called kindling, makes future treatment and withdrawal much harder.

13.4 Food Addiction

Food addiction is a disease similar to drug or alcohol addiction in which a chemical reaction in the brain is triggered by a certain behavior. With food addiction, the behavior that triggers the reaction can be eating a particular food or a particular amount of food. This addiction manifests itself in the uncontrollable cravings that one has for excessive eating and typically involves eating salty, sugary or carbohydrate rich foods for satisfaction.

The cravings are so strong that the addict cannot control them and in many cases, food addiction will lead to a deteriorated quality of life. Physical, emotional, social and spiritual happiness and well-being are all affects of food addiction. Once an individual who is addicted to food eats and experiences the "high" or pleasurable state that they feel when they are done eating, they will quickly feel the need to eat more or to eat again to feel that feeling. Tolerance can build as an individual eats more and this can lead to a desire to eat even when they are already full. In fact, tolerance can result in an individual's need to consume more and more food with less and less satisfaction from their eating over time. Because of the tolerance that builds, scientists believe that food addiction plays an important role in obesity and in the struggle to lose weight.

13.4.1 *Effects to Addicts*

Left untreated, compulsive overeating can lead to serious medical conditions. For example, compulsive overeating usually results in weight gain and obesity, although it is not the only cause thereof. In addition, compulsive overeating could potentially lead to high cholesterol, diabetes, heart disease, hypertension, sleep apnea, and major depression. Additional long-term side effects of the condition include kidney disease, arthritis, bone deterioration, and stroke. In severe cases, compulsive overeating can result in death. Other negative effects may include the amount of money that is wasted on food and the feelings of low self-esteem that usually accompany bingeing.

Scientists are still working to figure out and fully understand every facet of food addiction but there have been some treatments which have been proven to be effective at helping people to come out on top of their addiction. Many argue that food addiction is actually more complicated than certain types of drug or alcohol addiction simply because people can refrain from using drugs or alcohol but they cannot completely refrain from eating. This means that for those who do suffer from food addiction, there will always be the presence of food in their lives which can cause potential relapse.

Food addiction treatment typically consists of behavioral therapy, nutrition counseling, education and social support. If an addiction to food is primarily the result of an emotional disorder such as anxiety or depression, psychological counseling and medication to treat the mental illness can often reduce the adverse addiction to food.

Nutritional counseling is often effective at helping those who are addicted to food to at least learn about the foods that are better for them so that they can eat healthier. Nutritionists can help those with a food addiction to learn how to cook healthier meals, learn about the foods that they can indulge on and learn about the foods that they can safely eat to make them feel full for longer. Healthy eating habits can become a normal part of everyday life for recovering food addicts with the help of some nutritional counseling, therapy and support.

13.4.2 *Introduction of TCM Therapy to Food Addiction*

The Traditional Chinese Medicine approaches to consider food addiction or obesity is to figure out each patient's individually presenting illness mechanisms resulting in the accumulation of phlegm, dampness and turbidity, which means treating based on the patient's particularly manifesting characteristics. For professional TCM, there is no one-size-fits-all treatment. In clinical practice, the physicians first identify the patient's unique Chinese medical patterns using inquiry and examination and then base the treatment upon that pattern discrimination using the treatment principles as the bridge between these two. Food addicts typically present with other disease diagnoses. Some other conditions commonly associated with obesity are high cholesterol, diabetes, hypertension, osteoarthritis, back or knee pain, depression, and heart disease. Depending on the condition and whether or not it shares one or more disease mechanisms with the food addiction, we may also (and usually) choose to treat these as well. When such comorbidities are taken into consideration, therapy tends to be even more personalized. Food addiction can significantly lead to obesity, heart disease and so forth. Generally, there are several steps about treating these problems with TCM: (1) Differentiate the patterns; (2) State the treatment principles for that/those; (3) Erect a treatment plan using acupuncture and/or Chinese herbal medicine and diet and lifestyle recommendations based upon those treatment principles; (4) Adjust the pattern differentiation and treatment plan according to the patient's change over the course of treatment. We will take an example to demonstrate the main patterns of overweight and obesity recognized to day with professional TCM. For didactic purpose, they are presented as individual, discrete patterns. However, in real life, patients tend to present with multi-pattern combinations. Therefore, the treatment protocols under each pattern are given as examples will typically need to be modified with additions and subtractions [67].

13.4.3 *An Example of the Patterns of TCM*

- **stomach heat and damp blockage (wei re shi bi, 胃热湿闭)**

1. **Signs and symptoms:** Obesity, head distention, dizziness, rapid hungering after meal, heaviness of the limbs which caused them to be indolent, thirst, a predilection for drinking, constipation, a red tongue with slimy, slight yellow fur, and a slippery, slightly rapid pulse.
2. **Note:** In this case, being overweight is due to food addiction or just overeating in turn due to stomach heat. The stomach takes in more food and drink than the spleen can move and transform. The excess then becomes dampness and phlegm.
3. **Treatment principle:** clear heat and eliminate dampness.
4. **Acupuncture-moxibustion:** Nei Ting (St 44); Jie Xi (St 41); Ying Ling Quan (Sp 9); Feng Long (St 40); Zhong Wan (Cv 12); Tian Shu (St 25); Da Chang Shu (Bl 25).
5. **Formula explanation:** Nei Ting and Jie Xi heat from the stomach. Ying Ling Quan seeps dampness. Zhong Wan and Feng Long transform phlegm. Tian Shu and Da Chang Shu free the flow of the stools, clear and precipitate.

6. **Additions and subtractions:** for head distention and dizziness, add Tai Yang (M-HN-9) and Feng Chi (GB 20). For thirst, add Cheng Jiang (CV 24) and Zhao Hai (Ki 6). For constipation, add Zhi Gou (TB 6) and Zhao Hai.
7. **Chinese herbal formula:** Fang Feng Tong Sheng Tang (Ledebourieslla Sagely Free the Flow Decoction).
8. **Ingredients:** Shi Gao (Gypsum Fibrosum),12 g; Hua Shi (Talcum), 12 g; Fang Feng (Radix Ledebouriella), 9 g; Zhi Zi (Fructus Gardeniae), 9 g; Lian Qiao (Fructus Forsythiae), 9 g; Jing Jie Sui (Herbal Schizonepetae), 9 g; Huang Qin (Radix Scutellariae), 9 g; Jie Geng (Radix Platycodi), 9 g; Bai Zhu (Radix Atractylodis Macrocephalae), 9 g; Bai Shao (Radix Alba Paeoniae), 9 g; Dang Gui (Radix Angelicae Sinensis), 9 g; Chuan Xiong (Radix Chuanxiong), 3–6 g; Ma Huang (Herbal Ephedrae), 3–6 g; Bo He (Herbal Menthae Haplocalycis), 3–6 g; Da Huang (Radix Et Rhizoma Rhei), 3–6 g; Mang Xiao (Natri Sulfas), 3–6 g; Gan Cao (Radix Glycyrrhizae), 3–6 g; Sheng Jiang (uncooked Rhizoma Zingiberis), two to three slices.
9. **Formula explanation:** Shi Gao, Hua Shi, Zhi Zi, Lian Qiao, Huang Qin, Jing Jie Sui, Bo He, Da Huang, and Mang Xiao all clear heat. Hua Shi also seeps dampness, while Bo He and Jing Jie Sui resolve the exterior and move and rectify the qi. Ma Huang strongly resolves the exterior. It, along with Jing Jie Sui and Bo He, are windy, acrid medicinals which upbear and out-thrust yang. From a western biomedical point of view, Ma Huang stimulates and increase the basal metabolic rate. Da Huang and Mang Xiao also discharge heat, free the flow of the stools, and relieve constipation. Bai Zhu fortifies the spleen and dries dampness. Jie Geng transforms phlegm and also guides the other medicinals to the upper half of the body. Thus, this formula is targeted to treat central obesity. Bai Shao, Dang Gui, and Chuan Xiong nourish and quicken the blood. By nourishing the blood, the help prevent the attack and draining medicinals from damaging the righteousness. By quickening the blood, they are able to prevent phlegm and dampness from engendering stasis. Sheng Jiang aids Jie Geng in transforming phlegm. It also eliminates dampness and harmonizes the stomach. Along with Gan Cao, it harmonizes and regulates all the other medicinals in the formula. With the exception of Bai Zhu, Dang Gui and Bai Shao, all the ingredients in this formula are draining, and this formula as a whole is strongly attacking and draining. It should only be used in patients with a replete constitution. If there is no constipation, delete Da Huang and Mang Xiao or use only with care.

In the book Chinese Medicine & Healthy Weight Management: An Evidence-based Integrated Approach, Aiyana Juliette also gave the appendix, pointing out main foods classified according to their thermal nature that is important in TCM. Then, Yang supplements help to warm the spleen and kidney and move the qi, blood and body fluids, preventing the negative effect that food addiction contributes to [67]. Yang supplements are showed in following tables, all of which will help our readers to make further understanding on eating problems in the point of TCM.

Basil	Fennel	Raspberry
Beef	Fenugreek seed	Rosemary
Cayenne	Garlic	Sage
Chestnut	Dried ginger	Savoy lettuce
Chive	Jasmine tea	Shrimp & prawn
Cinnamon	Lamb	Star anise
Clove	Lobster	Thyme
Crayfish	Nutmeg	Walnut
Dill seed	Pistachio	

The movement of qi is stimulated by the acrid flavor. Foods which move the qi are called qi-rectifiers in TCM. Some commonly eaten qi-rectifies include:

Basil	Coriander	Peppermint
Caraway	Dill	Radish
Cardamom	Garlic	Spearmint
Carrot	Jasmine tea	Star anise
Cayenne	Marjoram	Tangerine peel
Chive	Mustard leaf	Turmeric
Clove	Orange peel	

Cold conditions are generally improved by warming foods. In chronic cases, warm, sweet and acrid foods are used to warm us steadily. In acute cases, warm or even hot foods are combined with stronger acrid-flavored to drive out the cold. Warming foods include:

Anchovy	Dill	Nutmeg	Sweet potato
Basil	Fennel	Oat	Sweet rice
Bay leaf	Garlic	Onion	Trout
Black pepper	Ginger	Peach	Turnip
Coconut	Kohlrabi	Quinoa	Vinegar
Cayenne	Lamb	Rosemary	Walnut
Cherry	Leek	Scallion	Wine
Chestnut	Mussel	Shrimp	
Chicken	Mustard leaf	Spelt	
Coriander	Mutton	Squash	

In addition, dampness results from the body's failure to move and transform fluids. It is treated in TCM by two main methods, one can use acrid, aromatic, warm foods to dry or transform dampness or bland foods to seep dampness. Effective dampness-eliminating foods include:

Adzuki bean	Garlic	Onion
Alfalfa	Green tea	Parsley
Anchovy	Horseradish	Papaya
Amaranth	Jasmine tea	Pumpkin
Barley	Kidney bean	Radish
Buckwheat	Kohlrabi	Rice bran
Celery	Lemon	Rye
Corn	Mackerel	Scallion
Cranberry	Marjoram	Turnip
Daikon radish	Mushroom	Umeboshi plum
Eel	Mustard leaf	

13.5 A Short Conclusion

As this literature has showed, TCM might be a good alternative solution for several types of addiction, as well as certain complex chronic disease, such as cancer and diabetes, because its effectiveness has gradually gained the support of evidence-based medicine. According to the consensus of experts from both TCM and modern medicine circles, the future of TCM herbs will largely depend on its safety and efficacy. As the multi-component therapeutic strategies and practices are ongoing, mixture and combination could be a feasible direction in facilitating TCM herbs modernization to answer the criticisms of underlined components, uncontrolled quality and undermined toxicity. Furthermore, compared with the complicated and elusory formula, the safety and efficacy of combination would definitely be much easier to answer.

References

1. Lu L, Shepard JD, Scott Hall F, Shaham Y (2003) Effect of environmental stressors on opiate and psycho-stimulant reinforcement, reinstatement and discrimination in rats: a review. Neuro Sci Biobehav Rev 27:457–491
2. Shaham Y, Shalev U, Lu L, De Wit H, Stewart J (2003) The reinstatement model of drug relapse: history, methodology and major findings. Psychopharmacology 168:3–20
3. Wang J, Fang Q, Liu Z, Lu L (2006) Region-specific effects of brain corticotropin-releasing factor receptor type 1 blockade on foot shock-stress-or drug-priming-induced reinstatement of morphine conditioned place preference in rats. Psychopharmacology 185:19–28
4. Guo M, Wu CF, Wang JH, Pei G (2004) Effects of ginsenosides on the actions of morphine. Zhongguo Zhong Yao Za Zhi 29:299–301

5. Xu T, Jin XL, Cao HM (2001) Pharmacological action of tetra-hydropalmatine: recent progress. Chin J Clin Pharm 10:58–616
6. Su JK, Zheng YS (1998) Application of traditional Chinese medicine in detoxification. Chin J Integr Tradit West Med – Angiocardiopathy 14:58–60
7. Li LJ, Xing XF, Shao HX (2003) Preparation of traditional Chinese herbs on addiction detoxification: recent progress. Chin Med 34:20–22
8. Li SC, Li B, Cheng DG, Li F (2005) Studies on treatment of traditional Chinese herbs on opiate addiction. J Beijing Univ Tradit Chin Med 28:84–88
9. Liang Y, Cao HB, Mu JP, Wang JH (2003) Recent progress of traditional Chinese herbs in opiate addiction detoxification on clinical and experimental studies. Chin Mag Drug Abuse Prev Treat 9:40–45
10. Duan JW, Li DH, Li YG, Zhang WJ, Li DY, Duan JY (1999) Traditional Chinese herbs in opiate addiction detoxification: recent progress. Chin J Inform Tradit Chin Med 6:8–12
11. Gao XM, Shong SL, Bai XJ (2001) Traditional Chinese medicine on addiction detoxification. Chin Mag Drug Abuse Prev Treat 7:3–9
12. Tang Y (2004) Traditional Chinese herbs on detoxification. J Pract Tradit Chin Med 20:268–269
13. Zheng YS, Su JK (1999) Current situation, problem and recommendation of traditional Chinese medicine on detoxicity. Chin J Inform Tradit Chin Med 6:19–20
14. Chen M, Jin G (1996) Role of d-tetrahydropamatine in the inhibition of dopamine and serotonin or synaptosomes in the rat brain. Chin J Physiol Sci 12:25–30
15. Liu Y, Liang J, Yan L et al (2005) Effects of l-tetrahydropalmatine on locomotor sensitization to oxycodone in mice. Acta Pharmacol Sin 26:533–538
16. Ge X, Zhang H, Xu Z et al (1999) Experimental study of tetra-hydroprotoberberines inhibiting morphine withdrawalsyndromes. Chin J Drug Depend 8:178–184
17. Yang Z, Shao Y, Li S et al (2008) Medication of l-tetrahydropalmatine significantly ameliorates opiate craving and increases the abstinence rate in heroin users: a pilot study. Acta Pharmacol Sin 29:781–788
18. Miller L (1998) Herbal medicinals. Arch Intern Med 158:2200–2211
19. Gillis C (1997) Panax ginseng pharmacology: a nitric oxide link. Biochem Pharmcol 54:1–8. KTBotanicals.com [online]. Available from URL: www.ktbotanicals.com/xuan-xia-qudu-jia-onang-100-capsules-p-7994.html
20. Attele A, Wu J, Yuan C (1999) Ginseng pharmacology: multiple constituents and multiple actions. Biochem Pharmacol 58:1685–1693
21. Takahashi M, Tokuyama S (1998) Pharmacological and physiological effects of ginseng on actions induced by opioids and psychostimulants. Methods Find Exp Clin Pharmacol 20:77–84
22. Shi J, Xu G, Liu T et al (2008) A comparative clinical study of the effects of the traditional Chinese medicine Jinnui capsules and lofexidine on acute heroin withdrawal symptoms. Am J Drug Alcohol Abuse 34:792–800
23. Harris I (2000) Regulatory and ethical issues with dietary supplements. Pharmacotherapy 20(11):1295–1302
24. Li Z, Xu N, Wu C et al (2001) Pseudoginsenoside-F11 attenuates morphine-induced signalling in Chinese hamster ovary-mucells. Neuroreport 12:1453–1456
25. Seo J, Lee J, Lee W et al (2008) Inhibitory effects of ginseng total saponin on up-regulation of cAMP pathway induced by repeated administration of morphine. Arch Pharm Res 31:167–170
26. KTBotanicals.com [online]. Available from URL: www.ktbotanicals.com/xuan-xia-qudu-jia-onang-100-capsules-p-7994.html
27. Hao W, Zhao M (2000a) A comparative clinical study of the effect of WeiniCom, a Chinese herbal compound, on alleviation of withdrawal symptoms and craving for heroin in detoxification treatment. J Psychoactive Drugs 32:277–284
28. Ward J et al (2011) Herbal medicines for the management of opioid addiction. CNS Drugs 25(12):999–1007
29. Jiang WY (2005) Therapeutic wisdom in traditional Chinese medicine: a perspective from modern science. Trends Pharmacol Sci 26:558–563

30. Zhang L (2001) Electro-acupuncture and auricular seed press in the treatment of alcoholism: an 18 case report. Shanghai J Acu-mox (Shanghai Zhenjiu Zazhi) 20(1):S30
31. Cochrane J, Chen HH, Conigrave K, Hao W (2003) Alcohol use in China. Alcohol Alcohol 38(6):537–542
32. Wang DY (2003) Basic Chinese traditional medicine and acupuncturology. In: Wang DY (ed.) (pp. 12–13). Peking University Medical Press, Beijing
33. Zhang XL (2005) New perspective on the damp pattern. Chin J Inform TCM (Zhongguo Zhongyiyao Xinxi Zazhi) 12(5):90–91
34. Qu JH, Wang ZP (2008) Huanglianwendan plus-minus decoction in the treatment of alcohol dependence: 30 cases report. Shangxi J TCM (Shangxi Zhongyi) 29(1):8–9
35. Wang BX (2004) Pharmacology and clinical implementation of modern Chinese medicine (Xiandai Zhongyao Yaoli Yu Linchuang). In: Wang BX (ed) (pp. 31–39). Tianjin Science & Technology Translation & Publishing Co, Tianjin
36. Wang CH, Yang ZL, Duan SY, Cheng BL, Zhang XL (2005) Clinical investigation on intervened course of acute severe alcoholism with infusion of Jiawei Gehua Jiecheng Tang into stomach. Prac Clin Med 6:30–33
37. Wing MK, Bert LV (1998) Daizin and its antidipsotropic analogs inhibit serotonin and dopamine metabolism in isolated mitochondria. Proc Natl Acad Sci 95:2198–2203
38. Liu CM, Li T (2003) Pharmacological study of Jie-jiu Oral Liquid. J Changchun College TCM (Changchun Zhongyiyao Daxue Xuebao) 19(3):81–82
39. Liu YJ, Du H, Liu M, Liu YZ, Wu H (2002) Tranditional Chinese medicine Jie-jiu oral liquid in the treatment of alcohol dependence: a 36-case report. J Tradit Chin Med (Zhongyi Zazhi) 43(11):839–840
40. Liu YJ, Du H, Lu YQ (2003) Traditional Chinese medicine Jie-jiu oral liquid in the treatment of alcohol dependence: a 146-case report. Chin Gener Pract (Zhongguo Quanke Yixue) 2(8):95–96
41. Lin RC, Guthrie S, Xie CY, Mai K, Lee DY, Lumeng L, Li TK (1996) Iso avonoid compounds extracted from Pueraria lobata suppress alcohol preference in a pharmacogenetic rat model of alcoholism. Alcohol Clin Exp Res 20:659–663. Suppression of alcohol intake after administration of the Chinese herbal medicine, NPI-028, and its derivatives. Alcohol Clin. Exp. Res. 20: 221–227
42. Overstreet DH, Lee YW, Wezvani AH, Pei YH, Criswell HE, Janowsky DS (1996) Suppression of alcohol intake after administration of the Chinese herbal medicine, NPI-028, and its derivatives. Alcohol Clin Exp Res 20:221–227
43. Wing MK, Oscar L, Lisa K, Bert LV (1996) Potentiation of the bioavailability of daidzin by an extract of Radix Puerariae. Proc Natl Acad Sci 93:4284–4288
44. Rezvania AH, Overstreet DH, Marina P, Massi M (2003) Plant derivatives in the treatment of alcohol dependency. Pharmacol Biochem Behav 75:593–606
45. Arolfo MP, Overstreet DH, Yao L, Fan P, Lawrence AJ, Tao G, Keung WM, Vallee BL, Olive MF, Gass JT, Rubin E, Anni H, Hodge CW, Besheer J, Zablocki J, Leung K, Blackburn BK, Lange LG, Diamond I (2009) Suppression of heavy drinking and alcohol seeking by a selective ALDH-2 inhibitor. Alcohol Clin Exp Res 33(11):1935–1944
46. Sun FL (2004) Researches on alcohol abstinence using transitional Chinese medicine. J Chin Clin Med (Zhongguo Linchuang Yixue) 5(9):69–71
47. Dou JJ, Gao HJ, Liu BS (2003) The effect of Guadi wine in the treatment of alcohol dependence. Chin J Med (Zhongguo Yikang) 2(36):48
48. Shang YH, Zhao YH, Gao SH (2005) A clinical control study on the effect of Guadi capsule in the treatment of alcohol abstinence. Chin J Psychiatry (Zhonghua Jingshenke Zazhi) 31(3):S195
49. Linde K, Ramirez G, Mulrow CD, Pauls A, Weidenhammer W, Melchart D (1996) St John's Wort for depressions – an overview and meta-analysis of randomised clinical trials. BMJ 313:253–258

50. Merikangas KR, Mehta RL, Molnar BE, Walters EE, Swensden JD, Augilar-Gaziola S (1998a) Co-morbidity of substance use disorders with mood and anxiety disorders: results of the international consortium in psychiatric epidemiology. Addict Behav 23:893–907
51. Swensden JD, Merikangas KR, Canino GJ, Kesler RC, Rubio-Stipec M, Angst J (1998) The co-morbidity of alcoholism with anxiety and depressive disorders in four geographic communities. Compr Psychiatry 39:176–184
52. Grant B, Harford TC (1995) Comorbidity between DSM-IV alcohol use disorders and major depression: results of a national survey. Drug Alcohol Depend 39:197–206
53. Markou A, Kosten TR, Koob GF (1998) Neurobiological similarities in depression and drug dependence: a self-medication hypothesis. Neuropsychopharmacology 18:135–174
54. Ciccocioppo R, Panocka I, Colombo G, Gessa GL, Massi M (1999) Antidepressant-like effect of alcohol revealed in the forced swimming test in Sardinian alcohol-preferring rats. Psychopharmacology 144:151–157
55. Rezvania AH, Overstreet DH, Yang Y, Clark JE (1999) Attenuation of alcohol intake by the extract of Hypericum perforatum (St John's Wort) in two different strains of alcohol preferring rats. Alcohol Alcohol 34:699–705
56. Viglinskaya IV, Overstreet DH, Koshevkaya OP, Badishtov BA, Kampov-Polevoy A, Seredin SB (1995) To drink or not to drink: tests of anxiety and immobility in alcohol-preferring and alcohol-nonpreferring rat strains. Physiol Behav 57:937–941
57. Feng C, Amir HR, Andrew JL (2003) Autoradiographic quantification of neurochemical markers of serotonin, dopamine and opioid systems in rat brain mesolimbic regions following chronic St. John's wort treatment. Naunyn-Schmiedeberg's Arch Phramacol 367:126–133
58. Muller WE, Rolli M, Schafer C, Hafner U (1997) Effect of Hypericum extract (LI 160) in biochemical models of antidepressant activity. Pharmacopsychiatry 30(Suppl 2):102–107
59. Nathan PJ (2001) Hypericum perforatum (St John's Wort) a nonselective reuptake inhibitor? A review of the recent advances in its pharmacology. J Psychopharmacol 15:47–54
60. Singer A, Wonneman M, Muller WE (1999) Hyperforin, a major antidepressant constituent of St John's wort, inhibits serotonin uptake by elevating free intracellular Na+. J Pharmacol Exp Ther 290:1363–1368
61. He YS, Fu JY (1999) The action of Ningshen Wendan decoction in the central neurotransmitters of mice. Chin J Inf TCM (Zhongguo Zhongyiyao Xinxi Zazhi) 6(4):25–26
62. Wang P, Shi XM (2002) The in uence of Jiawei Wendan prescription on EAAs in hippocampi, Cortex and striate body of the brain of SAM-P/10 Chin. J Hosp Pharma (ZhongguoYiyuan Yaoxue Zazhi) 22(11):645–648
63. Cao GY, Zhang JY, Zhang GQ, Zhou HS, Deng CJ (2007) A control study comparing herbal Jiejiu decotion and furazolidone in the treatment of alcohol dependence. Hebei Med (Hebei Yiyao) 28(11):1196–1197
64. The State Pharmacopoeia Commission of People's Republic of China (2005) Pharmacopoeia of the people's republic of China, vol 1. Chemical Industry Press, Beijing
65. Zou F, Xu P, Diao HW (2008) The therapeutic effect of TCM Jiejiu decoction on the alcohol dependence. Chin J Prac Chin Modern Med (Zhonghua Shiyong Zhongxiyi Zazhi) 21(13):1158–1160
66. Zou L, Guang TJ (2009) Discuss on the elements affecting the therapeutic effects of decoctions in TCM. Hunan J TCM (Hunan Zhongyi Zazhi) 25:100–102
67. Juliette A (2007) Chinese medicine & healthy weight management: an evidence-based integrated approach. Blue Poppy Enterprises, Inc., Boulder

Chapter 14
Nutrition Support Therapy

Li-Jun Xiao and Ran Tao

Abstract In most addictions, serious nutritional deficiencies of major proteins, fats, vitamins and minerals exist which prevent their capability to digest carbohydrates efficiently. This review aims to point out some treatment approaches in nutrition management for alcohol addiction, drug addiction, food addiction, Internet addiction and sex addiction, according to existing literatures.

Keywords Nutrition intervention • Dietary behaviors • Nutritional deficiency • Omega 3 fatty acids • Nutrition education • Metabolic systems • Disorders of eating • Nutrition assessment

14.1 Introduction

Historically, addiction was considered as an illness of poor characteristics and was not systematically addressed by the medical and academic societies until twentieth century. Drug addiction and abuse including opiates, methamphetamine, cannabis, and alcohol have already become a major public health problem [1, 2], so have the other types of addictions we've already mentioned above such as the Internet addiction, mobile phone addiction, food addiction, sexual addiction, etc. development in scientific circle has uncovered that there is a correlation between health and diet for various genders, ages and ecological conditions. The daily nutrients which are necessary for the human body to grow and sustain normal function of life are namely carbohydrates, fats, protein, vitamins, minerals, and water. Studies have reported poor diets with overweight and obesity among people in recovery from substance and non-substance addiction. In most of these addiction, serious nutritional deficiencies of major proteins, fats, vitamins and minerals exist which prevent their capability to digest carbohydrates efficiently. For instance, physical and biochemical changes that occur from drug and alcohol use also cause nutritional deficiencies and imbalance [3]. Therefore, this part aims to point out a couple of treatment

L.-J. Xiao • R. Tao (✉)
Department of Psychological Medicine, PLA Army General Hospital, Beijing 100700, China
e-mail: bjptaoran@126.com

X. Zhang et al. (eds.), *Substance and Non-substance Addiction*, Advances in Experimental Medicine and Biology 1010, DOI 10.1007/978-981-10-5562-1_14

approaches in aspects of food and nutrition management according to existing literatures.

14.2 Substance Addiction(Uncategorized)

14.2.1 Background

Poor diets and high rates of obesity have been reported among people in recovery from substance addiction. Poor diets in this population may be related to a lack of nutrition knowledge and food preparation skills as well as food environments in treatment facilities that do not support healthy eating behaviors. Residential treatment facilities provide a unique environment to promote healthy eating and build food preparation skills that could be transferred to independent living.

14.2.2 The Nutrition Intervention

The RHEALTH intervention was implemented in all six sites. The educational and environmental components included weekly nutrition and food preparation classes for participants and policy changes in the house food environment to increase opportunities for healthy eating. The latter were made in collaboration with staff at each site. Both intervention components focused on increasing fruit and vegetable consumption and reducing the consumption of total energy, total fat, and added sweets. Two-hour weekly classes were conducted by the primary investigator for six consecutive weeks in each site. The classes engaged residents in active learning, practice and sharing of food preparation knowledge, skills and strategies, and tasting easy, inexpensive and healthy dishes. The program, adapted from previous interventions, was guided by a learner-centered approach. Topics included calorie, fat and fiber content of foods, food labels, increasing fruits and vegetables, reducing portion size, fats and sweets including sweetened beverages, and diet-related weight management. Weekly challenges encouraged participants to try and track new dietary behaviors and win supermarket gift cards. The environmental component targeted food policy changes to provide healthy foods at intervention sites through menu development, food procurement, food availability, and access. Each site was asked to: limit fried foods; use more non-meat dishes; purchase lower fat milk; substitute water and 100 % fruit juices for sweetened drinks; provide more fruits and vegetables, include at least one vegetable or fruit salad with low-fat dressings at each dinner; purchase fewer sweet snacks; and provide healthy snacks. Details about the environmental component are available elsewhere.

14.2.3 A Short Conclusion

Results suggest that men in residential treatment facilities may benefit from dietary interventions, but these interventions need to consider addiction and treatment history. Because people in recovery from substance addiction tend to have poor dietary patterns and are at an increased risk for chronic health conditions, additional studies are warranted to address dietary concerns in this population. These results add to the growing evidence that environmental factors impact dietary behaviors which subsequently affect obesity as well as support the importance of skill development in healthy eating demonstrated by previous successful programs with diverse groups. Treatment facilities have tremendous potential for providing healthy food options and skill development to improve resident dietary behaviors [4].

14.3 Alcohol Addiction

14.3.1 Adverse Effects of Alcohol Abuse on Nutrition

NIAAA has noted the complications in absorption, digestion and the effective functioning of nutrients within the body of those abusing alcohol, which may also have lifestyle factors that may lead to poor nutrition. Others have reported that those in recovery from alcohol and other drugs have an increased craving for carbohydrates, which may be a result of a reduction in serotonin [5].

Besides the primary illness and environmental factors, alcohol addiction, similar to another kind of dependency, is being considered as the causative factor of malnutrition problems. Alcohol and drug dependence people due to distorted eating behavior are predisposed not only to eating disorders (anorexia, bulimia) but also to eating abuse (obesity, overweight, habitual eating).

Wilkens Knudsen A and the fellows described nutrient intake, nutritional status and nutrition-related complications in a Danish population of outpatients with alcohol dependency. That was a cross-sectional study with a 6-month follow-up enrolling persons with alcohol dependency (n = 80) admitted to a hospital-based outpatient clinic. Body mass index, the waist-to-hip ratio and handgrip strength (HGS) were measured, a 7-day food diary was collected, and biochemical testing was conducted. Dual-energy X-ray absorptiometry was performed to determine body composition and bone mineral density (BMD). Results of this experiment was profound: patients with alcohol dependency have an altered nutritional status and risk of complications, as evidenced by osteopenia/osteoporosis and reduced muscle strength. Treatment at an outpatient clinic improved the variables related to liver function, but no change was observed in nutritional status over time. These findings suggest that specific screening and targeted treatment regimens for nutritional deficits could be beneficial .

14.3.2 Active Effects of Certain Nutritional Ingredient on Alcohol Abuse

Other researchers are inclined to stress the basic importance of certain nutritional ingredient such as omega 3. Omega 3's have been shown to have an impact on the production of dopamine and serotonin [5, 6], chemicals in the brain that are both affected by depression and substance abuse [7]. Studies by Buydens -Branchey et al. have shown that those with cocaine addiction have higher levels of omega 6's than omega 3's [8]. An increase in the level of omega 3 in substance users has been found to decrease anger, anxiety, and aggression [9, 10]. Buydens-Branchey and colleagues have also found that adding omega 3's into the diet reduced relapse vulnerability for those with substance use problems [11]. The research that has been done to date on the effectiveness of omega 3's in the treatment of substance use disorders is scant; however, as was mentioned previously, positive effects have been noted for many of the co-occurring disorders found with substance use disorders.

14.3.3 A Short Conclusion

Throughout history, a great many people and institutions have tried to help alcoholics and addicts. Currently, there are thousands of different programs in the United States trying to help those people who have a social or personal problem with drugs or alcohol. Though there have been a range of researches in the area of wellness and nutrition related to substance use treatment, little is known about what typical programs may be doing in this area.

The role of nutrition as a risk factor during drug addiction is in provoking disease pathogenesis, for example, alcohol addiction's normally linking to liver injury is well known. Nutritional deficiency may increase the risk of cell damage by augmenting excitotoxicity, reducing energy production, and lowering the antioxidant potential of the cells. One of the risk factors for brain developmental disorders is nutritional deficiency. Role of nutrition in drug addiction is well studied for alcohol addiction but not for other types of substance addiction.

14.4 Drug Addiction

14.4.1 Adverse Effects of Drug Addiction on Nutrition

Although designer drugs, such as amphetamine and its analogs, are popular and considered safe by the addicts, a couple of adverse effects have been associated with their use, such as the serotonin syndrome, hepatotoxicity, neurotoxicity, and psychopathology [12]. Abused drugs have adverse effects on the mind and on the body itself. They weaken the immune system and affect nutrition [13]. The effects on nutrition are related to behavioral changes as well as direct effects on the energy balance-related signaling systems. Alcohol-and drug-dependent subjects have distorted eating behavior that predisposes them to eating disorders (anorexia, bulimia, obesity, habitual eating). Many alcoholics are malnourished since alcohol and its metabolism prevent the body from properly absorbing, digesting, and using those nutrients. These nutritional deficiencies are not as well studied in other types of drug addiction. Drug addicts suffer from calorie and protein malnutrition with over 90 % of them being underweight [14] and 74 % of them showing clinical signs of nutrient deficiency with significantly lower hemoglobin and serum total protein levels [15].

Treatment with antipsychotics has been related to increasing risk of developing diabetes and ketoacidosis [16, 17]. The role of nutrition together with other risk factors such as level of education, poverty, heredity, and environmental factors must be taken into account in assessing whether a drug addict will develop the metabolic syndrome. The strategies available, in particular, the nutritional approaches to protect the drug addicts from the metabolic syndrome and other diseases have to be paid attention to.

14.4.2 Active Effects of Certain Nutritional Ingredient on Drug Addiction

As the principal type of substance addiction, drug addiction includes the dependency of nicotine, cocaine, opiates or designer drugs. Due to their special traits, drug addiction is mainly treated via non-nutritional approached, by substances like methadone, and morphine, or maintenance therapy, detoxification therapy and so forth. Apart from the compounds mentioned above, a number of dietary antioxidants, such as coenzyme Q10, lipoic acid, resveratrol, melatonin, polyphenols (green tea, curcumin), and flavonoids (quercetin, isoflanones, and catechins), as well as other compounds, such as omega 3 fatty acids, may also be useful in drug abuse and in the prevention of the metabolic syndrome; however, further studies are necessary.

There is a delicate balance between the metabolic systems producing energy and the intrinsic cellular protective mechanisms. Drug addiction can be detrimental to that balance. This may explain the higher risk of developing the metabolic syndrome in drug abuse condition [18].

The correction of metabolism as well as the mineral, vitamin, specific metabolic cofactors, and supplemental compounds may reinforce the balance. Nutrition education is an essential component of substance abuse treatment programs and can enhance substance abuse treatment outcome [19].

14.5 Food Addiction (FA)

14.5.1 What Is Food Addiction?

As we have showed, food addiction is a disease similar to drug or alcohol addiction in which a chemical reaction in the brain is triggered by a certain behavior. Eating helps to build strong bones and muscles, replenish vitamins and minerals and food is a vital source to promote life. Unfortunately, for some people, food causes an uncontrollable craving that manifests itself as an addiction and leads to excessive consumption of sugars and other foods which has physical, emotional and social consequences.

Until recently, many did not believe that there was a condition in which people could actually become addicted to food but recent scientific research has confirmed that food addiction is possible and does happen. Experiments in animals and in humans have shown that in some cases, the reward and pleasure centers that are triggered when using certain drugs can also be activated with food. The probability of such addiction is highly likely with foods that are rich in sugar, fat or salt but other foods can also play into an addiction as well.

In some cases, an individual's decision to consume large amounts of food is followed by excessive exercising, vomiting or use of laxatives to eliminate or reduce the number of calories that were eaten. This is known as bulimia. In other cases, excessive eating is followed by instances of limiting food for days or even weeks at a time which is a form of anorexia. Both of these eating disorders are characterized by an addiction too food in some manner and can lead to extreme weight loss.

There is an overwhelming consensus that obesity, diabetes or disorders of eating are normally generated by food addiction in modern society. Therefore, food addiction neuroscience is revolutionizing our understanding of how obesity evolves and sustains itself through time. This may include how weight is either gained or lost in specific eating disorders such as anorexia, bulimia, binge eating, etc. Equally important, food addiction may constitute the missing link that helps to explain the often unpredictable and nonlinear interplay between select macronutrients (e.g., high sugar and high sugar/high fat) and maladaptive psychological adaptations to food.

14.5.2 Three Main Treatments for Weight That Incorporate Food Addiction

In their penetrative study, Richard Shriner and Mark Gold [22] introduced three important models that serve to integrate food addiction with dietary/nutritional science and the study of psychological adjustments that underlie eating disorders. The first model (a Tripartite Model) incorporated metabolic, addictive and behavioral (i.e., relationship) drivers of weight; the second model (a stress/weight matrix called SWEAM, which stands for: Stress, Weight, Eubaric, Allobaric and Matrix) helps to illustrate how food addiction, through the processes of cueing and craving, had a significant impact on obesity and other disorders of eating; the last model was a simple but powerful neurochemical map for weight. It diagramed how key macronutrients create the thermic energy drivers that stimulate gut peptides, neurometabolic transmitters and endocrines which then traveled to the brain to stimulate either net weight gain or loss. Finally, they suggested three main treatments for weight that incorporate food addiction management: (1) Metabolic interventions. The central aim in the dietary or macronutrient intervention of obesity and diabesity is to understand lower weights are often associated with higher fatty acid oxidation (FOX), lower insulin, higher glucagon, increased ketones, etc. Those diets that can instrument some of these strategic nonlinear metabolic changes will most likely lead to more successful nutrition balance. (2) Avoiding addictive foods, valuing fellowship and relapse prevention. Overall, from previous studies [20], foods higher in dopamine and mu-opioid agonism (i.e., ones that stimulate the production of dopamine or opioids), such as high sugar colas, desserts and cafeteria style foods, may constitute the best nutritive candidates for food addiction, which need to be avoided by people who are addicted to them. Besides, the sequence or timing of ingestion may also be of importance, which means that diets should encourage continuous eating patterns of ingestion and not periodic abstinence. And, in order to achieve and maintain food sobriety over time, social networking and fellowship appear to be of tremendous value [21]. (3) Relationship intervention. Patients in their Living with Food program learn how to cook less addictive and less diabesigenic meals via the food kitchen and group discussion formats. As that article states, they learn the value of fellowship support networks and attend aftercare patient run groups; they learn how to more effectively deal with, express, and reintegrate the five Flow Emotions (especially anger, guilt and empathy). This defuses their need to use food as a substitute for not feeling valued by others.

14.5.3 Need to Identify More Specific Nutrition Intervention Strategies

Food addiction shows a pervasive and enduring pattern of both food perception (how we view and feel about food) and food-related behavior (how we go at procuring and ingesting food) which biases our relationship with food in harmful, non-resilient and unhealthy ways [23]. While others focus on the implications and treatments of substance use disorder, eating disorder and food addiction, with the sharp increase in substance use disorders (SUDs), mental health professionals are seeing increased numbers of patients with both SUD and ED. While the interaction between SUDs and EDs is not fully understood. In a recent textbook, Brewerton and Dennis explored links and correlations between EDs, SUDs, and addictions across genetic, neurobiological, and behavioral domains, and advocated for an integrated treatment approach [24]. Though food addiction has been well described in the eating disorder and obesity researches, incorporating the concept of FA into the spectrum of disordered eating has been difficult for eating disorder treatment experts for a couple of reasons. The Disordered Eating Food Addiction Nutrition Guide (DEFANG) was developed by David A. Wiss and Timothy D. Brewerton for clinical practice at treatment facilities for FA, SUDs, EDs, and related disorders, such as post-traumatic stress disorder (PTSD). The aim is to plot patient symptoms onto a diagram (outside of the circle, inside of the square) in order to craft effective, individualized intervention strategies (see Fig. 14.1). They conclude, further research on brain structure and function would help better model the complex interaction between EDs, SUDs, and addictions. Currently, there is no consensus on how to most effectively treat FA; therefore, efforts to identify more specific nutrition intervention strategies are clearly needed. A more recent research relies on the daily

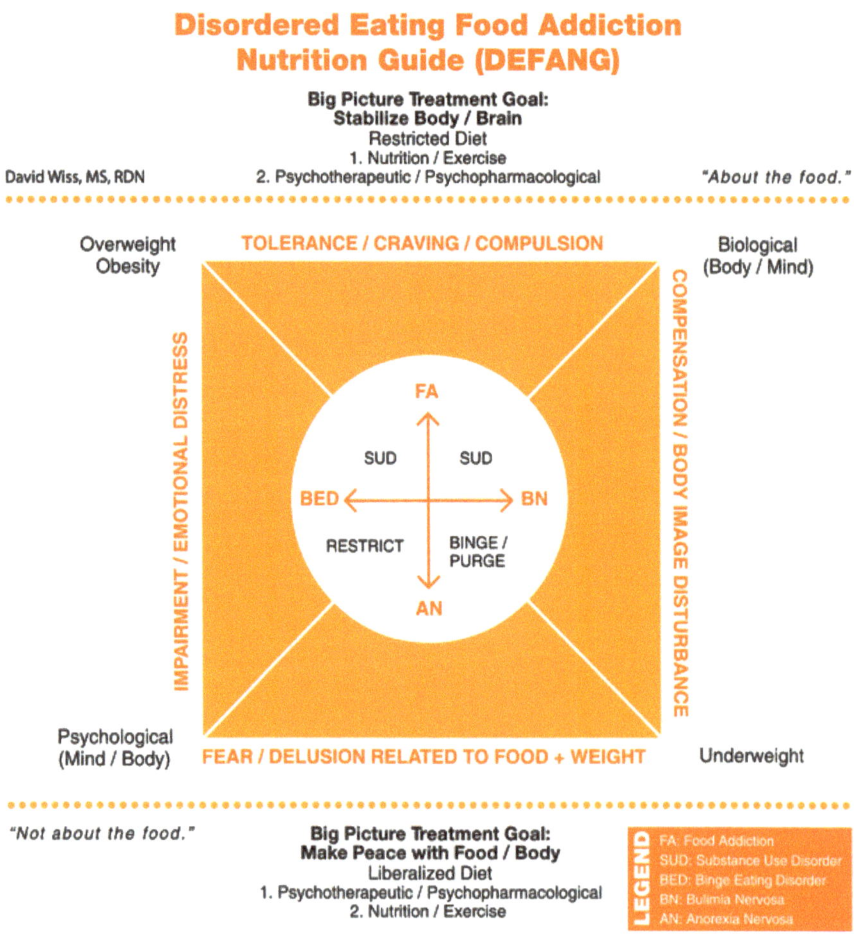

Fig. 14.1 A conceptual framework for individualized nutrition interventions designed to promote sustainable eating disorder recovery

dietary nutrition data and the substance use measures in the 2007–2008 National Health and Nutrition Examination Survey of America, aiming to address the impact of nutrition on alcohol and drug use problems in a nationally representative sample of US adults. The findings generally show that macronutrients increase the odds of substance use and micronutrients decrease the odds of substance use, especially among females. In addition, nutrient imbalance is a particularly strong predictor of substance use for both males and females. Depression partially accounts for the relationship between dietary nutrition consumption and substance use. In conclusion, nutrition represents a promising extension of the biosocial perspective in substance use disorders [25].

14.5.4 A Short Conclusion

Compulsive overeating is treatable with nutritional assistance and medication. Psychotherapy may also be required, but recent research has proven this to be useful only as a complementary resource, with short-term effectiveness in middle to severe cases.

Many eating disorders are thought to be behavioral patterns that stem from emotional struggles; for the individual to develop lasting improvement and a healthy relationship with food, these affective obstacles need to be resolved. Individuals can overcome compulsive overeating through treatment, which should include talk therapy and medical and nutritional counseling. Such counseling has been recently sanctioned by the American Dental Association in their journal article cover-story for the first time in history in 2012: Given "the continued increase in obesity in the United States and the willingness of dentists to assist in prevention and interventional effort, experts in obesity intervention in conjunction with dental educators should develop models of intervention within the scope of dental practice." Moreover, Dental appliances such as conventional jaw wiring and orthodontic wiring for controlling compulsive overeating have been shown to be "efficient ways in terms of weight control" in properly selected obese patients and usually no serious complications could be encountered through the treatment course.

14.6 Internet Addiction (IA)

In general, it is also called compulsive Internet use (CIU), Internet overuse, problematic computer use, or pathological computer use, problematic Internet use, or Internet addiction disorder. Yet in the most recent version of the DSM-5, Internet Gaming Disorder is the latest term to describe this problem.

In contrast to alcohol dependency, which is recognized as an independent clinical picture, Internet addiction seems rather to be a novel syndrome within the context of known mental disorders, even though there are phenomenological parallels to substance-related dependency and to the impulse-control disorders.

14.6.1 Relationship Between Diet and IA

Poor nutrition and physical inactivity were shown to be significantly associated with PIU. Adolescents who spend longer hours online potentially navigate towards unhealthier foods. It is postulated that online gamers drink high-caffeinated energy drinks and eat high-sugar snacks to increase alertness for online gaming. Subsequently, these factors could make online gamers more inclined to sedentary behaviors compared to non-gamers. Moreover, there is an extensive loyalty among gamers, particularly those who displace food, personal hygiene and physical activity, with playing online games. This could pose serious health-risks and may lead to severe psychosomatic symptoms [26].

14.6.2 A Useful Research

Though the nutritional approaches haven't yet been systematically introduced in treatments of Internet addiction, relationship between diet and IA were frequently explored before. Yeonsoo Kim' team performed a study to examine lifestyle patterns and dietary behavior based on the level of Internet addiction of Korean adolescents. Data were collected from 853 Korean junior high school students. The level of Internet addiction was determined based on the Korean Internet addiction self-scale short form for youth, and students were classified as high-risk Internet users, potential-risk Internet users, and no risk Internet users. The associations between the students' levels of Internet addiction and lifestyle patterns and dietary behavior were analyzed using a chi-square test. Irregular bedtimes and the use of alcohol and tobacco were higher in high-risk Internet users than no risks Internet users. Moreover, in high-risk Internet users, irregular dietary behavior due to the loss of appetite, a high frequency of skipping meals, and snacking might cause imbalances in nutritional intake. Diet quality in high-risk Internet users was also worse than in potential-risk Internet users and no risk Internet users. They demonstrated in this study that high-risk Internet users have inappropriate dietary behavior and poor diet quality, which could result in stunted growth and development. Therefore, the results of this study suggest that children should be educated as to what a balanced diet and optimum physical activity routine is to remain healthy. Furthermore, the government should take an active role in designing and evaluating Internet addiction-related health intervention strategies. Given the likely adverse effects of Internet addiction on adolescents' development because of poor dietary behavior, it is critical to raise awareness about Internet addiction. Close attention should be paid to students at risk of Internet addiction, as well as students at low risk to prevent them from becoming addicted to the Internet.

14.7 Sex Addiction

Sexual addiction, also known as sex addiction, is a state characterized by compulsive participation or engagement in sexual activity, particularly sexual intercourse, despite negative consequences. Proponents of a diagnostic model for sexual addiction, as defined here, consider it to be one of several sex-related disorders within an umbrella concept known as hypersexual disorder. In clinical diagnostics, the term sexual dependence may also refer to a conceptual model that is used to assess people who report being unable to control their sexual urges, behaviors, or thoughts. Related models of pathological sexual behavior include hypersexuality, erotomania, nymphomania, satyriasis, Don Juanism (or Don Juanitaism), and paraphilia-related disorders.

14.7.1 Main Therapy and Nutritional Approach

Goodman presented a psychotherapeutic stage model integrating pharmacotherapeutic, behavioral, and psychodynamic approaches, which might be beneficial to treatment of sex addiction. In the first stage (initial behavior modulation), individuals who engage in addictive sexual behavior learn how to modulate their behavior through a combination of inner motivation, psychological support, and affect-regulating medication; the second stage (stabilization of behavior and affect) addresses the question of relapse prevention, with a distinction between high- and low-risk forms of sexual behavior. Patients are taught to engage in "healthier" forms of sexual behavior. However, since sex addiction has been sometimes recognized as a category of non-substance addiction, we consider that integrating approaches involving the above treatments and health plan or nutrition plan are needed, too. Lisa L. Kirkland and the partners provide the hospitalist with an overview of screening, assessment, and development and implementation of a nutrition care plan in the acutely ill hospitalized patient, which can be utilized in the area of sex addiction in some extent. In this process, Nutrition Screening identifies patients with nutritional deficits who may benefit from further detailed nutrition assessment and intervention [27]; Nutrition Assessment should be performed by a dietitian or nutrition consult provider in patients whoare at risk for malnutrition to characterize and determine the cause of nutritional deficits [28]; the formal nutrition assessment of the at-risk patient derives the information needed for the development of a nutrition care plan. This plan guides the provision of nutrition therapy, the intervention, the monitoring protocols, evaluation, and reassessment of nutrition goals or termination of specialized nutrition support [28]; the development stage includes dietary modifications, enteral nutrition, and parenteral nutrition.

14.8 Conclusion

From the point of view of the nutritionist, it is satisfactory to see that adequate nutrition can abolish a dangerous habit, which becomes an addiction and doubtless has an unfavorable influence on the psychological state of the individual and the cultural evolution of a whole population. One might hope that this analysis of the biological background of a great social problem will lead to its abolition by means of better nutrition.

Substance abusers have nutrition deficits before commencing treatment. Improvement in diet and nutrition during treatment can prevent resumption of substance abuse in many patients.

Individualized nutrition counseling within comprehensive nutrition education programs was found to significantly improve the 3-month success rate of substance abuse treatment units. Further, dietary interventions have been shown to reduce the recidivism experienced by multiple driving-under-the-influence subjects.

Nutrition education, particularly with a substance abuse treatment focus provided within a group setting, is associated with positive substance abuse treatment outcomes and should be included as a component of substance abuse treatment. There is opportunity for dietitians to promote and deliver nutrition services, especially nutrition education services to patients enrolled in substance abuse treatment programs. Dietitians should develop viable nutrition education services to offer to residential and outpatient substance abuse treatment programs existing in private, public, and government health care settings.

References

1. Habil MH, Said MA, Rashid RA, Sulaiman AH, Peters H, Zahari MMA et al (2010) Substance abuse violence behavior. Int J Addict Sci:1–5
2. United Nations Office on Drugs and Crime (2010) World drug report
3. Lyle MH (2006) The reclassification of sugar as a drug. Lethbridge Undergrad Res J 1(1):1–5
4. Cowan JA, Devine CM (2013) Diet and body composition outcomes of an environmental and educational intervention among men in treatment for substance addiction[J]. J Nutr Educ Behav 45(2):154–158
5. Wrase J, Reimold M, Puls I, Kienast T, Heinz A (2006) Serotonergic dysfunction: brain imaging and behavioral correlates. Cogn Affect Behav Neurosci 6(1):53–61
6. Heinz A, Mann K, Weinberger DR, Goldman D (2001) Serotonergic dysfunction, negative mood states, and response to alcohol. Alcohol Clin Exp Res 25(4):487–495
7. Volkow N, Wang G, Fowler J, Tomasi D, Baler R (2012) Food and drug reward: overlapping circuits in human obesity and addiction. Curr Top Behav Neurosci 11:1–24

8. Buydens-Branchey L, Branchey M, McMakin DL, Hibbeln JR (2003) Polyunsaturated fatty acid status and aggression in cocaine addicts. Drug Alcohol Depend 71:319–323
9. Buydens-Branchey L, Branchey M (2006) N-3 polyunsaturated fatty acids decrease anxiety feelings in a population of substance abusers. J Clin Psychopharmacol 26(6):661–665
10. Buydens-Branchey L, Branchey M, Hibbeln JR (2008) Associations between increases in plasma n-3 polyunsaturated fatty acids following supplementation and decreases in anger and anxiety in substance abusers. Prog Neuro-Psychopharmacol Biol Psychiatry 32:568–575
11. Buydens-Branchey L, Branchey M, Hibbeln JR (2009) Low plasma levels of docosahexaenoic acid are associated with an increased relapse vulnerability in substance abusers. Am J Addict 18(1):73–80
12. Maurer HH (2004) Chemistry, pharmacology, toxicology, and hepatic metabolism of designer drugs of the amphetamine (ecstasy), piperazine, and pyrrolidinophenone types: a synopsis. Ther Drug Monit 26:127–131
13. Daniel EL (1991) Nutritional implications in recovery from substance abuse. Empl AssistQ 7:1–7
14. Santolaria-Fernandez FJ et al (1995) Nutritional assessment of drug addicts. Drug Alcohol Depend 38:11–18
15. Nazrul Islam SK et al (2001) Serum vitamin E, C and A status of the drug addicts undergoing detoxification: influence of drug habit, sexual practice and lifestyle factors. Eur J Clin Nutr 55:1022–1027
16. Timar O, Sestier F, Levy E (2000) Metabolic syndrome X: a review. Can J Cardiol 16:779–789
17. Kohen D (2004) Diabetes mellitus and schizophrenia: historical perspective. Br J Psych Suppl 47:S64–S66
18. Weitzman M et al (2005) Tobacco smoke exposure is associated with the metabolic syndrome in adolescents. Circulation 112:862–869
19. Grant LP, Haughton B, Sachan DS (2004) Nutrition education is positively associated with substance abuse treatment program outcomes. J Am Diet Assoc 104:604–610
20. Hoebel B, Avena N, Bocarsly M, Rada P (2009) Natural addiction: a behavioral and circuit model based on sugar addiction in rats. J Addict Med 3:33–45
21. Doheny K (2012) Group programs for weight loss may work best. Healthday report
22. Shriner R, Gold M (2014) Food addiction: an evolving nonlinear science. Nutrients 6(11):5370–5391
23. Shriner R (2011) Food as a bariatric drug. Curr Pharm Design 17:1198–1208
24. Brewerton TD, Dennis AB (2014) Eating disorders, addictions, and substance use disorders. Springer, Heidelberg
25. Schroeder RD, Higgins GE (2016) You are what you eat: the impact of nutrition on alcohol and drug use. Subst Use Misuse:1–15
26. Tony D, Vladimir C, Birgitta F et al (2016) Pathological internet UEuropean adolescents. Int J Environ Res Public Health 13(3):1–17
27. American Society for Parenteral and Enteral Nutrition (A.S.P.E.N.), Board of Directors and Clinical Practice Committee (2012) Definition of Terms, Style, and Conventions used in A.S.P.E.N. Board of Directors-Approved Documents
28. Ukleja A, Freeman KL, Gilbert K et al (2010) Standards for nutrition support: adult hospitalized patients. Nutr Clin Pract 25(4):403–414

Chapter 15
Psychotherapy

Ri-Hui He and Ran Tao

Abstract This chapter focuses on psychotherapy of substance and non-substance addiction (see Cognitive Behavioral Therapy in Chap. 16) and introduces the latest advances, mainly in the mindfulness-based relapse prevention, PITDH, and points out that complete elimination of psychological addiction is hopefully to become the target and core of the psychotherapy of addiction disorder. This chapter also introduces methods and progress of various types of substance and non-substance addiction.

Keywords Psychotherapy • Motivational enhancement therapy • Behavioral enhancement therapy • Psychological addiction elimination technology • Mindfulness-based relapse prevention • UCS memory retrieval-extinction paradigm • PITDH

15.1 Introduction on Psychotherapy of Addiction

Generally speaking, addiction therapy is a long process, psychological behavior therapy is an important part of addiction therapy, and its main goal is to improve the understanding of addiction, improve therapy compliance, prevent relapse, rebuild a healthy lifestyle, maintain long-term withdrawal state, or even be completely healed.

The goal of psychological behavior therapy is different based on different therapy and rehabilitation stages of patients. Early treatment is mainly to increase the motivation of treatment, improve self-confidence and self-efficacy of patients; therapeutic rehabilitation at middle and later periods is mainly to correct a variety of psychological and behavioral problems induced by abuse of addictive substances or addiction behavior, help patients learn a variety of psychological skills, improve the

R.-H. He
RiHuiAddiction and Mental Disorders Medical Center, Guangzhou 510000, China

R. Tao (✉)
Department of Psychological Medicine, PLA Army General Hospital,
Beijing 100700, China
e-mail: bjptaoran@126.com

X. Zhang et al. (eds.), *Substance and Non-substance Addiction*, Advances in Experimental Medicine and Biology 1010, DOI 10.1007/978-981-10-5562-1_15

ability to resist addictive substances or behavior, establish a healthy lifestyle, and prevent relapse [1].

The field of addiction medicine has developed a number of treatment methods and strategies on psychological behaviors for drug addicts. Psychological behavior therapy can be divided into motivational intervention, cognitive behavioral therapy, behavior intervention, aversion therapy, addiction eliminating technology, etc., according to different theoretical basis, and individual treatment, group therapy, family therapy, etc., according to the form of psychological behavior treatment. These methods can be used alone or in combination for different treatment forms and treatment sites, and are the basic methods for the treatment of various substance addictions. These methods can be applied to the psychological treatment of non-substance addiction after being properly improved. Therefore, although the psychotherapy mentioned in this chapter is mainly derived from the clinical research of substance addiction, they also provide a good reference for non-substance addiction.

15.2 Common Psychotherapy of Addiction

15.2.1 Motivational Intervention

Most addicts do not have a strong "treatment motive", so it needs special treatment skills for those addicts who lack "treatment motivations". Motivational intervention is just developed based on this characteristic. Motivational intervention adopts a number of interview techniques to help patients recognize their current or potential problems, understand their own ambivalence, strengthen their motivation to change their addictive behavior and to help change their addictive behaviors. Motivational intensive therapy suggests that the intrinsic motivation of addict is the real motivation and a key factor for the change.

Motivation intervention is a psychological treatment technology mainly based on the theory of behavior change stage proposed by American psychologist Di Clemente. According to the theory of change stage, rehabilitation of substance addiction and non-substance addiction is a long-term process that goes through various stages, and can be divided into the following six periods according to the intrinsic motivation of addicts: precontemplation, contemplation, preparation, action, maintenance and relapse. In the recovery process of addicts, therapists can take many strategies to influence patients to change their attitudes, awareness, emotions and behavior, to help them successfully get through the six stages, and ultimately recover.

Motivation intervention is a visitor-centered counseling model, can expose and resolve the ambivalence that occurs while using addictive substance or having addictive behaviors, and ultimately causes the change of ambivalence. This method is particularly effective for visitors who are hesitant or at the thinking stage. Motivation intervention is mainly achieved through the use of motivational interviewing (MI). Motivational conversation is a kind of psychological counseling strategy and technique, which is a way of interpersonal communication with visitors. The therapist firstly needs to establish a trust and cooperative treatment relationship with the visitor. The basic principles of motivational conversation: expressing empathy; presenting a gap; avoiding controversy; resolving resistance; supporting self-confidence and so on.

15.2.2 Cognitive Behavioral Therapy

Cognitive behavioral therapy for addicts is based on the identification and alteration of irrational cognition of patients to reduce or eliminate undesirable emotions or behaviors (such as substance addiction or behavioral addiction); the main purpose of the therapy is to change the cognitive process that causes addicts to cope with undesirable behaviors, to intervene in a series of events that lead to substance addiction and non-substance addiction, to help patients effectively cope with psychological cravings for addictive substances and behaviors, and to develop various skills keeping away from addictive substance and behaviors. The most widely used cognitive behavior therapy is relapse prevention, which is designed to help patients strengthen self-control to avoid relapse of substance addiction. But cognitive behavioral therapy of non-substance addiction is also extremely important to prevent relapse.

According to the theory of social learning, Marlatt et al. proposed a cognitive-behavioral model of relapse in 1985, suggesting that the cognitive and coping patterns of substance addicts in high-risk situations determine the likelihood of relapse.

Relapse prevention is based on the theory of cognitive behavioral therapy in psychology, and the main goal is to change the misconception of relapse by patients, in order to change the behavior of relapse. The patient should learn various skills to deal with high-risk situation under the guidance of consultant by allowing patients to identify their own high-risk situation of relapse, so as to improve their self-efficacy, learn to establish a new life-style replacing substance addiction or addiction behaviors, and ultimately to prevent relapse and maintain a long-term withdrawal. Prevention of relapse is the process allowing patients to learn new cognition and behaviors, can be used in individual or group treatment to emphasize the patient's participation and repeated practice. Prevention of relapse is suitable for patients with strong therapeutic motivation, and needs to be used in combination with other psychological and behavioral intervention methods, such as motivational intervention, so that patients can maintain treatment motivation, and better cooperate with the therapy. Prevention of relapse is a professional job, so consultants need to receive relevant training, and consultants and visitors need to cooperate with each other to develop their therapy goals. If group therapy is used, consultant should participate in group activities as counselors or coordinators, with positive psychological interaction with team members.

Early prevention of relapse can be mainly used to strengthen the treatment motivation with adoption of motivational intervention and establishment of a good relationship with addicts; a variety of skill training are taken at latter stage, and many technologies and strategies are used in the course of treatment, such as identification of adverse cognition, correction of absurd belief, self-supervision, assigned job scoring, self-confidence training, relaxation training and some social problems (such as looking for a job, maintaining work skills, using leisure time and financial skills, etc.). These technologies and strategies are not static, but should be changed according to the actual situation of patients, to take emphasis on repeated practice and practical use.

The process of preventing relapse is to teach patients how to deal with and respond to real or potential relapse of high-risk situations; to help them understand various psychological processes that lead to relapse: in addition to specific behavioral exercises, it also stresses the changes of lifestyle and establishment of social support network. In recent years, relapse has been considered as a normal phenomenon in the process of rehabilitation, a process of patient learning and accumulating experience to get complete recovery. Prevention of relapse training can help patients repeat the behavior correction, and rehabilitation is a spiral progress. There are multiple relapses in the process of rehabilitation, but it ultimately moves toward the goal of complete abandonment of addiction behaviors. Cognitive behavioral therapy will be elaborated in the following chapter.

15.2.3 Behavioral Enhancement Therapy

Behavioral enhancement therapy is based on Skinner's principle of "enhanced contingency", that is, when an act is strengthened at the time of establishment, it will increase the chance of another occurrence in the future. The contingency management (CM) is one of the most commonly used behavioral interventions for the treatment of substance addiction. A large number of studies have confirmed that the contingency management technique can help the substance addicts improve the therapy compliance and increase the therapeutic effect. This method has formed a complete set of theoretical system and operational techniques, and has been widely used in many international institutions of addiction treatment. It needs to be further studied for taking behavior intervention to treat non-substance addiction.

CM aims to treat substance addiction with operational instrumental conditioning and learning theory, that is, take use of the combination of reward (positive reinforcement) and punishment (negative reinforcement) to systematically manage a certain target behavior of substance addict (such as not to use addictive drugs, but take regular treatment and other behaviors for the benefit of long-term rehabilitation) under the instructional treatment environment, make substance addicts maintain long-term target behavior, change the original adverse behavior (such as addiction substance use) and promote long-term rehabilitation of patients.

Behavioral intervention can be carried out with many treatment methods and forms. At present, voucher method and goldfish bowl drawing method are more mature CM methods. A number of studies on behavioral intervention have shown that CM technique has a sustained effect on a variety of substance addictions, such as opioids, cannabis and tobacco.

It should be emphasized that behavioral intervention is a targeted intervention mainly aims at recurrent adverse cognitive, emotion and other undesirable behavior caused by a certain target behavior of patients (more emphasis on drug abuse behavior or addiction behavior). Behavioral intervention needs to be combined with other medications or psycho-behavioral therapy to increase the therapeutic effect.

15.2.4 Mindfulness-Based Relapse Prevention

Mindfulness-based relapse prevention (MBRP) is a new type of psychological intervention technology, initially proposed by Sarah and her colleagues from the psychological research team at Addiction Behavior Research Center, University of Washington in 2010. MBRP is a combination therapy of mindfulness meditation and cognitive behavioral therapy. It is mainly used to prevent relapse in the rehabilitation period of addicts, with the aim to improve self-awareness of triggering factors, inertial model and self-reaction by means of mindfulness exercises, help addicts adopt present experience, and overcome deep-rooted, often disastrous habitual thinking and behavior patterns. MBRP is suitable for those who are hospitalized or have taken outpatient treatment, with motivation of taking maintenance treatment, and also willing to change the way of life to get happiness and rehabilitation.

MBRP is mainly carried out in group form, and includes eight stages, 2 h for each stage, and each stage includes 20–30 min of meditation exercise. The main contents are formal mindfulness meditation exercises (including body scan, meditation, imaginary meditation and tadasana meditation, mindfulness yoga, etc.), informal mindfulness meditation (including mindfulness meditation in daily life: such as breathing techniques, healthy lifestyle and so on) and family exercises (including daily tracking tables, mindfulness meditation exercise, body scan, etc). For family exercises at each stage, it needs to record your daily exercise, any ideas, encountered obstacles and gains of experiences through the daily tracking tables. The first three stages of MBRP focus on identifying automatic response and high-risk situations, practicing to integrate mindfulness meditation into daily-life activities. The next three stages emphasize the acceptance of the current experience and the application of mindfulness meditation exercises to prevent relapse. The last two phases include training the patient's self-care, getting supportive system, and having ability of work-life balance.

Previous studies have shown that MBRP has a significant effect on treating patients' cravings and negative emotions, reducing drug use or addiction behavior, and preventing relapse. Regular mindfulness meditation exercises can not only help patients reduce the "automated" and reactive behaviors, but also help patients be more aware of their own experiences and choices. But it is not easy to have long-term adherence to mindfulness meditation exercises, and difficulties and obstacles often occur. Therefore, it needs to treat them well, and begin to take practice and exercise from here and now [2].

15.2.5 Aversion Therapy

Aversion therapy, also called counter conditioning, is a powerful tool for treating alcohol or other drug addiction. The aim is to reduce or eliminate the "pleasure memory" or desire of the drug, and at the same time form a disgust or avoidance response to the substance. Unlike punishment, the time of saliency for the use of substance is usually delayed, but aversion therapy relies on the immediate association of patient's vision, olfactory sensation, taste and behavior with his unpleasant or disgusting experience. Moreover, negative consequences are imposed on individuals for taking saliency, and on behaviors for aversion therapy, and negative consequences only appear with the use of drug behavior. This has a very important benefit to the patient's self-esteem. As the patient participates in a positive recovery activity, he or she immediately receives positive support to form a new way of behavior and thinking. As long as the patient relies on alcohol or medication, he or she will immediately feel the same discomfort as before. Thus, self-esteem is reconstructed by separating drugs from self. Meanwhile, training and conditional reactions must be repetitive, requiring appropriate testing to produce a sense of disgust, and to maintain and strengthen the sense of disgust, to prevent its attenuation. In clinical practice, several common methods are developed according to the principle of aversion therapy: chemical-induced nausea aversion therapy; induced electricity anorexia therapy; covert sensitization. Clinical practice has proven to be effective in the treatment of heroin addiction, alcohol addiction, nicotine addiction, cocaine addiction, amphetamine addiction, marijuana addiction, etc.

Intensive aversion therapy achieved better results. Four hundred thirty-seven of 600 alcohol, marijuana or cocaine addicts who took chemical and induced electricity aversion therapy were followed up for a year. The results showed 1 year complete withdrawal rate of 29.4 % for alcohol addicts without taking any intensive treatment, 50.5 % after receiving an intensive aversion treatment, 68.5 % after two intensive aversion therapies, and 80 % after more than two intensive aversion therapies.

Induced electricity aversion therapy almost has no unsafe side effect and it is found to be safe for patients with pacemakers and for pregnant women. The patient must have no medical contraindication for chemical aversion treatment, such as esophageal varices, severe coronary artery disease or active gastrointestinal abnormalities.

The rate of treatment in patients who seek aversion therapy in clinical practice is as high as those seeking other mature therapies. Both the US government and private medical institution recognize aversion therapy as an appropriate therapy for addictive disease.

Schick Shadel Hospital has developed new therapies for OxyContin addiction. The therapy makes full use of Naltrexone to counteract the psychoactive effects of oxycodone hydrochloride. First, oxycodone hydrochloride addicts receive detoxification treatment, and then take daily treatment of Naltrexone from the first morning. The treated subjects use oxycodone hydrochloride in the usual manner during vomiting therapy. The therapy has been widely accepted, and more and more patients take the initiative to take it.

15.2.6 Psychological Addiction Elimination Technology

The key anatomical structures of reward/saliency loop are the nucleus accumbens (NAc) and ventral tegmental area of the midbrain. The reward effect focuses on the relevant mechanisms of acute effect of addictive drugs or behaviors on neural networks. The strengthening effects of addictive drugs or behaviors are divided into positive effects (inducing individual euphoria, satisfaction, and thus enhancing drug seeking or addictive behavior) and negative effects (inducing individual and physical discomfort, and which can be avoided with intake of drugs or re-engaging addictive behavior). Reward/saliency loop theory can explain the occurrence and maintenance of obsessive-compulsive drug intake or addiction, but does not adequately explain the mechanism of relapse [13].

Recently, the study of memory/learning conditioning/habits loop better explains the relapse mechanism. The key anatomical structure of this loop is amygdala and hippocampus. Even after long term withdrawal, drug or addiction behavior-related clue stimulation can still recall the strong past experience, leading to relapse, which is the difficulty of addiction treatment. At present, addiction memory consolidation and conditioned reflex are important mechanisms leading to relapse after withdrawal. Addiction memory can persist, each use of addictive drugs or addiction behavior or contact with their related clues to stimulate can complete a re-consolidation process, so that addiction memory continues to strengthen, leading to gradual increase of carving to addictive drugs or addictive behavior, and eventually induce relapse. Intervention on addiction memory re-consolidation process can reduce the desire of addicts to prevent relapse. Neutral stimuli that have not previously induced drug addiction or addictive behavior are associated with intensifying factors (such as drugs), which can induce dopamine release in the striatum by conditioned reflexes, and the addicts will trigger carving to reward as receiving the stimulus again, and then produce a strong desire to seek medicine or addiction behavior impulse.

15.2.6.1 CS Memory Retrieval-Extinction Paradigm and UCS Memory Retrieval-Extinction Paradigm

The team led by Professor Lu Lin, Director of China Institute of Drug Dependence, Peking University published papers on Science and Nature Communications in 2012 and 2015, proposing CS memory retrieval-extinction paradigm and UCS memory retrieval-extinction paradigm. It is found that the memory manipulation paradigm can eliminate the addictive memory of addictive animals and human beings, reducing their psychological desire for addictive drugs and the risk of relapse [4, 5].

The root cause of mental dependence is the pleasant sensation for drug addiction to form a lasting and unusually strong addiction memory. They have found that as general memory, addiction memory is also a dynamic process of change [6], can be saved in the brain at about 6 h after formation under normal circumstances, and form a stable long-term memory. That's why most people get addicted to drugs after their initial exposure.

Addiction memory will persist once formed, which is the difficulty of addiction therapy. Addiction memory is a kind of pathological memory, which repeatedly connects pleasant sensation of drugs with drug-taking environment in the process of abuse. After drug addiction therapy, even if the physical drug addiction seems to have been withdrawn, the addict still has a strong sense of desire for drugs once coming to the previous drug-related environment. When the addict again encounters clues related to addiction memory, the original addiction memory is evoked and becomes unstable. This provides an opportunity for the removal, processing and updating of addiction memory. At this time, it needs to take repeated exposure and addiction-related stimuli to destroy previous addiction memory, and change the patient's awareness of drugs.

Addiction memory often contains an unconditional stimulus and a number of conditional stimuli. For heroin addicts, heroin is unconditional stimuli, while syringes, needles and other drug addiction tools can be seen as conditional stimuli. For heroin addicts, therapists firstly present syringe or drug-taking pictures to the drug addicts, and then repeatedly expose drug taking pictures or tools to addicts after they have a desire for drugs, so as to destroy the addiction memory of drug tools or pictures. The addict will no longer have carvings when he or she is once again exposed to the drug tools or pictures, to get purpose of eliminating psychological addiction, and there is no need to drug combination therapy [7]. This is CS Stimulus retrieval-extinction paradigm proposed by the researchers.

On this basis, team led by Lu Lin has successfully developed a more effective and more widely used "UCS memory retrieval-extinction paradigm". They found in animal experiments that if you firstly give small dose of drugs to addictive animals, and then take intervention, you can erase all addiction emotional memories associated with the addictive substance. Due to the restrictions on relevant provisions, it has not yet been verified in drug addicts, but it has been verified in tobacco addicts.

Tobacco addiction is essentially nicotine addiction. Nicotine can change the brain's neural pathways in plasticity, forming of a strong and lasting nicotine addiction memory, so that nicotine addicts continue to crave smoking. This desire will weaken or even destroy their determination to quit smoking. Therefore, the key to the success of clinic smoking cessation is to eliminate the pathological addiction memory of nicotine addicts and to reduce their cravings.

On March 1st, 2017 Prof. Lu Lin's research group published a paper on JAMA Psychiatry titled Effect of selective inhibition of reactivated nicotine-associated memories with propranolol on nicotine craving [8]. The study successfully transferred animal studies to findings in clinical trials. First, in the nicotine addiction memory model of animals, it was found that after administering low-dose nicotine as an unconditional stimulus to evoke addiction memory, giving Propranolol in enhanced time window can effectively eliminate all the nicotine addiction memories of animals. Further study in the smoking addicts has found that administration of propranolol in the enhanced time window resulted in the elimination of smoking-related memory and reduced the psychological craving induced by various smoking-related hints. The groundbreaking results of Prof. Lu Lin's research group has brought new hope for addicts who find it difficult to quit smoking. Hopefully, the results may tackle the medical and social conundrum that a long-term effective treatment is yet to be found for mental and psychological problems related to pathological memory caused by nicotine addiction etc.

These findings have also been repeated and validated by their international counterparts in other addictive animal models and human beings. It is worth mentioning that a randomized controlled clinical study on smoking cessation has also been published on the same issue of JAMA Psychiatry. It conducted by researchers from the United States using CS retrieval-extinction paradigm proposed by the Prof. Lu Lin's research group. Also on the same issue is a review article titled *Behavioral and Pharmacological Strategies for Weakening Maladaptive Reward Memories: A New Approach to Treating a Core Disease Mechanism in Tobacco Use Disorder* [9], which commented that these studies have changed the long-held viewpoint that pathological addiction memory is hard to eliminate once it is formed and found the treatment to eliminate the kernel pathogenesis of substance addiction and they are a milestone in treating smoking addiction and other substance addictions.

15.2.6.2 Program Implantation Technology Under Deep Hypnosis, PITDH

With many years of clinical practice, the author (Ri-Hui He, similar hereinafter in this chapter) introduced PITDH technology under deep hypnosis for efficient elimination of psychological addiction. In the author's opinion, the mental dependence or psychological addiction of addictive disease is actually a pathological conditioned reflex at the subconscious level, which means pathological addiction memory proposed by Prof. Lu Lin's research group. According to the current research, the author has categorized the operating mechanism into three stages: stimulating arouse, impulse generation and behavioral implementation.

First of all, patients will immediately think of addictive substance or behavior when feeling bored, upset, stressful, and even happy or seeing some conditioned stimulus associated with addiction like injection needle, which is the step of stimulus retrieval. The addicts will immediately feel the positive emotions of excitement, pleasure when thinking of substance abuse or addictive behavior, which is the aforementioned drug or addictive behavior to evoke a strong past experience. Pursuit of pleasure is the instinct of animals, so the patient will have the impulse of substance abuse or engaging in addictive behaviors. This is the step to produce impulse. The longer the addiction time, the stronger the pleasure sensation of addiction is, and the stronger the impulse is.

For example, patients with heroin abuse will feel discomfort, once stop taking heroin, or immediately think of heroin, once feel worry or bored. And they will think of the relaxed, comfortable and pleasant feeling after taking heroin, and then have the desire and impulse of re-abuse of heroin.

It can be seen that stimulus arousal and impulsive production is a series of conditioned reflex that occur at the subconscious level, without the need of rational thinking. Impulse is produced, followed by the possibility of rational thinking, such as taking into account of the source of money, drug sources, drug abuse and other hazards. At this time, impulse and reason will conflict with each other in the patient's brain. If the impulse overcomes the reason, the patient will eventually go to buy heroin and become heroin abuser. This is the third step of "psychological addiction", behavior implementation. If the reason overcomes, the patient will not have abuse until the production of next impulse.

The patient can get extremely strong pleasant sensation caused by addictive substance and behavior, especially in early addiction, and rational thinking is very little. Moreover, once again abuse or engage in addictive behavior, it is easy to produce psychological "abstinence violation effect (AVE)", and it is difficult to have the courage to withdraw. For addictive disorder, the traditional psychological treatment is mainly cognitive behavior therapy, which is carried out at the level of rational thinking, needs to repeatedly detect and reflect, so it is inefficient and unstable. Even the popular MRBP mainly carried out in the state of meditation only takes the heart and body in a mild state of relaxation, and is not ideal to eliminate psychological addiction.

In contrast, deep hypnosis is much more efficient. Hypnosis is to give patients a high degree of concentration, and temporarily put down rational thinking follow the hypnotic language and guide. In applying PITDH technology, the hypnotist begins the "implant procedure" when he is in deep hypnosis, which eliminates the conditioned reflex of the patient to original substance or behavior, and creates a new conditioned reflex. For example, the patient should firstly think of listening to music as feeling upset or boring, have negative emotions such as nausea or fear when thinking of addictive substance or behaviors; the patient turns around and walks away, even sees addictive substances and behavior, and feels happy for his or her restraint.

Studies have shown that the brain's learning and memory efficiency in the state of deep hypnosis is 30–50 times to that at the level of consciousness. In other words, each "program implantation" is equivalent to 30–50 times of rational thinking of patients. After many "implantation", the patient's psychological addiction is getting weaker until the patient has instinctive resistance to addictive substances and behavior., If the treatment process is relatively simple, it takes about 5–10 h; if it is carried out in deep, or at systematic design, it takes about 20 h.

The author takes this method to eliminate psychological addiction of addicts after rapid recovery of body by UROD or of behavior addicts after emotional stability that the patient can quickly recover or even be completely cured. With clinical application of 3 years, more than 100 patients have been treated, and 3 years-conduct rate is of 82 %.

15.2.7 Individual Therapy

Individual therapy takes intervention on the symptoms and related problems of substance addiction and non-substance addiction, and focuses on the current content and structure of rehabilitation program, putting emphasis on the behavioral changes of patient and allowing the patient to learn the skills and tools needed for rehabilitation. Individual therapy or consultation can be used in different theoretical treatments, such as motivational intervention, cognitive therapy, behavioral therapy, psychological therapy and so on.

Individual therapy is generally two times a week, 45 min each time. If it is difficult to do so, it should at least twice a week at the initial stage of therapy. Individual therapy is generally 6 months for a course, a total of 36 times of treatment. Individual therapy is generally 36 weeks for a course, the first 1–12 weeks for the early treatment, two times a week; 13–24 weeks for the active therapy period, once a week; 25–36 weeks for the consolidated therapy period, once a month.

Individual therapy is one of the most basic forms of addiction treatment, and widely used. It has the flexibility to meet the individual needs. Individual therapy can combine with group therapy and family therapy to play a good effect.

15.2.8 Group Therapy

As individual therapy, group therapy is an important psychological behavioral therapy for addicts. Many basic skills of individual counseling or treatment also apply to group therapy, but group therapy has different characteristics and uses a number of different techniques.

Group therapy has the following effects: reduce the common loneliness, boredom and other emotions of the addicts; provide active peer support and maintain the pressure of personal integrity; inspire hope and confidence after observing the rehabilitation of others, and the patient realizes that "if the other can do, the author also can do."; the group can also provide support and encouragement besides of the group; help learn how to deal with substance abuse or addiction and other problems in the group, such as interpersonal relationships, work, family and so on. Group therapy also provides much useful information for new rehabilitation members. Group therapy provides feedback on the values and abilities of the members, and corrects bad behavior and cognition; group therapy provides a family-like environment in which members can learn how to get along with other members; when a team member encounters difficulties, the group can provide incentives, coaching, support and strength; members learn social psychological skills in the collective environment, to replace drug abuse or addiction behavior; have effective control on substance abuse and addictive behaviors through the collective force of the members. In addition, some rules are set by group therapy to cultivate the sense of responsibility and discipline of substance dependents. Group therapy also has effect on related issues of addiction, such as depression, anxiety, loneliness, shame, sick personality.

15.2.9 Family Therapy

Family therapy for addicts is to treat addictive patients in a family environment, taking the whole family as a treatment unit, and treating family-related relationships as the focus of treatment. Therapist helps addicts solve problems by thinking with the whole family member (or some family members) or discussing with other family members. Family therapy is one of the most important therapy mode of drug abuse in Europe and the United States. It is considered that family therapy is one of the most promising therapy modes for adolescent drug abuse and related behaviors.

The development direction of contemporary family therapy is "integrated mode", which is characterized by "multi-system" and "multi-dimension", focusing on the entire living environment of drug abusers, assessing the entire social life network affecting drug abuse, and adopting comprehensive ways to intervene on multiple systems.

With the development and application of evidence-based medicine in recent years, many studies have evaluated the family therapy, and the family therapy based on evidence-based medicine is as follows:

1. Behavioral contract: The goal of therapy is to let all family members understand problems that arise in the family, such as drug addiction or addiction behavior problems of family members are the matter for whole family, and promote patient to get rehabilitation by creating an environment without addictive drug or addictive behaviors and helping family members address emotional problems induced by addictive behavior.
2. Marital behavior therapy: Taking couple as the object of addiction treatment, using social learning theory to handle the problem of addiction and family functions, focusing on the intervention on relevant factors of current drug addiction and behavior addiction, and changing cognitive emotional state causing drug abuse and behavioral addiction.
3. Brief strategy family therapy: mainly for adolescent addicts. Young people lack of personal development skills are likely to have addictive behavior, strict family structure can increase the risk of adolescent addiction, and family and cultural conflicts are risk factors for addiction behavior.
4. Multidimensional family therapy: MDFT is developed by Dr. Liddle in 1985. Treatment objects are young people with drug abuse and other behavioral problems. MDFT mainly intervenes in the four aspects of generation and maintenance of adolescent drug abuse: adolescents; parents; family environment and family relationships; and family external systems related to adolescents and parents, such as schools, judicial system, companions and social support networks.

5. Multi-family therapy community: helping family members change the dysfunctional family rules, roles and alliances to assist the families of addicts to take withdrawal and rehabilitation, consolidate withdrawal state. After reaching the withdrawal, to help family members establish a sense of intimacy, maintain the core functions of healthy family, especially for the addictive family members playing a buffer and balance role.
6. Multi-system therapy: understand and deal with addictive problems from a broader system. The primary goal is to involve family members in the therapy, understand the barriers to family participation and develop strategies to overcome these obstacles, and assess advantages and needs of each system and their relationships, intervene on risk factors and protective factors affecting family in a large range, and determine the role of family members during therapy.
7. Behavioral family therapy: combine individual interventions within the framework of solving family problems to help each family member set up treatment goals.
8. Cognitive behavior family therapy: combine principles and techniques of traditional family therapy with those of behavioral family therapy, and cognitive behavior family therapy takes addictive behavior as a conditional behavioral response.
9. Network therapy: treatment objects not only include core family members, but also other members caring for the patient, such as companions, friends, colleagues, teachers and others supporting network. The network usually includes three to four members. Therapists encourage these subjects to help patients stop using addictive drugs or taking addictive behaviors, support patients to stay in a withdrawal state, and network treatment helps patients make full use of community resources to support their rehabilitation. During the initial phase of therapy and the relapse phase, network members can help cope with the tendency of patients to deny substance abuse or addiction behavior, and they can also provide therapist with a way to communicate with recurrent patients and how to help them. Effective intervention only requires network members to provide advice during treatment. The importance of relationship between the patient and his or her optional network member and his ability to respond to the efforts of member are powerful tools to ensure compliance.
10. Problem-solving therapy: different from the traditional expert guidance, it stresses cooperation between therapist and patient, seeking to solve the problem of treatment to change drug abuse or addictive behavior.

Family therapy begins after the patient's therapy, and involves participation of core family members, spouse of addicts (marriage treatment), fellow siblings, all family members or major social support personnel. The therapists guide them how to face addicts and help them get recovery, including encouraging families to support the practice of addicts, providing family members with attitudes towards drug addiction or addictive behavior, requiring families to urge addicts to participate in treatment or self group therapy, supporting the addicts to adapt to society and work, guiding them how to maintain marital relations and mutual communication, resolve differences, improve interpersonal relationships, contact with addictive behavior, etc. Therefore, it plays a positive role for the addicts and their family rehabilitation [10].

15.2.10 Therapeutic Community

Modern therapeutic community (TC) of addiction is a highly effective therapy that has helped drug addicts and alcohol addicts successfully recover in the past few decades, and the withdrawal rate for many years is higher than 90% in some well-qualified therapeutic communities. It is not clear why this therapy is so effective and how it works when many other addiction therapies are ineffective. Modern TC or "concept TC" of addiction is a combination of self-help and social-support that are suitable for addictive patients and patients with comorbid mental disorders.

There are two kinds of comments on TC: TC is too expensive and demanding. Due to the high cost, only the one who has received several therapies and get ineffective results can accept TC. In addition, the cost of TC is only a small part of the fee for closed treatment, usually about one-third of each household's annual consumption. In the TC project, this cost is easily compensated by a low relapse rate or a recidivism rate [10].

15.2.11 Twelve Steps for Recovery

Self-help or support groups are non-negligible in rehabilitation of addiction. For rehabilitation group, Alcoholics Anonymous (AA) and its basic ides shall be mainly introduced. Of course, other programs also will be involved, such as Al-Anon, Alateen, Narcotics Anonymous (NA), Alcoholics Anonymous (AA), Adult Children of Alcoholics (ACOA) etc. Although support groups and twelve steps are led by companions but not professionals, these programs are still the significant part for rehabilitation of many addicts. Therefore, it is crucial for professionals of addiction medicine to know the basic principle of 12 steps [10].

15.2.11.1 Alcoholics Anonymous

The recovering alcoholics Bill Wilson and his companion Bob Smith created AA after meeting their physician in 1935. The conference of AA always begins with reading AA Core Knowledge, in which many reasons that why AA will be effective are included. The following is the Core Knowledge of AA:

> A.A. is a fellowship of men and women who share their experience, strength and hope with each other so that they may solve their common problem and help others to recover from alcoholism. The only requirement for membership is a desire to stop drinking. There are no

dues or fees for A.A. membership; we are self-supporting through our own contributions. A.A. is not allied with any sect, denomination, politics, organization or institution; does not wish to engage in any controversy, neither endorses nor opposes any causes. Our primary purpose is to stay sober and help other alcoholics to achieve sobriety.

The specific content for twelve steps is as follows:

1. We admitted we were powerless over alcohol – that our lives had become unmanageable.
2. Came to believe that a Power greater than ourselves could restore us to sanity.
3. Made a decision to turn our will and our lives over to the care of God as we understood Him.
4. Made a searching and fearless moral inventory of ourselves.
5. Admitted to God, to ourselves, and to another human being the exact nature of our wrongs.
6. Were entirely ready to have God remove all these defects of character.
7. Humbly asked Him to remove our shortcomings.
8. Made a list of all persons we had harmed, and became willing to make amends to them all.
9. Made direct amends to such people wherever possible, except when to do so would injure them or others.
10. Continued to take personal inventory and when we were wrong promptly admitted it.
11. Sought through prayer and meditation to improve our conscious contact with God as we understand Him, praying only for knowledge of His will for us and the power to carry that out.
12. Having had a spiritual awakening as the result of these Steps, we tried to carry this message to alcoholics, and to practice these principles in all our affairs.

Twelve steps are the theoretic core of AA, which has been introduced in detail in the book *Twelve Steps and Twelve Traditions*, which has been regarded as one of most classic textbooks by members of AA. Twelve steps have been successfully applied to solve other life issues, such as narcotic drugs, cocaine, gambling, sex, shopping and food addiction etc.

Many specific groups have been established with the development of AA. If patients report to the group that he/she is not feeling well in the AA group, he/she may be transferred to specific groups for help, such as groups for female, young person, faggotry and lesbian, African-American or other races/nationalities. International Doctor Group of AA (IDAA), especially, it is a group based on AA but with service objects such as physician, psychologist, dentist, veterinarian, educator and any other persons with doctoral degree.

Help-based relation is the basic concept of AA. The helper who has identical gender with the new members and has stayed in AA at least for 1 year will be the guide for twelve steps and model-like tutor. The help-based relation can last for the whole life. Member who has been totally recovered will be asked to help a new member at a certain time. In a 10 years' follow-up research, after being helper, 91 % previous alcoholics have stable personal integrity.

15.2.11.2 Family Support Group

It is estimated that an alcohol addict has a profound effect on an average four individuals (usually family members). Because alcohol and drug addiction are becoming increasingly complex, a concept emerged that addiction is family disease. Based on such understanding, the 12-step support team for the family members of the addicts was able to develop rapidly. The focus is on helping family members, rather than drugs or alcohol addicts. The earliest family support association for alcohol addicts was initiated by Loisw (wife of Bill w). Wives of alcohol addicts realized that they have been impacted by alcohol addiction and applied twelve steps into their lives. Founded in 1957, Alateen is a part of family groups of family support association for alcohol addicts. Alateen aims at teenagers and follows the steps and traditions of family support association for alcohol addicts. Every Alateen group shall be assigned a member from family support association for alcohol addicts as helper for providing guidance and keeping stability of the group.

15.2.11.3 Role of Physician

In order to help addicts, physicians need to be familiar with rehabilitation support group, including twelve steps. Research of MATCH projects showed that the acceptance of patients for twelve steps can be improved if trained professionals participate in the group with positive and non-forced way. The best way for referral patients taking part in the twelve steps program is to ask the alcoholics to contact with members of AA or other twelve steps groups.

15.2.11.4 Twelve Steps in the Therapy Project

Twelve Steps Facilitation (TSF) in hospital can be regarded as continuous treatment but not independent single intervention. After patients discharged from community therapy, twelve steps are effective adjuvant therapies to other subsequent interventions. Since most inpatients are at the early stage of rehabilitation and the therapy project has short period, most TSFs shall focus on the first three to five steps.

15.2.12 Social Intervention for Addiction Related Problems

Sociological theories of substance abuse assumes that substance abuse is a social phenomenon, mostly caused by or related to cultural, social and economic reasons. From the point of view of socialization theory in sociological theory, substance abusers can be regarded as persons who have socialization process disorder. They are losers with social maladjustment and cannot be restrained and controlled by the society. What they need is to be socialized again (re-socialized) so as to help them receive the influence from society and belief and values from social groups, at the same time, learn life skills, production skills as well as code of conduct, to fit the social environment.

The social intervention process of substance abuse also is the combined and interactive process of personal cognitive, behavioral and environmental factors of substance abuser, which is a comprehensive system project and common responsibilities of whole society, with many aspects involved. Thus, while social intervention does not fall into the category of traditional psycho-behavior therapy, it is mentioned here to emphasize the importance, so that the medical professionals can have a greater awareness of the need of social intervention in addiction rehabilitation [11].

15.3 Common Psychotherapy for Substance Addiction

15.3.1 Individual Psychotherapy for Alcohol Addiction

All above mentioned psychological therapy methods for addition are applicable to treat alcohol addiction, in which aversive therapy has especially notable effect on patients addicted to drunkenness.

The method is described as follows: in general, the patient shall take nothing but cleaning liquid orally 6 h prior to the therapy during the duration of therapy to reduce the craving. The patient shall enter the therapy room after fully understanding therapeutic processes. In the therapy room, there are various alcoholic beverages put on the shelf along the wall and advertisements about wine pasted on the wall. Those are designed to associate most of the visual stimulations of the patient with visual cues of drinking alcohol. Then, the patient shall be given emetine oral liquid and water as well as electrolyte solution as emetic. Before the patient feels nausea, the nurse shall provide the patient with his/her favorite alcoholic beverage in accordance with medical order immediately and let the patient smell the alcoholic beverage and sip it. The nurse shall guide the patient to keep the wine in his mouth and fully enjoy the taste. After that, the patient shall spit the wine into the water tank. The above process allows the patient to fully feel the feeling brought by his/her favorite alcoholic beverage in the aspects of vision, smell and taste before he feels nausea. Immediately following that, the patient shall feel nausea and vomit, which turn "smell, taste and vomit" into "smell, taste and swallow". The swallowed alcoholic beverage turned into emetic within a short period, so that only a little alcohol can be absorbed. Once a course of treatment completes, the patient shall return to the ward and be given a cup of alcoholic beverage mixing with emetine oral liquid and tartar emetic, which will slowly induce nausea for 3 h. Every patient shall take five courses of treatment on average, once every 2 days, 10 days in total [10].

15.3.2 *Individual Psychotherapy for Opioids Substance Addiction*

1. Treatment for opioids receptor stimulant Milestone researches in the field showed that patient with opioids substance addiction will have more remarkable improvement by receiving methadone maintenance treatment and professional psychotherapy at the same time compared to drug consultation. Researches showed that although patients who receive methadone maintenance treatment and do not have severe psychiatric symptoms can be improved whether by professional psychotherapy or drug consultation, for those patients who have severe psychiatric symptoms, the improvement only can be achieved through receiving professional psychotherapy, which transpires that psychotherapy is the best choice when lack of resources.
2. Contingency management Various studies tried to reduce the application of illegal drug of patients who received methadone maintenance treatment with contingency management. In these studies, the patient who has a particular target behavior (such as negative for drug urinalysis, to achieve a specific treatment goal or participating a treatment conference) shall be given a reinforcing stimulus (reward). For example, allowing patients to bring methadone home for medication when reducing drugs as contingent reward, which may be applicable to all patients received methadone maintenance treatment. In the same way, contingency management system can be used to patients of opioids substance addiction who received methadone maintenance treatment when they stick to non-addict lifestyle. A reward (such as cinema ticket or sports goods) can be provided to intensify their substance withdrawal behavior, which has been testified that can effectively reduce the application of illegal opioids and cocaine.

Treatment of opioids antagonist (naltrexone) is superior to methadone maintenance treatment in many aspects, including no drug addiction, no worry that patients use the drug for other purposes and slight adverse reaction. In addition, it can save time for both professionals and patients since it doesn't need go to the clinic every day or almost every day like methadone maintenance treatment does. However, naltrexone hasn't been made full use of while there is a little naltrexone treatment and hasn't been fully used, which is mainly subject to the poor patient compliance, especially for the early stage. Nevertheless, to enhance the obedience and effect of naltrexone treatment, contingency management provide solutions, such as providing vouchers to patients who have good obedience.

3. Aversion therapy: some studies form a special aversion therapy method through displaying aversive stimulus and cognitive image when heroin is not used. They ask patients to describe only when they produce a strong mental image by imagination. In the second stage of the treatment, addicts are asked to imagine appropriate social behaviors, including employment, education or non-drug recreation behavior. At that time, the potential meaning of verbal exposition shall be measured. At the beginning, addicts can imagine positive thinking related to the

application of heroin quickly while feel obviously blocked when imagining beneficial non-drug behaviors. Subjects are arranged in the half-way houses prepared for Heroin abusers for group treatment and relaxation therapy, as well as aversion therapy. Induction stimuli also can be used when addicts imagining about narcotics for aversive stimulus [10].

4. Psychological addiction eliminating technology:

   After patients with opioid addiction have been able to eliminate the addiction of opioid after drug withdrawal and UROD treatment, the PITDH technique eliminates the psychological addiction of opioid addicts. In the meantime, the patients are further implanted with programs. When they think about and even see opioids addictive substances, the treatment will not only make them feel nausea, fearful and dreadful through aversive conditioning but also feel their responsible to have positive behaviors. The technology has been mature. For patient who positively cooperates, it takes about 20 h to totally eliminate the psychological addiction.

15.3.3 Individual Psychotherapy for Cocaine or Methylamphetamine Addiction

1. Contingency management As mentioned above, providing vouchers to patients who are negative in urinalysis to enhance drug withdrawal shows that this method has higher receptivity, obedience and withdrawal rate comparing to standard 12 step consultation process. In fact, the withdrawal shall basically remain when vouchers being replaced by incentive with low value and voucher system have been proved for lasting effect. However, although the program has been proved effective by solid evidence, they are rare for clinical therapy program. Recently, a CM program with low cost has been developed, which may provide solutions for putting these effective methods into clinical program. A research finished recently indicated that: a variable enhancement program, which is to increase the intensified stimulus while decrease the probability of getting stimulus can effectively maintain individual psychotherapy and reduce substance abuse. Since only a proportion of behaviors can obtain intensified stimulus, its cost will be much lower than standard voucher system.
2. Cognitive behavioral therapy (CBT) Another effective behavior therapy for cocaine abuse is CBT. Aiming at helping patients to master the individual coping strategy, CBT uses this as replacement of substance abuse to achieve drug withdrawal. It seems like that CBT is more effective to patients who have severe cocaine addiction or comorbidity. Moreover, effect of CBT shall last longer, even make patients reduce the application of cocaine after the end of the treatment.
3. Manualized disease-model approaches: manualized disease-model approaches have been proved effective by randomized clinical trials, one method of which is

Twelve Step Facilitation (TSF). Contents of TSF, mainly emphasize acceptance and surrender (for example, to help the patient realize that it is possible to get rid of drug addiction with the help of others), are consistent in twelve step therapy of AA though there is no correlation between it and AA or CA. Another important goal of the therapy is to improve the enthusiasm of patients for participating in self-help group in addition to withdrawal from various psychoactive drug substances. Its effect is comparable to that of TSF regarding cocaine use reduction.

4. Aversion therapy: in the research of chemical aversion therapy for cocaine abuse, articaine has been researched and developed as the artificial replacement of cocaine from tetracaine, mannitol and quinine. Patient shall feel nausea when he/she smells the replacement with emetine (ipecamine). A trial has been designed to assess aversion therapy for cocaine abuse of volunteers who participated in the substance abuse treatment programs of Augusta VA medical Center, the results of which show that drug-free percentage of participants who accepted emetic therapy is 57.9 % after therapy for 6 months through follow-up, significantly higher than that of control group at 26.5 %.
5. Psychological addiction eliminating technology: for methylamphetamine addicts, mental intervention is more significant for recovery due to slight reaction of physical withdrawal. The writer implanted psychological addiction eliminating technology to patients of opioids addiction who are recovered through UROD therapy under deep hypnosis program, the psychological addiction of methylamphetamine addicts can be eliminated. In the meantime, the patients have been implanted further program that will not only make them feel nausea, fear and dread through aversive conditioning when think about even see methylamphetamine, but also feel their responsibility to have positive behaviors. The technology has been well developed. For patient positively cooperated, it takes about 15 h to totally eliminate the psychological addiction.

15.3.4 Individual Psychotherapy for Marijuana Addiction

Treatment researches concerning marijuana abuse and addiction are relatively few up to now. Yet no effective pharmacotherapy has been found; only control studies of behavioristics have been conducted, but few in number. A recent study compared the effect of two behavioristic methods, one is twice motivation interviews and another is nine interviews for motivation treatment and coping skills. Results showed that both methods can significantly reduce the application of marijuana through 9 months' follow-up visit, but the latter has better intervention effect than the former and the difference remains among the 9 months' follow-up period. Another research showed that adopting CM and vouchers incentive to coping skill and motive enhancement also can improve the treatment effect of marijuana addicts.

Induction electric aversion therapy has been used in clinical practice for marijuana addicts. The basic principle of it is similar to that of alcohol therapy except using different smoke pipe, drug paraphernalia and visual image. Artificial marijuana, replacement and marijuana fragrance shall be used in the therapy. After being treated for 5 days, it is said that the withdrawal rate for 1 year can reach to 84 % combined with Self-Management Skills Group's therapy three times per week.

15.4 Common Psychotherapy for Psychological Addiction

15.4.1 Psychotherapy for Gambling Addiction

Currently, no randomized controlled trial can support the efficacy of psychodynamic psychotherapy, Gamblers Anonymous and self-withdrawal contract to gambling addiction, yet three conclusions can be drawn. Firstly, gambling addicts can be effectively treated. Especially for cognitive-behavioral therapy, it shows good effects to improve gambling addiction. Secondly, the gambling addiction intervention just starts to receive rigorous empirical test comparing to the treatment for other addictive behaviors. Thirdly, few studies has devoted to the development of simple treatment of gambling addiction. Simple outpatient therapy has been successfully applied to other addictive behaviors, however, it hasn't been given enough attention in the field of gambling addiction.

Effects of psychological treatment methods, such as cognitive therapy, behavioral therapy, cognitive-behavioral therapy, brief intervention and motivational interviewing have been proved by researches. Effect of combined CBT and intervention to improve treatment compliance or CBT and mapping-enhanced treatment is superior to CBT alone. (See "Cognitive Behavioral Treatment" for details).

Comparing to workbook group or waiting list group, group of workbook with motivation intervention has lower gambling probability after 6 months and less financial loss.

Self-treatment and Gamblers Anonymous: for gambling addicts, self-oriented therapy manual or reading therapy are the most effective method with minimum strength. Although these methods may have less effect than cognitive behavior therapy made by therapist, they fill the gap of gradient therapy methods.

According to research findings, marriage situation of gambling addicts usually is terrible, which means the joint participation by husband and wife shall have certain effect.

The author implanted psychological addiction eliminating technology to ten gambling addicts with PITDH technology by three stages, combining with family or marriage treatment, without drug therapy and made follow-up up to now, no one gambles again. The therapy is promise to totally withdraw gambling addiction after being further improved.

15.4.2 Psychotherapy for Internet Addiction

The interventions to internet addiction at home and abroad mainly include psychotherapy and comprehensive intervention, treatment made only by medicine is rare. Psychotherapy mainly consists of cognitive-behavioral therapy, motivational interviewing, Mindfulness Based Cognitive Therapy, group therapy, family therapy, multi-mode psychotherapy and "eight stages and three parts therapy" etc.

The following aspects are involved for research when adopt cognitive-behavioral therapy for the treatment of internet addiction: learning to practice management strategy; identify factors that can trigger internet addiction, such as specific internet using behaviors, emotional state, maladaptive-cognition and life events; learning how to manage emotions and control the impulse to use the Internet, such as muscular relaxation or breathing relaxation training; improving interpersonal communication and interaction techniques; cultivating alternative activities that can meet psychological needs.

Research made by domestic and foreign experts has proved that group counseling is an effective treatment method for internet addiction, which can effectively control internet behavior and help internet addicts get rid of internet addiction. Supportive therapy, cognitive therapy, relaxation training method, behavior training, psychodrama and role play, family psychotherapy, solution focused therapy, communication analysis therapy and Gestalt therapy etc. can be adopted in group counseling, which is an integrated psychological intervention.

The family based intervention is necessary to the treatment of internet addiction, whatever being a factor of nosetiology or therapeutic factor. In the family therapy of youth internet addiction, Young thought it is crucial to pass the method for treatment of internet addiction to the whole family. System standpoint not only plays the theoretical basis of different family therapy school, but the theoretical cornerstone of family based intervention to internet addiction. Education and training provided to parents of internet addicts by above mentioned research should include the following: strengthen the awareness of internet addiction symptoms, learn to identify children's emotional state, make effective communication among family members, learn solutions to problems and methods to control emotions and behaviors as well as understand children's psychological growth process.

Multi-mode psychotherapy involves various psychotherapy and different treatment forms mentioned above, including group therapy, individual psychotherapy and family therapy as well as school based intervention. In effect, any single therapy method shall not be recommended for the psychogeny therapy, but should combine with various psychotherapy methods, not only for internet addiction. Apparently, different methods can be complementary when one method is not enough. Therefore, in every therapeutic interview, different treatment methods are usually intersected and overlapped.

Based on neuro-psychological mechanism, diagnosis and treatment practice of internet addiction, the group led by Professor Tao Ran gradually researched and formulated an "eight stages and three parts therapy" for psychological intervention of internet addiction.

The general goal for "eight stages and three parts therapy" is to stimulate the motivation of internet addicts to seek for help by integrating various psychological treatment technologies and to objectively describe their physical and mental symptom, explain reasons, predict and control symptoms, thus to achieve the goal that make individual take control of internet use and improve personality.

"Eight stages and three parts therapy" actively integrates time series with space structure of treatment and perform in practice in a dynamic and open way. Wherein, the eight stages include: stimulate motivation based on withdrawal stress; objective description of symptoms to explore and research; find attribution and evaluate and identify; make inquiries about life meaning and make plans; awaken latent power and let what past stay past; adjust thinking model to reconstruct cognition; strengthen positive habits and control addiction; maintain psychological circle and grow together; three parts refer to the relative proportion and organic bond of individual, family and group therapy during treatment process [12, 13].

15.4.3 Psychotherapy for Sex Addiction and Sexual Preference Addiction

Traditional psychodynamics and psychoanalytic therapy have been proved invalid to sexual preference addiction through several decades' verification [14].

One third psychotherapy used for treatment of sexual preference addiction can prevent relapse, which is to help the patient maintain the improvement made in behaviors. Physician shall train the patient to identify the reinforcement process of abnormal sexual behavior period and let them practice early intervention strategies to cope with the cycle progress of abnormal sexual behavior, to reduce the risk of abnormal sexual behavior.

The most common method is cognitive-behavioral therapy, which treats the addicts by teaching them some skills to gradually withdraw and prevent relapse. Twelve-step therapy is conducive to the participation of addicts and has no environmental limitation. It also can be used for self-help group treatment, including anonymous sexual addict group and anonymous sex addict group etc.

Zhong Youbin from China created a cognitive comprehend therapy for sexual preference addiction. Unlike psychoanalysis attaching great importance on memory, finding crucial reason in childhood or specific event of early anxiety, the therapist shall discuss and analyze with patients the absurdity of symptoms. The treatment needs heuristic conversation and repeat discussion with patients and make them get to know and understand their childish pathological emotions and behaviors. The root cause of illness is from childhood, but currently, we should survey and value those emotions and behaviors again with views and attitudes of adults. Most patients will suddenly be enlightened and symptoms shall be relieved and disappear. Before the therapy, patients know nothing about the childish symptoms, but they can see the ignorant emotions, ideas and behaviors through the therapy, which can be a comprehension of psychoanalytic type. Cognitive comprehend therapy can make patients comprehend and be cured only within a short time by a few times' interviews, which greatly shorten the therapy time [15]. The clinical experience of the writer manifests that the therapy has obvious effect to moderate sexual preference addicts while poor to severe patients.

The writer implanted psychological addiction eliminating technology to 12 sexual preference addicts (6 fetish addiction, 4 foot fetishism and 2 pedophilia) with PITDH technology, combining with cognitive comprehend therapy, without psychiatric drug therapy for 15–20 h and made follow-up up for 2 years to now. Only one fetish addict has fetish behavior occasionally, others never showed sexual preference.

15.4.4 Psychotherapy for Food Addiction

Psychotherapy provides basis for the treatment of food addiction, which shall include psychological education, nutrition management, cognitive behavioral strategies and motivation enhancement, to comprehensively solve the problems brought by food addiction to patients. Solid research evidence has proved the effect of cognitive behavior therapy to bulimia nervosa, at the same time, a few research evidence also shows the effect of interpersonal therapy to bulimia nervosa. Moreover, cognitive behavior therapy and interpersonal therapy have been proved to be effective to the treatment of f bulimia. Recently, researches show that dialectical behavior therapy can be used to treat bulimia. In addition, preliminary evidences indicates that the participation of family members could be beneficial to teenager's BN.

The writer implanted psychological addiction eliminating technology to two patients with bulimia nervosa and 1 patient with bulimia with PITDH technology. It has been found that the intrinsic excitatory reflexes of patients on food can be quickly eliminated, so does the overeating behavior. Notably, when implanted new conditioned reflex, the patient is calm rather than feeling disgusting when he/she sees food. Bulimia has been cured only through three times' psychological intervention and bulimia nervosa for eight to ten times' psychological intervention. The writer made follow-up for 1 year, no relapse appears. Certainly, large-scale clinical practice is needed to confirm the effect of psychological addiction eliminating technology implanted with PITDH technology to food addiction.

15.4.4.1 Cognitive-Behavioral Therapy for Bulimia Nervosa

Results of a large number of randomized controlled clinical studies consistently show that CBT has the same effect with controlled treatment or has an advantage over controlled treatment. One significant feature for CBT when applied for treatment of BN is fast-changing behaviors of patients. (See "Cognitive Behavioral Treatment" for details).

15.4.4.2 Interpersonal Psychotherapy for Bulimia Nervosa

Interpersonal psychotherapy (IPT) refers to a psychological treatment method with short course of treatment and time limit, which improved by Fairburn (1997) for the treatment of BN. The course of treatment for IPT generally is 3–5 months and 12–20 times by three stages like CBT.

There are some IPT based application been developed. Two important researches for BN outpatient psychotherapy make comparison of IPT and CBT, the result of which shows that IPT group has significant improvement at the end of the therapy while the remission rate, improvement effect of bulimia and withdrawal behaviors is inferior to CBT group. Difference of the curative effect has been after 1 year's follow-up, the main reason of which is the condition of patients in IPT group continues to improve while that of CBT group stops. Since there is no significant difference between IPT and CBT in follow-up studies, studies made by Fairburn et al. found that the effect of IPT is significantly better than behavioral therapy group, thus IPT has been widely believed effective to BN at present [16].

15.4.4.3 Outpatient Psychotherapy for Binge Eating Disorder (BED)

Preliminary research evidence indicates that dialectical behavior therapy (DBT) is another effective method to BED while weight management behavior may be applicable to those overweight BED patients.

As a comprehensive therapy plan, DBT is the representative of the third generation of behavior therapy based on cognitive and behavioral principle and supplemented by mindfulness strategy created by Zen Buddhism. DBT clearly defines a framework that can availability change therapeutic goals adapt to delinquent conducts and find a balance by identifying patients' views and treatment focuses that hard to accept changes.

Telch et al. improved and verified group DBT therapy designed for BED patients. That therapy can reduce overeating behaviors with effect and change the maladaptive attitude to feeding, body type and weight. The effect of DBT still needs follow-up control study in a long term while preliminary research results indicate it's effective to BED. Although these preliminary research results need to be testified by studies on large sample, the research results show that psychological intervention methods including the third generation of psychotherapy have a wide application prospect in the therapy of eating disorder.

References

1. Hao W (2016) Addiction medicine: theory and practice. People's Medical Publishing House, Beijing
2. Earleywine M, Maisto SA, Connors GJ et al (2012) Psychological treatment of addiction disorder. Kewa Zhang, Yan Bao, Peilian Chi translation. China Light Industry Press, Beijing
3. Chen L, Wang X (2015) Research progress of deep brain stimulation for the treatment of drug addiction. Chin J Contemp Neurol Neurosurg 15(10)
4. Luo Y-X, Xue Y-X, Liu J-F et al (2015) A novel UCS memory retrieval-extinction procedure to inhibit relapse to drug seeking. Nat Commun 14(6):7675
5. Xue YX, Luo YX, Lu L et al (2012) A memory retrieval-extinction procedure to prevent drug craving and relapse. Science 336:241–245
6. Lee JL, Di Ciano P, Thomas KL, Everitt BJ (2005) Disrupting reconsolidation of drug memories reduces cocaine-seeking behavior. Neuron 47:795–801
7. Jing J (2016) Is it possible for the brain to be wiped out? China Science and technology network. 2016-6-24 http://www.stdaily.com
8. Xue YX, Deng JH, Chen YY et al (2017) Effect of selective inhibition of reactivated nicotine-associated memories with propranolol on nicotine craving. JAMA Psychiat 74(3):224–232
9. Kamboj SK, Das RK (2017) Behavioral and pharmacological strategies for weakening maladaptive reward memories: a new approach to treating a core disease mechanism in tobacco use disorder. JAMA Psychiat 74(3):209–211
10. Christopher A. Cavacuiti (2004) Essentials of addiction medicine. Wei Hao, Tieqiao Liu, the main translation. People's Medical Publishing House, Beijing
11. Barlow DH (2004) Clinical handbook of psychological disorders. China Light Industry Press, Beijing
12. Tao R, Ying L (2007) Stress. Intervention addiction (physical and mental five. Comprehensive treatment) and intervention research on internet addiction, vol 7. Shanghai people's Publishing House, Beijing, pp 153–163
13. Tao R (2007) Analysis and intervention of internet addiction. Shanghai people's Publishing House, Shanghai
14. Shi J (2002) Addiction medicine. Science Press, Beijing
15. Zhong Y (1999) Cognitive insight therapy. Guizhou Education Press, Guiyang. (in Chinese)
16. Hales RE (2010) Textbook of psychiatry. Mingyuan Zhang translation. People's Medical Publishing House, Beijing

Chapter 16
Cognitive-Behavioral Therapy

Hong An, Ri-Hui He, Yun-Rong Zheng, and Ran Tao

Abstract Cognitive-behavioral therapy (CBT) is the main method of psychotherapy generally accepted in the field of substance addiction and non-substance addiction. This chapter mainly introduces the methods and technology of cognitive-behavior therapy of substance addiction, especially in order to prevent relapse. In the cognitive-behavior treatment of non-substance addiction, this chapter mainly introduces gambling addiction and food addiction.

Keywords Cognitive-behavioral therapy • Relapse prevention

16.1 Introduction

Cognitive-behavioral therapy (CBT) is based on cognitive theory and behavioral theory, which reflects the work of its pioneers, namely Ellis and Bandura. The therapy emphasizes the importance of thought and feeling, including how individuals feel and explain life events, which is a decisive factor in action. CBT also attempts to help patients realize maladjustment, teach them how to heed, seize, monitor, interrupt the "cognitive-affective-behavioral chains" and eventually adapt to it, in order to achieve the highest goal [1].

CBT is a structured, short-term, well-targeted, psychological treatment that focuses on the current problems of substance addiction, and helps addicts to identify, evade and respond to the factors inducing substance addiction, in order to maintain integrity,

The original version of this chapter was revised: The order of the authors are corrected. The erratum to this chapter is available at https://doi.org/10.1007/978-981-10-5562-1_18

H. An • Y.-R. Zheng
English Department, Tianjin University of Technology and Education, Tianjin 300222, China

R.-H. He
RiHuiAddiction and Mental Disorders Medical Center, Guangzhou 510000, China

R. Tao (✉)
Department of Psychological Medicine, PLA Army General Hospital,
Beijing 100700, China
e-mail: bjptaoran@126.com

X. Zhang et al. (eds.), *Substance and Non-substance Addiction*, Advances in Experimental Medicine and Biology 1010, DOI 10.1007/978-981-10-5562-1_16

prevent relapse. The efficacy has been verified. The actual curative effects of CBT in the substance addiction has also been valued and adopted in psychotherapy of non-substance addiction. However, frankly speaking, CBT has no confirmed efficacy in treating psychological addiction, which accounts for the development of other psychological addiction elimination technology, such as MBRP and PITDH etc.

CBT of substance addiction combines the behavioral theory (classical conditioned reflex and operant conditioned reflex), social learning theory (the decisive effect of observational learning, the influence of role models, and cognitive anticipation on behavior) and foundations of cognitive theory (thinking, Cognitive schemas, beliefs, values, attitudes and attribution).

According to behavior theory, substance addiction is an acquired behavior pattern through learning and reinforcement. On the other hand, learning and changing behavioral reinforcement can modify addictive behavior. Addictive substance produces a strong physiological effect in the human body, that is, positive effects, and reduce anxiety, relaxation as a negative reinforcement effect, which leads to the continuous development of addictive behavior. Studies have shown that addicts are prone to the urge and desire for alcohol or addictive substance addiction. Treatment measures include identifying inducing factor, managing impulses and cravings. The treatment techniques based on behavioral theory have cue exposure treatment, contingency management (CM) and Community Reinforcement Approach (CRA).

According to cognitive theory, cognition, emotion, behavioral interact with each other. Cognitive factors function as mediation in emotional and behavioral responses. Some scholars have found that some distorted cognition such as catastrophic, rigid, over-generalized and irrational cognitions are related to substance abuse disorder. The treatment is mainly aimed at the specific problems of the current life of the addicts, who are encouraged to reflect on what they do, formulate plans and change the maladjusted cognition, emotions and behaviors [2].

According to social learning theory, substance addiction is influenced by multiple factors of observational learning, imitating role models (such as parents, siblings and companions), social reinforcement, anticipation effect etc. Individuals acquire addictive behaviors by observation and imitation. Substance addiction behavior as a negative coping style may result from the lack of appropriate coping skills of the addicts and the low self-efficacy. Relapse prevention mode is the treatment for low self-efficacy and negative coping style.

16.2 Content and Operational Points of Substance Addiction CBT

Substance addiction CBT includes three core elements: functional analysis, coping skills training and prevention of relapse patterns [3].

Functional analysis is the use of questionnaires, interviews and role-playing methods to identify the factors of substance addiction, including factors of cognition, emotional and physiological status, society, environment or situation.

Functional analysis also includes the assessment of addicts in coping ability and self-efficacy. A comprehensive assessment of the influencing factors of substance addiction can help develop individualized intervention programs and treatment plans.

Coping skills training is the core of substance addiction CBT. It includes training for substance addiction, such as: dealing with craving, rejection of drugs, and also general coping skills training, such as: communication skills, emotion management. It has skills training for addicts themselves (such as identifying negative automatic thoughts, regaining the fun of life, solving problems, developing contingency plans, etc.) as well as interpersonal skills training (such as developing intimate relationships and social support networks, engaging with important others, general social skills training, etc.). It is necessary and effective to choose individualized training, which is one of the advantages of CBT.

Prevention of relapse patterns include functional analysis, identification of high-risk situations and coping skills training. Bandura argues that the most significant and lasting way to improve self-efficacy is skill training. The therapist should help the addicts perceive negative attribution patterns and corresponding mood changes, gradually modify distorted cognition and negative attribution, and reduce feelings of helplessness and loss of control. Relapse prevention also includes preparing for the possibility of relapse and the developing a coping plans, so as to reduce potential harm and prevent development into comprehensive relapse. Developing a balanced, healthy lifestyle is equally important for relapse prevention.

CBT is typical highly structured, organized in a goal-oriented way, with relatively short treatment time. Every session has a clear goal, and the discussion is also closely related to the topic of drug abuse. The key to progress of reaching the goal rests with the close monitoring and positive attitude on the part of the experts.

Operation points of substance addiction CBT are as follows. The obligations of the therapist, the expectations and responsibilities of the addicts and the time and number of sessions must be clarified. Those who are absent or late are required to notify by telephone in advance. During the treatment, the patient is required to provide a urine test and keep the conduct. Typically, the treatment mainly comes in two forms: individual (60 min each time) and group (90 min each time) treatment for about 12–16 times. If short-term treatment can not achieve a stable effect, it is necessary to extend the course of treatment. Treatment can be performed under different settings, such as outpatient, inpatient, follow-up care, etc. Since functional analysis, skills training and relapse prevention are more effective when repeatedly practiced in daily life and assessed, they are conducted usually in the clinic.

Operation points of substance addiction CBT are as follows. The obligations of the therapist, the expectations and responsibilities of the addicts and the time and number of sessions must be clarified. Those who are absent or late are required to notify by telephone in advance. During the treatment, the patient is required to provide a urine test and keep the conduct. Typically, the treatment mainly comes in two forms: individual (60 min each time) and group (90 min each time) treatment for about 12–16 times. If short-term treatment can not achieve a stable effect, it is necessary to extend the course of treatment. Treatment can be performed under different settings, such as outpatient treatment, inpatient treatment, follow-up care, etc. Since functional analysis, skills training and relapse prevention are more effective when repeatedly practiced in daily life and assessed, they are conducted usually in the form of outpatient treatment.

The most widely used CBT for substance prevention is relapse prevention, developed by Marlatt et al. using CBT technology. The main goal is to change the misconception of relapse by patients, in order to change the behavior of relapse. The patient should learn various skills to deal with high-risk situation under the guidance of consultant by allowing patients to identify their own high-risk situation of relapse, so as to improve their self-efficacy, learn to establish a new life-style replacing substance addiction or addiction behaviors, and ultimately to prevent relapse and maintain a long-term withdrawal. Prevention of relapse is the process allowing patients to learn new cognition and behaviors, can be used in individual or group treatment to emphasize the patient's participation and repeated practice. Prevention of relapse is suitable for patients with strong therapeutic motivation, and needs to be used in combination with other psychological and behavioral intervention methods, such as motivational intervention, so that patients can maintain treatment motivation, and better cooperate with the therapy. Prevention of relapse is a professional job, so consultants need to receive relevant training, and consultants and visitors need to cooperate with each other to develop their therapy goals. If group therapy is used, consultant should participate in group activities as counselors or coordinators, with positive psychological interaction with team members [1].

Early prevention of relapse can be mainly used to strengthen the treatment motivation with adoption of motivational intervention and establishment of a good relationship with addicts; a variety of skill training are taken at later stage, and many technologies and strategies are used in the course of treatment, such as identification of adverse cognition, correction of absurd belief, self-supervision, assigned job scoring, self-confidence training, relaxation training and some social problems (such as looking for a job, maintaining work skills, using leisure time and financial skills, etc.). These technologies and strategies are not static, but should be changed according to the actual situation of patients, to take emphasis on repeated practice and practical use.

The process of preventing relapse is to teach patients how to deal with and respond to real or potential relapse of high-risk situations; to help them understand various psychological processes that lead to relapse: in addition to specific behavioral exercises, it also stresses the changes of lifestyle and establishment of social support network. In recent years, relapse has been considered as a normal phenomenon in the process of rehabilitation, a process of patient learning and accumulating experience to get complete recovery. Prevention of relapse training can help patients repeat the behavior correction, and rehabilitation is a spiral progress. There are multiple relapses in the process of rehabilitation, but it ultimately moves toward the goal of complete abandonment of addiction behaviors.

The principle and basic techniques of CBT will be introduced as follows using Relapse Prevention Therapy (RPT) as an example.

Compared with other therapies, RPT has a more rigorous structure and mode. More teaching and training methods are used. The therapist plays a more active role of the mentor. Individual or collective therapy can be adopted to prevent relapse.

16.2.1 Structure and Mode of Therapy

The course of PRT is generally 3–6 months, conducted once a week, 1 h each time. There are many tasks for each treatment, such as reviewing the last skill exercise, briefly discussing the problems encountered since the last treatment, skill training, feedback on skills training, skill training during the therapy, putting forward a plan for next week. Each session is generally divided into three stages, 20 min for each stage, which is what is called the 20/20/20 rule.

1. At the beginning stage:

   In the first 20 min, the therapist mainly tries to collect information about the patient's past status, general functional level, drug use, craving status, skills training experience. He also assesses the current status of the patient, urine test results, solve the patient's current problems and discuss homework etc. At this stage, although it is the therapist that guides the patient and observes his response, but the patient often does more talking.

2. The intermediate stage: the therapist spends the second 20 min in introducing and discussing a certain skill, including introducing the topic, explaining the relationship between the content of the topic and the current situation of patient, and making sure that the patient gets the message by observing his response. At this stage therapist does more talking, but what he introduces must be combined with the specific circumstances of the patient, and the therapist should provide some examples to make himself well understood.
3. The final stage: in last 20 min of the therapy, the patient plays the leading role by agreeing to a specific skill proposed by the therapist to work on next week, making a detailed plan, and understanding the high risk of relapse next week that he might take next week and ways to handle such situations, which further includes arranging skills to practice next week, analyzing the high-risk situations that might occur next week, and making next week's specific plans and so on.

16.2.2 The Main Content of Relapse Prevention

The process of RPT is mainly to make the substance addicts learn to identify the high-risk situations that leads to their relapse and to change the misconceptions that trigger relapse, learn the effective way to deal with high-risk situations, improve self-efficacy, prevent relapse and recover from the addiction. The main contents are as follows.

1. Establish a good relationship between therapists and patients and enhance the latter's motivation for therapy. In the initial stage of therapy, the main purpose for therapists is to establish a good relationship with substance addicts, adopt motivational enhancement interview technique to enhance the patient's motivation for treatment and adhere to their commitment. It is done to reduce the patient's resistance and conflict towards changing their behavior. Therapists should present potential problems to the patient in simple language and give them enough information on the hazards and consequences. Through helping substance addicts analyze their decision and understanding the long-term and short-term consequence of addictive substance abuse, patients can think independently and make decisions to receive therapy.
2. Identify and monitor high-risk situations. Identify their own high-risk situations based on High Risk Situation (HRS) List developed by professionals and rate the risks according to it. Conduct self-monitoring every day to identify potential dangers (such as bad moods, friend's risky invitation, etc.). According to the recording or video in the high-risk situations, patients describe their own cognitive and behavioral response to assess how much confidence they have to resist temptation (marking of self-efficacy), and assess their coping skills in high-risk situations.
3. Deal with high-risk situations. For a variety of specific high-risk situations, therapists should help patients shape appropriate coping behavior with cognitive and behavioral methods,. Common methods are as follows. Control stimuli and reduce exposure to high-risk situations, such as discarding drugs and tools carried on the patients. Avoid situations related to bad behaviors like the place where the addicts abuse certain substance. Practice how to act decisively and refuse temptation of friends through role-playing. Stop relapse fantasy; teach patients to identify relapse fantasy accompanied by cravings; say "stop!" out loud or silently to oneself to interrupt the fantasy. Take along "craving tips" to build self-control when feeling a craving.
4. Respond to cravings. By discussing cravings with substance addicts, counselors instruct patients to learn how to deal with psychological cravings, such as helping substance addicts to understand and use the principle of reflex so as to reduce the patient's craving for drugs. Repeated combining conditional and unconditioned stimulations can produce conditioned response. On the contrary, the absence of it can gradually reduce conditioned response until it finally disappears. Trainers try to help patients understand and recognize conditional cravings in order to identify the conditioned situation of their cravings, avoid exposure to these situations, effectively cope with these cravings and reduce conditional cravings.
5. Cognitive restructuring (CR) changes irrational cognition. Irrational cognition on the part of the addicts leads to substance abuse and relapse. In the rehabilitation process, one time or occasional drug use is common, for which convalescent

patients should have correct understanding. Consultants need to explore whether patients have wrong way of attribution for occasional drug use in order to avoid abstinence violation effect (AVE) which leads to a relapse. In the therapy, therapists usually deal with attribution and emotional response after one time drug use using cognitive restructuring technology. They reshape internal, stable, general attributional styles into external, temporary, special ones. For instance, occasional drug use is similar to the mistakes made in the learning process. Patients can restructure occasional drug use as a one-time mistake rather than a complete failure so that they still have a second chance to study in the correct way. Patients are encouraged to consider occasional drug use as mistakes and distinguish the experiences of mistakes from failure. Cognitive restructuring includes attributing occasional drug use to external, special, controllable factors. (Controllable factors should be specified. Active avoidance can be adopted to control them). Occasional drug use can be finally avoided and followed by abstinence. The only goal of relapse prevention is to prevent the next occasional drug use or relapse.

6. Learning a variety of psychological skills. Substance addicts are more likely to have a relapse and unable to cope with pressure, manage and adjust their mood effectively because of the absence some psychological skills. In addition, substance addicts also lack the skills to solve problems and interpersonal relationships, which are detrimental to their rehabilitation and adopting a new lifestyle. Consultants can guide the substance addicts to learn these psychological skills and methods, including problem-solving skills, emotional adjustment techniques, stress coping strategies, interpersonal and communication skills in order to reduce relapse and promote rehabilitation. Technology, coping strategies, interpersonal and communication skills to reduce relapse and promote rehabilitation.
7. Improve self-efficacy. In the process of therapy, it is necessary to reinforce the positive changes of substance addicts, affirm their efforts and achievements, enhance their confidence and self-efficacy, so that they will take positive actions to maintain the state of withdrawal.
8. Build a social support system. Many patients insist that they are fully able to control their own recovery process, which is actually wrong. Patients must work hard to learn to establish an external social support system, which is an important part of the behavior management program. The support system is comprised of a group of people who have close relationships with patients, including their parents, spouses, friends, colleagues and doctors. They have to learn to support, remind and politely confront the patients and work together to help them keep the conduct. It is necessary to make the members of the support group know in detail the specific high-risk situations, and understand the existence of a sequence from occasional drug use to relapse, and that the support system is seen as the first defense line of intervention.
9. Develop an alternative, healthy lifestyles in place of addicted behaviors. A healthy lifestyle is very important for patients to maintain long-term conduct. In the therapy, patients are encouraged to participate in some alternative activities (such as meditation, relaxation or running) which is helpful to improve the old

way of life. If these alternative activities become "what they really want", they will become a healthy "addictive" behavior. A healthy "addiction" must meet the following five requirements. It should be able to operate independently; it should can be easy to operate; it should have short-term and long-term benefits for individuals; people can participate in stably and have a sense of progress after a period of time; and the operator will have a sense of accomplishment.

In summary, the process of relapse prevention is to teach patients how to deal with and respond to real or potential high risk of relapse. It helps patients understand various psychological processes that lead to relapse. In addition to specific behavioral exercises, it also emphasizes the establishment of social support networks. In recent years, relapse is seen more a normal phenomenon in the process of rehabilitation, in which patient learn and gain experience until they fully recover. Relapse prevention training can help patients repeat behavior modification. Rehabilitation is a spiral process, in which patients may have multiple relapses, but ultimately they are making progress toward the goal in abstaining addiction behavior [2].

In CBT of behavioral addictions, Ladouceur and his colleague developed cognitive-behavioral therapy, which possibly has most extensive application for gambling addiction. There are mainly five parts in their therapy: provide training to gambling addicts in aspect of games of chance; correct their cognitive bias; provide skill training for problem solving; provide social skills training and prevent relapse. The treatment may need about 17 times, 1 h each time. Gambling addicts who received therapy have obviously improvement than those who didn't. The therapy scheme of Ladouceur for individual is available to both young and adult gambling addicts.

Although some antidepressants are proved effective to BN, the control study result of CBT supports the effect of CBT. Reliable results have been obtained, compared to other psychotherapies (including supportive psychotherapy, behavioral therapy, interpersonal psychotherapy, support-expressive psychotherapy) to prove that CBT is the most effective method to treat BN. Currently, Oxford CBT mode established by Christopher Fairburn is the most reasonable cognitive-behavioral therapy for bulimia nervosa with most wide application [4].

Two summaries regarding CBT treatment control study to bulimia nervosa draw a conclusion that CBT is applicable to most patients. CBT shows better effect comparing to antidepressant drug alone, self-surveillance and supportive psychotherapy as well as behavioral therapies excluding cognitive therapy. Result of 1-year follow-up of CBT treatment indicates that it can remain effect better than antidepressant drug therapy. A combination of antidepressant drug therapy and CBT is more effective than drug therapy alone or CBT. CBT seems having better effect than antidepressant drug therapy in a long term [4].

References

1. Gabbard GO (2010) Gabbard's treatments of psychiatric disorders. Zhao Jingping, the main translator. People's Health Publishing House, Beijing
2. Hao W (2016) Addiction medicine: theory and practice. People's Medical Publishing House, Beijing
3. Shujun P, Min Z, Jiang D et al (2011) CBT for substance addiction. J Psychiatry 24(1):69–72
4. Barlow DH (2004) Clinical handbook of psychological disorders. China Light Industry Press, Beijing

Part V
Summary and Prospect

Chapter 17
Summary and Prospect

Zhengde Wei, Xueli Chen, and Xiaochu Zhang

Abstract In this chapter, the main content is to summarize the similarities and differences between substance and non-substance addictions in several aspects, involving definition, mechanisms, diagnosis and treatment. We try to display the complete picture of addictions in a brief but comprehensive way. Mechanism includes molecule and neural circuit, genetics, neuroimaging and cognitive psychology; diagnosis includes diagnostic criterion, diagnostic scales, biochemical diagnosis and new diagnostic techniques; treatment includes drug therapy, physical therapy, traditional Chinese medical therapy, nutrition support therapy, psychotherapy and cognitive-behavioral therapy. This chapter also covers some prospect which will induct future studies on addiction. We aim at providing the researchers and graduate students with better understanding of substance and non-substance addictions.

Keywords Substance addiction • Non-substance addiction • Comparison

Addiction is a global problem which causes millions of lives yearly and leads to great damage. There are mainly two types of addition, substance addiction (e.g., nicotine, alcohol, cannabis, heroin, stimulants, etc) and non-substance addiction (e.g., gambling, computer gaming, internet, etc.).

Z. Wei • X. Chen
Key Laboratory of Brain Function and Disease, Chinese Academy of Sciences, School of Life Sciences, University of Science & Technology of China, Hefei, Anhui 230027, China

X. Zhang (✉)
Key Laboratory of Brain Function and Disease, Chinese Academy of Sciences, and School of Life Sciences, University of Science & Technology of China, Hefei, Anhui 230027, China

School of Humanities & Social Science, University of Science & Technology of China, Hefei, Anhui 230026, China

Centers for Biomedical Engineering, University of Science & Technology of China, Hefei, Anhui 230027, China
e-mail: zxcustc@ustc.edu.cn

X. Zhang et al. (eds.), *Substance and Non-substance Addiction*, Advances in Experimental Medicine and Biology 1010, DOI 10.1007/978-981-10-5562-1_17

It is important to highlight that attempts at understanding and explaining addiction have been made for centuries. It is, however, just five decades ago, with the growth of science and technology that more interest has been observed in this field. What is addiction? We discuss this question as tackled from three different perspectives (biological, psychological and social-cultural perspectives) and demonstrate how each perspective, if independently conceived and propounded falls short in adequately addressing the questions.

American Society of Addictive Medicine defines addiction as a primary, chronic disease of brain reward, motivation, memory and related circuitry. Addiction is thus characterized by inability to constantly abstain, diminished behavioural control, craving, diminished recognition of significant problems with one's behaviour and dysfunctional emotional response. Addiction affects reward system of the brain, including the nucleus accumbens and midbrain.

A psychological perspective of addiction views addiction as a consequence of psychological maladjustment and functional impairment. The addicted persons find it difficult to resist the urge and consequently fulfil it in order to avoid the negative experiences that follow from withdrawal. The psychological perspective holds that the disruption of the reward system in the brain acts as a major barrier for the normal decision process. This perspective shows impulsivity plays a significant role in the early stage of addiction while maladaptive learning processes and habit formation lead to the development of compulsive behaviour in the later stage of addiction. The two should not be taken as parallels. Dealing with impulsivity may help prevent addiction, but dealing with compulsivity may help in treating addiction.

Both sociologists and psychologists study the influence of groups (families, organizations, societies, cultures) on individual behaviour. From a sociological perspective, addiction is a harmful behaviour that affects both individuals and groups. Thus, we should study addiction within the context of the society in which it occurs. Studies have also shown that addicted persons that receive proper social and family support, find it easier to recover from addiction than those who lack social and family support. From the social cultural perspective, addiction's compulsive nature cannot be resolved by social support alone, but by directed and deliberate therapeutic intervention. Recent neuroscience evidence has shown that some people are genetically more vulnerable to addiction than others, while some are vulnerable as a result of non-social psychological traits.

Based on the reviews, addiction is a crosscutting phenomenon, thus, it should be explained from different perspectives to present a holistic picture of what it is and its developmental process. Each perspective, separately, has limitations.

Studies of addiction focus on a wide range of underlying mechanisms. This field has an ultimate goal to develop a comprehensive strategy for addiction treatment. Based on existing evidence, both types of addiction cause negative consequences to the person's physical, mental, social and financial well-being, the common mechanisms underlying which consist with reward neural system dysfunction and some gene transcription factors However, there are also some different properties of these two types of addiction.

This book focuses on the similarities and differences between the substance and non-substance addiction involving mechanism, diagnosis and treatment.

17.1 Definition of Substance and Non-substance Addictions

Drug addiction is a neuropsychiatric disorder characterized by a recurring desire to continue taking the drug despite harmful consequences [15]. The non-drug addiction (or non-substance addiction) is a behavioral addiction and covers gambling disorder, food addiction, internet addiction, and mobile phone addiction. Their definitions were similar to drug addiction but differ from each other in some specific domains.

Addiction was largely seen as a moral failure in will-power until improved models based on brain function were explored. While early use of a drug may indeed be by choice, the neurobiological changes that occur with continued use, particularly to the prefrontal cortex among other regions related with executive function, compromise inhibitory control which when coupled with physiological and psychological craving for the drug lead to uncontrolled drug use [15, 16]. The transition from DSM-4 to DSM-5 saw the preference for the word 'dependence', as a euphemism for addiction, reportedly as an attempt to help destigmatize addicted patients [38]. This however, resulted in confusion amongst clinicians, where 'dependence' in a DSM-sense was really 'addiction', yet dependence was known as the normal physiological adaptation to the repeated use of a drug or medication [38]. Thus, it is important to highlight that pharmacological dependence is characterized by tolerance and/or withdrawal symptoms that arise from the continued exposure of the central nervous system to a drug. This is distinct from addiction, which is characterized by compulsive drug-seeking behavior.

Gambling disorder, used to be called as pathological gambling (PG), a widespread activity around the world, involves risking something of value in the hopes of obtaining something of greater value [12]. The PG was first defined as an impulse control disorder. As the growing scientific literature on PG reveals common elements with substance use disorders, PG was moved to the category "Substance-Related and Addictive Disorders (SAD)". Moreover, in the DSM-5, PG had been renamed as "gambling disorder". Gambling disorder is thus the only non-substance related disorder or behavior addition in the SAD category until now.

The idea of food addiction (FA) was firstly proposed in the mid-50s [41], while scientists found that eating behaviors of obese and overeating individuals are addiction-like, so they tried to use an addiction model to explain the abnormal eating patterns in obese and overweight people as well as in patients with Eating Disorders (EDs) [54]. It has been defined as a chronic and relapsing condition caused by the interaction of many complex variables that increase cravings for certain specific foods in order to achieve a state of high pleasure, energy or excitement, or to relieve negative emotional or physical states [20].

Internet addiction was firstly defined as an impulse control disorder that does not involve poisoning. Chinese researcher Tao Ran defined it as: individuals overuse of the internet caused by a mental and behavioral disorders; the performance of the re-use of the internet have a strong desire to stop or reduce the withdrawal of the use of the internet reaction; at the same time may be associated with mental and physical symptoms [44]. These two definitions are both comprehensive description of internet addiction and have been used widely.

Mobile phone addiction (MPA) is a term used to describe the uncontrolled use or overuse of mobile phone. Despite there is no uniform definition of MPA in psychological circles, however, a more consistent view is that MPA, together with pathological gambling and internet gaming addiction, can be grouped into the spectrum of behavioral addiction.

17.2 Comparison Between Substance and Non-substance Addictions in Mechanisms

17.2.1 Similarities and Differences in Neurobiology

A variety of evidence has suggested important commonalities in the neurobiology of substance addiction and non-substance addiction which include several neurotransmitter systems. Here we review the similarities and differences between substance addiction and non-substance addiction with a focus on dopamine (DA), serotonin (5HT), opioid, glutamate and norepinephrine (NE) systems.

Dopamine has become characterized as the ‘pleasure’ neurotransmitter in the human brain that produces reward [21, 52, 57] and there is a prevailing view that the dopamine system has a central role in addiction. Both substance and non-substance addiction has been shown to be associated with dopamine release, suggesting a role for dopamine in rewarding and reinforcing effects of addiction. However, some differences have been shown between them. Decreased striatal dopamine D2 and D3 receptor availability has proven a contribution to substance addiction in cocaine [49], methamphetamine [25, 47, 55] and alcohol dependence [18, 19, 33, 51, 53]. And higher striatal dopamine receptor availability has been reported to protest against alcoholism in high-risk individuals [50]. Unlike with substance addiction, there appeared to be no marked differences in baseline D2/D3 receptor availability in individuals with gambling problems compared with healthy subjects [5, 7]. Non-substance addiction may also differ from some substance addictions with respect to dopaminergic response to particular manipulations.

As for serotonin, significantly lower the cerebrospinal fluid levels of the serotonin metabolite 5-hydroxyindoleacetic acid (5-HIAA) and similar response after administration of serotonin agonists were found in gamblers [37] and alcohol dependence [13, 42], which indicates an important role of serotonin in both substance and non-substance addiction. But, differences in the contribution of serotonin in non-substance addiction and substance addiction may also exist. Tryptophan depletion, which reduced central serotonin levels and altered serotonin neurotransmission, significantly reduced the number of decisions made to chase losses in simulated gambling [6], while among those with a positive family history of alcoholism, tryptophan depletion impaired performance on the behavioral inhibition task [10, 26].

Preclinical evidence indicates that opioid receptors are distributed widely in the mesolimbic system and are implicated in the hedonic aspects of reward processing [1, 39]. Opioid function may influence dopamine release in the mesolimbic pathway extending from the ventral tegmental area to the nucleus accumbens or ventral striatum [43]. Clinical trial results with opioid antagonists for substance addiction, particularly alcohol and opiate dependence, share similarities with those for gambling disorder. Further, opioid receptors have been shown to mediate or modulate other rewarding or motivational behaviors where many 'natural' rewarding stimuli such as palatable food-seeking, social behavior and maternal reward have a opioid receptor-mediated component [32].

A persuasive body of preclinical evidence has indicated that glutamate transmission and glutamate receptors play a critical role in drug reward, reinforcement and relapse. Pharmacological manipulation of the glutamate system may be an effective treatment for addiction. Norepinephrine is critically involved in mediating stimulant effects including sensitization and reinstatement of drug seeking. Noradrenergic function has been linked to risk seeking behavior in gambling disorder, which shares some similarities with substance addiction. Further study is needed to investigate the relationships between noradrenergic system and non-substance addiction.

Evidence supports the involvement of dopamine, serotonin, opioids, glutamate and norepinephrine in both substance addiction and non-substance addiction while differences may exist with respect to the contributions of these systems. Further studies are still needed to understand the similarities and differences more precisely, which can facilitate the understanding of neurobiology underlying these addictions as well as the development of treatment across substance addiction and non-substance addiction.

17.2.2 Similarities and Differences in Genetics

There were similar symptomatology manifestations and higher co-morbidity in substance and non-substance addictions, suggesting that there may be a common pathogenesis between them. Genetic findings provide new view for addictions. The investigation of the similarities and differences in genetic mechanism between substance and non-substance addiction may extend the understanding mechanisms of addiction and promote more distinct criteria for diagnosing both substance and non-substance addiction.

Thanks to the development of genetics and neuroscience theory and technology, and lower cost of the experiment, more and more addiction related genes were found in dopamine system, 5-HT system, the endogenous opioid system, norepinephrine system. Existing researches in which gene of multiple systems related with substance addiction were also correlated with non-substance addiction suggested parallels between non-substance addiction and substance addiction. DA is an important neurotransmitter in the brain, and is closely related to spontaneous activity, euphoria, etc., which plays an important role in occurrence, maintenance and relapse in addiction. The dopamine receptors and dopamine transporter gene are all associated with different addictive pathways or behaviors more or less. The 5-HT transporter gene, specifically 5-HTTISLC18A2, were associated with both substance and non-substance addiction with higher s allele. Moreover, the 5-HT receptor gene is also related to addictive disorders. Other genes, such as μ opioid receptor (OPRM1) gene, Catecholamine-O-Methyltransferase (COMT) gene, and Monoamine oxidase A (MAOA) gene, have been proved in addiction.

However, a few studies revealed the difference between them, which were mainly related to 5-HTT and brain-derived neurotrophic factor (BDNF) gene. Based on previous studies on 5-HTT, S allele seems to be related with substance and non-substance addictions. However, inconsistent results were found in PG patients and shopping addiction. Some studies even got the opposite results. Evidence from animal and clinical studies have shown that BDNF activity was related with the pathogenesis of substance addiction, which may owe to the effects of BDNF Val66Met polymorphism on BDNF secretion. However, a study of 210 Caucasian women has shown that the 66Met homozygous carriers had more frequency and severity of binge eating [35].

Just like substance addiction, non-substance addiction is also affected by multiple genes [4, 17, 30]. Consequently, the effect of a single gene may be sometimes not significant so that many experimental results are difficult to repeat verification. Moreover, the current evidence suggests that non-substance addiction may be also affected by both disease-causing genes similar with substance addiction and environmental risk factors, e.g. poor family or other social environment. Therefore, the researchers should pay their attention to investigating the influence of environmental factors and mechanisms in the future. Furthermore, the effect of the epigenetics may also be a way to interact with the environment, so we believe that it will be an interesting topic on investigating the relationship with specific genes and non-substance addiction by interactions of gene-environment or gene-gene.

17.2.3 Similarities and Differences in Neuroimaging

With the development of imaging technology in the past three decades, imaging studies have provided information on the neurobiological effects of drugs, revealed neurochemical and functional changes in the brains of both drug-addicted and non-substance addicted subjects, and yielded important insights into individuals' subjective experiences and behaviors, which provide new evidences for the mechanisms underlying addiction and give clinicians opportunities to assess addiction, assign patients to appropriate care interventions, and monitor response to therapy [48]. These imaging techniques include computed tomography (CT), structural magnetic resonance imaging (MRI), functional magnetic resonance imaging (fMRI), magnetic resonance technology (MRS), positronemission tomography (PET), single photon emission computed tomography (SPECT) and magneto encephalography (MEG).

The brain imaging techniques have been well used to identify the mechanism of substance addiction and changes of the regions in brain. For substance addiction, especially opioids, cocaine, ketamine, alcohol and nicotine, a great number of imaging studies have illustrated the structural and functional changes on the white matter in reward and craving circuits areas, such as bilateral amygdala and nucleus accumbens [29, 31, 45] and the gray matter in decision-making and learning circuits regions, such as prefrontal cortex, cingulate gyrus, insula and temporal lobe, for instance, among opioids addiction [22, 27, 36]. Additionally, brain imaging techniques are also used to explore the structural and functional changes on brain caused by non-substance addiction such as internet addiction disorder (IAD) and pathological gambling (PG). According to the neuroimaging studies, like substance addicts, subjects with behavioral addiction experience the similar structural and functional damages on brain regions related to reward, decision-making and emotion processes [11, 40, 56, 58]. The neural basis of non-substance addiction has been more extensively studied and is better established compared to other forms of "addiction" (e.g., drug addictions).

However, with relatively fewer neuroimaging studies in non-substance addictions or direct comparison on the two groups, the differences between these two kinds of addictions are not well-identified where current results are not quite inconsistent. Although there are still some variations among the results of these studies and unidentified mechanisms for the non-substance addiction, most of the neuroimaging studies reveal similar neurobiological mechanism between substance and non-substance addiction. In the future research, more studies need to be conducted in order to deeply figure out the similarities and differences in neurobiological mechanism between substance and non-substance, and provide more evidences to establish the diagnostic criteria and treatment guidelines, particularly for non-substance addiction.

17.2.4 Similarities and Differences in Cognitive Psychology

The essential feature of substance addiction is a cluster of cognitive, behavioral, and physiological symptoms indicating that an individual continues using the substance despite significant substance-related problems, while similar to substance addiction in physiological and psychological changes, non-substance addiction is triggered by environmental cues. Even though all kinds of addiction affect brain and body's biochemical processes [3], it is not enough to attribute a certain addiction to biological and genetic predisposition [9]. We also need to take personal cognition, personality and social factors into consideration.

In the comparisons in psychology, we focus on differences and similarities in sensation seeking, inhibitory control, attentional bias, intertemporal choice and environmental factors between substance and non-substance addiction. We conclude that substance addicts have high sensation seeking scores, suggesting that personality traits and genetic factors significantly influence the formation of substance addiction; for non-substance addiction, only correlational – not causal – relationships may be drawn, indicating that sensation seeking does not play a major role in the formation of non-substance addiction. With regard to inhibitory control and attentional bias, substance addicts have significantly impaired basic cognitive functions in comparison to non-substance addicts. In non-substance addicts, intertemporal choice can effectively predict the severity of their addiction, suggesting that non-substance addicts' higher cognitive functions are relatively unaffected. Finally, environmental factors significantly influence non-substance addiction relative to substance addiction. Substance addiction and non-substance addiction share common cognition, personality, environmental factors, but the effects caused by substance addiction are profound and often permanent due to brain damage, genetic factors, and decline in cognitive function. Non-substance addiction (e.g. internet addiction) is a result of a variety of factors such as family, personality, environment and development (adolescence and early adulthood), indicating that the formation of substance addiction involves more endogenous factors while non-substance addiction involves more exogenous factors, implicating both nature and nurture.

Though several psychological sets are associated with addiction, the causality remains unclear. If the addiction results in psychological impairments, cognitive enhancement may improve patients' life quality. If changes in psychological sets lead to addiction, interventions at young age should decrease the risk of being addicted. Therefore, future studies should explore the causality of addiction and psychological deficits.

17.3 Comparison Between Substance and Non-substance Addictions in Diagnosis

17.3.1 Similarities and Differences in Diagnostic Criterion

Addiction is defined as a cluster of cognitive, behavioral, and physiological symptoms indicating that the individual continues addictive behavior despite significant addictive behavior-related problems. It is a syndrome centered on impaired control over a reward-seeking behavior leading to significant social impairment and withdrawal symptoms. In order to research this complicated syndrome, the first step we need to do is to diagnose what is addiction. From Goodman's "addictive disorder" to latest criteria in DSM-5, the diagnostic criteria of addiction are more and more scientific and effective. In general, DSM-5 by American Psychiatric association is the most authoritative tool to diagnose addiction nowadays. Though all ten drugs addiction can be diagnosed by a common diagnosis because of common direct brain reward system activation, the diagnostic criteria of different drug is different from each other more or less. Besides, behavioral addiction, such as gambling disorder, internet addiction, food addiction and hypersexual disorder, attracts great attention in recent years. Among them, gambling disorder is the only behavioral addiction that is now divided into non-substance-related disorder.

Though drug addiction is a human phenomenon, some of the behavioral characteristics of this syndrome can be satisfactorily modeled in animal. In this way, research in animals has contributed to the understanding of the neurobiological basis of drug addiction and of the brain reward system. Simple behavioral models of animals would greatly facilitate the understanding of circuit and molecular mechanisms that promote drug consumption, and could assist in the development of pharmacological therapies to target the compulsive drug intake, which remains intractable to human addiction. Vanderschuren and Ahmed have identified several ways in which those DSM-IV criteria can be researched in animal models, and then described the evidence that these symptoms can be observed in laboratory animals after repeated drug use [46].

Since animal models have contributed to the understanding of drug addiction, we can also apply animal models to behavioral addiction (gambling disorder specifically here) naturally. Several animal models have been used to model the gambling behavior, but none of them actually model the gambling disorder. The most important thing for an animal model of the gambling disorder is to build a valid paradigm based on the diagnostic criteria. Conditioned place preference (CPP) and drug self-administration are widely used approaches to model drug addiction, which would be applied to gambling disorder somehow. New animal models of addiction may promote to identify the neuropharmacologic and molecular mechanisms underlying the addictive-like behavior, developing new affective treatments for addiction by filling the gaps between preclinical and clinical studies.

17.3.2 Similarities and Differences in Diagnostic Scales

Diagnostic scale of addiction is a convenient and useful tool for addiction diagnoses in clinical, social work and research fields. These scales reflect diagnostic criteria of addiction, which includes impaired control, social impairment, risky use and pharmacological criteria, more or less. As tobacco and alcohol use are two major substance addictions which have harmful influence worldwide, we mainly focused on diagnostic scales of these two substance addiction. We also discussed diagnostic scales of non-substance addiction, including pathological gambling, pathological internet use, internet gaming addiction and so on.

According to the comparisons among substance scales, several differences involved. Interview and self-report are two formats that caused little variability. As for the context, every aspects relating to the substance might be contained in the scales, such as the quantity, frequency, purpose and consequence of substance use, the guilt, the help-seeking behavior. The application could be divided into three fields depending on the culture, law or age, which were also the barriers of the widespread use.

Non-substance addiction scales developed often depend on the case study, the literature or the diagnostic criteria of substance addiction. We found that the scales of pathological gambling are primarily based on the criteria of substance addiction, and the subsequent non-substance addiction scales are often based on the diagnostic criteria of substance addiction or pathological gambling. The reason is that no non-substance addiction scales are available for reference before the pathological gambling scale, and there are some common characteristics between pathological gambling and substance addiction, for example, they can both activate reward system. After the gambling scales were included in diagnostic criteria, the following non-substance will have the reference. As for the usage of the scales, the frequently-used scales have been widely used in the studies of differenct contries. Most of them are mainly for adults, and are also some scales for adolescents.

By discussions on addiction scales, some trends were proposed. First, new substance addiction scales will no longer limit the range to the drug itself, but include almost all the related substance, which highlights the integration trend of substance addiction scales. Second, substance addiction scales tended to be more concerned with psychometric properties and theoretical models. Third, the cross-cultural communication becomes more and more popular along with the social economic development, so cross-cultural applicability becomes an important feature and future direction in the addiction development. They all influence the design and development of substance and non-substance addiction scales by expanding the groups, computer-aided investigation, cross-cultural application, the arguments between the developing and mature models in theoretical and practice fields.

17.3.3 Similarities and Differences in Biochemical Diagnosis

Searching biochemical indices is a promising strategy for objective assessment. Biological specimens such as urine or blood sample, have a long history as a source for measuring health and remain an important tool for clinical diagnosis. The specimens not only are usually easy to collect, but also require minimal cooperation from the patients. It is also believed that addictive substances can induce biochemical reactions so that certain biomarkers of indulgent use can be found. On the other hand, addictive disorders are considered as diseases of multifactorial inheritance. It is likely that certain inheritable features carry susceptibility to one or more addictive disorders and responsiveness to treatment. These genetic biomarkers, if established, would help clinicians with more personalized therapeutic choices.

We classified those possible biomarkers into three categories: origin forms and metabolites of substances, markers from biochemical responses to certain addiction, and genetic and epigenetic biomarkers suggesting susceptibility to addiction.

Substance in its original or metabolite form can be found in specimens like blood, urine, sweat, saliva, hair, and nail. Their existence confirms contact of substance in a recent period. Importantly, The same abused substances can be detected in non-substance addiction, such as pathological gambling, sex addiction, and internet use disorder. Therefore, detection of substance use in non-substance addiction often indicate more severe degree of non-substance addiction.

Though the different mechanism underlying the substance or non-substance addiction, some biochemical reactions of our body to substance or non-substance addiction are similar. It is possible that the brain neurotrophic factors, immune function, and vascular integrity are changed by these behavioural patterns. There are three major modes of epigenetic regulation, namely histone acetylation and methylation, DNA methylation, and non-coding RNAs, which profoundly affect the expression of genes, the translation into proteins, cellular structures and functions, and ultimately the apparent phenotypes. Given technical convenience and affordability, changes of epigenetic regulation are among the most promising ones to be developed into biomarkers.

Genetic factors contribute to the formation of addictive disorders. Sibling and twin studies found the heritability of addictive disorders around 30–70 %. The genetic susceptibility to addiction comes from different aspects. (1) Inherited temperament, such as impulsivity, sensation-seeking, and novelty-seeking, may interact with environmental factors, thus increase the likelihood of substance exposure. (2) Deficiency in stress response contribute to the progress from misuse to dependence. These features may be shared by the most kinds of addiction, including non-substance addiction.

We observed a spectrum of biochemical changes in patients of addiction, which may reflect the inflammatory impairment that endanger the cardiovascular system. These markers include TNF, IL-6, homocysteine, alpha-synuclein, adipocyte-derived hormones, and several epigenetic markers. Some of these changes may be unspecific, but their existence calls more attention to study the inflammatory react and systematic impairment of cardiovascular and brain diseases caused by addiction.

17.3.4 Development of New Diagnostic Techniques

Addiction diagnose is a key part of addiction research and treatment. The traditional diagnoses on addiction heavily rely on the patient's self-report through using diagnose tools such as Diagnostic and Statistical Manual of Mental Disorder (DSM) and International Classification of Diseases (ICD). But, subjective reports are highly influenced by patient's memory and cooperation. Neuroimaging techniques are hoped to identify preclinical neural changes that predict subsequent addiction. However, there are several limitations when solely using the neuroimaging methods to diagnose addiction. In recent years, due to the significant development of neuroscience studies on addiction, several molecular, neurobehavioral, and neurological biomarkers have been found to potentially classify addicted persons. But, the biomarkers are often mixed and difficult to be replicated on individual level. Thus, there is an urgent need to find new methods to accurately classify individuals of addictions.

Machine learning (ML) is a data-driven procedure that learns construction of algorithms from training data and makes predictions on test data. During the last 10 years, ML methodology has been developed for the diagnoses of psychiatric disorder based on data from multiple domains, which is hoped to be a useful tool for addiction diagnose. Generally, there are four steps of ML processing for addiction diagnose: pre-processing, model training, cross-validation and clinical prediction. Recent studies have begun to employ ML in clinical usage, which support the idea that ML can be used to classify clinical disorders, separate patients with different disorders and predict the development of a disorder. ML has so far been used in some addiction classification studies and has been anticipated to be the most promising approach to classify vulnerable persons, addicted individuals, remitted patients, and chronic relapers.

There are several limitations when using ML to diagnose addiction. Firstly, the sample size of neuroimaging data is often too small. Secondly, the inter-subject and intra-subject variability are larger. Thirdly, the features of neuroimaging data are often too large for ML. Fourthly, the application of ML in diagnosing behavioral addiction is still scarce.

In conclusion, ML is a hopeful tool in diagnosing addiction. It has been found to contribute to classifying addicts and non-addicts, separating different types of addiction, and evaluating the effects of treatment. Future ML studies based on larger clinical dataset, prior knowledge, and information from multiple domains may help to improve the accuracy of diagnose. Moreover, the validity of ML-based diagnose need to be tested in more sub-types of substance addiction and behavioral addiction.

17.4 Comparison Between Substance and Non-substance Addictions in Treatment

17.4.1 Drug Therapy

The therapy of substance and non-substance addictions will be expected to transfer to elimination of psychological craving (i.e, addiction) based on physical rehabilitation, and the behavior is controllable after elimination of psychological addiction. Elimination of psychological addiction makes complete rehabilitation of substance and non-substance addictions possible, and neither addictions will be seen as chronic recurrent encephalopathy, which will change our understanding of this disease and the whole therapy system.

According to modern mainstream medicine, the goals and principles of substance and non-substance addiction therapy are similar with other chronic diseases. Three-level prevention and cure strategies should be adopted. Stress should be put on the prevention of diseases, early detection and early therapy. When the condition develops to addiction, comprehensive intervention should be carried out from the medical, psychological and social and other dimensions, and the main therapeutic target is to reduce the degree of addiction, prevent relapse, restore physical and psychosocial function, and reduce the harm. As for addicts who have strong desire for withdrawal, allowable economic conditions, and good social support system such as work and family, if the addiction rehabilitation therapy institutions have mature systematic therapy technology, especially mature technology to eliminate psychological addiction, the therapeutic target can be complete rehabilitation.

Based on the main goal of addiction therapy, 12 basic principles of substance and non-substance addiction therapy were established: (1) The prerequisite to therapy is that it should be harmless and shouldn't reduce patients' quality of life. It should be strictly prohibited to use electric shock therapy, electroconvulsive shock therapy and ablative procedure of cerebral nuclei when the patients are conscious; (2) The therapy should be personalized, humanized, efficient, integrated and multidimensional; (3) The therapy should be convenient and accessible; (4) The therapy should be flexible; (5) The therapy time should be adequate and follow-up visits should be on a long-term basis; (6) Great importance should be attached to psychological therapy, especially the elimination of psychological addiction; (7) Medication can be encouraged in substance addiction therapy but used with discretion in non-substance addiction therapy; (8) Aggressive treatment should be administered for comorbid psychiatric disorders; (9) Therapy is divided into three stages: acute withdrawal therapy, psychosomatic rehabilitation therapy, prevention of relapse and return to society; (10) Effective therapy does not require voluntary conduct; (11) Assessment of effects not only relies on monitoring explicit behavior, such as regular monitoring of the use of addictive substances or addictive behavior, but also on the patient's inner feelings and emotions; (12) Assessment and consultation should be undertaken on disease complications.

Addiction therapy is a long-term process, and can be divided into three stages: acute withdrawal therapy, psychosomatic rehabilitation therapy, prevention of relapse and return to society. The core content of addiction therapy includes diagnostic assessment, therapy plan and various comprehensive therapy measures. Nowadays, comprehensive intervention has gradually become the mainstream of addiction disorder therapy. In fact, the addiction disease therapy has showed systematized, comprehensive, effective and humanized trend in China at present.

17.4.2 Physical Therapy

Physical therapy has the evidence-based science knowledge to address a wide range of physical and psychological problems of addiction. Neuromodulation techniques is becoming more and more important in the treatment of addiction. Here, the efficacy of different neuromodulation techniques in addiction, such as transcranial direct current stimulation (tDCS), repetitive transcranial magnetic stimulation (rTMS), deep brain stimulation (DBS), is critically evaluated. Other physical therapy methods including Biofeedback, Physical Activity and Acupuncture are also presented.

Transcranial direct current stimulation (tDCS), as a non-invasive brain stimulation technique, induces plasticity via generation of sub-threshold, stimulation polarity-dependent alteration of membrane potentials modifying spontaneous discharge rates [24]. It is a promising tool in neuroplasticity research as well as a therapeutic instrument in neurological disorders [28]. Previous studies have demonstrated that the application of tDCS is helpful for substance disorders [14] and psychiatric diseases [24]. The main goals of tDCS application in behavioral addiction are all at once therapeutic, by modulating craving, impulsivity, executive functions and physiopathological, by enhancing the knowledge of neurophysiological basis of behavioral addiction.

Repetitive transcranial magnetic stimulation (rTMS), also as a non-invasive brain stimulation technique, has gained notable attention in neurologic and psychiatric research in recent years. It involves the use of a wire coil through which brief pulses of electrical current are passed, leading to the generation of magnetic fields that pass through the skull [2]. It can change the brain's neuronal activity depending on the target area and stimulation parameters, such as intensity and frequency. The rationale to use rTMS as a treatment for substance addiction and craving is that the DLPFC, which plays a major role in top-down inhibitory control mechanisms and reward mechanisms, is dysfunctional in these disorders.

Deep brain stimulation (DBS) has mainly been used in medication-refractory movement disorders, such as Parkinson's disease and essential tremor, in treatment of resistant obsessive-compulsive disorder and depression [23]. Comparing to traditional Radiofrequency Ablation, DBS has certain advantages because it is reversible, controllable, minimal-invasive, and has rapid recovery rate. Therefore, DBS has the potential to become one of the optimal physiotherapies for substance addiction (i.e. alcohol addiction). However, only limited researches and clinical trials have been done on the use of DBS in the treatment of substance. Researchers are expected to confirm the optimal stimulation place and parameters of DBS.

The neurofeedback training has been widely used in the treatment of many diseases and disorders [34]. Studies have shown that neurofeedback training is a good way to quit drug addiction whereas long-term use of the drug has a profound effect on the individual's EEG.

Yoga practices, including postures and meditation, direct attention towards one's health, while acknowledging the spiritual aspects of one's nature. There are growing number of clinical experiments and cases about substance addiction such as alcohol dependence and smoking showing the effect of yoga and mindfulness. Physical activity, and specifically exercise, as a potential non-pharmacological treatment for addiction, has been suggested as a potential treatment for drug addiction. More and more studies have revealed the direct efficacy of exercise as a prevention for addiction.

Acupuncture, a key component of Traditional Chinese Medicine, involves the penetration of the skin with thin metal needles, and is controlled by an appropriately trained practitioner or further stimulated by electrical stimulation. The current prevalent hypothesis for the use of acupuncture in the treatment of substance is the relationship between acupuncture and the Cascade Theory of Rewarding. Auricular acupuncture has been used extensively in substance abuse treatment programs, hospitals, and prisons throughout the USA and the world for the past 30 years.

17.4.3 Traditional Chinese Medical Therapy

Traditional Chinese medicine (TCM) is one of typical ethno-medicine that derives from the regular experiences of Chinese in the early age. It is also a systematic principle of thousands of years of clinical practice. Compared with modern medicine, experts believe that TCM has fewer side effects, is safe enough and has ideal effects in treating refractory chronic illness. Generally, TCM contains Chinese herbal medicine and acupuncture. It has been practiced in treatment of substance addiction and non-substance addictions.

TCM have efficacy in the rehabilitation of abnormal physical problems induced by chronic drug use, including improving immune function, increasing working memory, and protecting against neurological disorders.

TCM might be a good alternative solution for some type of addiction, as well as certain complex chronic disease, such as cancer and diabetes, because its effectiveness has gradually gained the support of evidence-based medicine. According to the consensus of experts from both TCM and modern medicine circles, the future of TCM herbs will largely depend on its safety and efficacy. As the multi-component therapeutic strategies and practices are ongoing, mixture and combination could be a feasible direction for facilitating TCM herbs modernization to answer the criticisms of underlined components, uncontrolled quality and undermined toxicity. Furthermore, compared with the complicated and elusory formula, the safety and efficacy of combination would definitely be much easier to answer.

17.4.4 Nutrition Support Therapy

The daily nutrients which are necessary for the human body to grow and sustain normal function of life are namely carbohydrates, fats, protein, vitamins, minerals, and water. Studies have reported poor diets with overweight and obesity among people in recovery from substance and non-substance addiction. In most of these addiction, serious nutritional deficiencies of major proteins, fats, vitamins and minerals exist which prevent their capability to digest carbohydrates efficiently. For instance, physical and biochemical changes that occur from drug and alcohol use also cause nutritional deficiencies and imbalance.

Poor diets and high rates of overweight and obesity have been reported among people in recovery from substance addiction. Poor diets in this population may be related to a lack of nutrition knowledge and food preparation skills as well as food environments in treatment facilities that do not support healthy eating behaviors. Residential treatment facilities provide a unique environment to promote healthy eating and build food preparation skills that could be transferred to independent living. Results suggest that men in residential treatment facilities may benefit from dietary interventions, but these interventions need to consider addiction and treatment history. Because people in recovery from substance addiction tend to have poor dietary patterns and are at an increased risk for chronic health conditions, additional studies are warranted to address dietary concerns in this population. These results add to the growing evidence that environmental factors impact dietary behaviors and subsequently obesity as well as support the importance of skill development in healthy eating demonstrated by previous successful programs with diverse groups. Treatment facilities have tremendous potential for providing healthy food options and skill development to improve resident dietary behaviors [8].

Compulsive overeating is treatable with nutritional assistance and medication. Psychotherapy may also be required, but recent research has proven this to be useful only as a complementary resource, with short-term effectiveness in middle to severe cases.

Poor nutrition and physical inactivity were shown to be significantly associated with IAD. Adolescents who spend longer hours online potentially navigate towards unhealthier foods. It is postulated that online gamers drink high-caffeinated energy drinks and eat high-sugar snacks to increase alertness for online gaming. But, the nutritional approaches haven't yet been systematically introduced in treatments of Internet addiction.

Since sex addiction has sometimes been recognized as a category of non-substance addiction, we consider that integrating approaches involving the above treatments and health plan or nutrition plan are needed, too. Lisa L. Kirkland and the partners provide the hospitalist with an overview of screening, assessment, and development and implementation of a nutrition care plan in the acutely ill hospitalized patient, which can be utilized in the area of sex addiction in some extent.

17.4.5 Psychotherapy

Generally speaking, addiction therapy is a long process, psychological behavior therapy is an important part of addiction therapy, and its main goal is to improve the understanding of addiction, improve therapy compliance, prevent relapse, rebuild a healthy lifestyle, maintain long-term withdrawal state, or even be completely healed.

The goal of psychological behavior therapy is different based on different therapy and rehabilitation stages of patients. Early treatment is mainly to increase the motivation of treatment, improve self-confidence and self-efficacy of patients; therapeutic rehabilitation at middle and later periods is mainly to correct a variety of psychological and behavioral problems induced by abuse of addictive substances or addiction behavior, help patients learn a variety of psychological skills, improve the ability to resist addictive substances or behavior, establish a healthy lifestyle, and prevent relapse.

The field of addiction medicine has developed a number of treatment methods and strategies on psychological behaviors for drug addicts. Psychological behavior therapy can be divided into motivational intervention, cognitive behavioral therapy, behavior intervention, aversion therapy, addiction eliminating technology, etc., according to different theoretical basis, individual treatment, group therapy, family therapy, etc., according to the form of psychological behavior treatment. These methods can be used alone or in combination for different treatment forms and treatment sites, and are the basic methods for the treatment of various substance addictions. These methods can be applied to the psychological treatment of non-substance addiction after being properly improved.

17.4.6 Cognitive-Behavioural Therapy

Cognitive-behavioral therapy (CBT) is based on cognitive theory and behavioral theory, which reflects the work of its pioneers, namely Ellis and Bandura. The therapy emphasizes the importance of thought and feeling, including how individuals feel and explain life events, which is a decisive factor in action. CBT also attempts to help patients realize maladjustment, teach them how to heed, seize, monitor, interrupt the "cognitive-affective-behavioral chains" and eventually adapt to it, in order to achieve the highest goal.

CBT is a structured, short-term, well-targeted, psychological treatment that focuses on the current problems of substance addiction, and helps addicts to identify, evade and respond to the factors inducing substance addiction, in order to maintain integrity, prevent relapse. The efficacy has been verified. The actual curative effects of CBT in the substance addiction has also been valued and adopted in psychotherapy of non-substance addiction.

CBT of substance addiction combines the behavioral theory (classical conditioned reflex and operant conditioned reflex), social learning theory (the decisive effect of observational learning, the influence of role models, and cognitive anticipation on behavior) and foundations of cognitive theory (thinking, Cognitive schemas, beliefs, values, attitudes and attribution).

Substance addiction CBT includes three core elements: functional analysis, coping skills training and prevention of relapse patterns. Operation points of substance addiction CBT are as follows. The obligations of the therapist, the expectations and responsibilities of the addicts and the time and number of sessions must be clarified.

The most widely used CBT for substance prevention is relapse prevention. The main goal is to change the misconception of relapse by patients, in order to change the behavior of relapse. The patients should learn various skills to deal with high-risk situation under the guidance of consultant by allowing patients to identify their own high-risk situation of relapse, so as to improve their self-efficacy, learn to establish a new life-style replacing substance addiction or addiction behaviors, and ultimately to prevent relapse and maintain a long-term withdrawal.

In CBT of behavioral addictions, cognitive-behavioral therapy was developed, which possibly has most extensive application for gambling addiction. There are mainly five parts in their therapy: provide training to gambling addicts in aspect of games of chance; correct their cognitive bias; provide skill training for problem solving; provide social skills training and prevent relapse.

CBT shows better effect comparing to antidepressant drug alone, self-surveillance and supportive psychotherapy as well as behavioral therapies excluding cognitive therapy. Result of 1-year follow-up of CBT treatment indicates that it can remain effect better than antidepressant drug therapy. A combination of antidepressant drug therapy and CBT is more effective than drug therapy alone or CBT. CBT seems having better effect than antidepressant drug therapy in a long term.

17.5 Conclusion

Based on the empirical evidence of similarities between substance and non-substance addictions, it shows that addictions share neurobiological and psychosocial factors that increase vulnerability. Thus, some researchers developed a syndrome model of addiction that involves both substance and non-substance addictions. However, addictions vary in distinctive expression and unique negative results (i.e. gambling debt in pathological gamblers or liver cirrhosis in alcoholics). Treating certain behaviors as potentially addiction is still an open question.

The purpose of comparing similarities and differences between substance and non-substance addictions is to increase knowledge of a insufficiently explored field, to take seriously the subjects who suffer from the lack of control over their behaviors, and to provide help for those who need it. Conclusively, the comparisons between substance and non-substance addictions are good for research in the area of addiction. We hope our book could help the ones who want to put their enthusiasm into this field.

References

1. Barbano MF, Cador M (2007) Opioids for hedonic experience and dopamine to get ready for it. Psychopharmacology 191:497–506
2. Barker AT, Freeston IL, Jalinous R, Jarratt JA (1987) Magnetic stimulation of the human brain and peripheral nervous system: an introduction and the results of an initial clinical evaluation. Neurosurgery 20:100–109
3. Berridge KC, Robinson TE (2003) Parsing reward (vol 26, pg 507, 2003). Trends Neurosci 26:581. 581
4. Blum K, Bailey J, Gonzalez AM, Oscar-Berman M, Liu Y, Giordano J, Braverman E, Gold M (2011) Neuro-Genetics of Reward Deficiency Syndrome (RDS) as the root cause of "Addiction Transfer": a new phenomenon common after bariatric surgery. J Genet Syndr Gene Therapy 2012:S2-001
5. Boileau I, Payer D, Chugani B, Lobo D, Behzadi A, Rusjan PM, Houle S, Wilson AA, Warsh J, Kish SJ, Zack M (2013) The D2/3 dopamine receptor in pathological gambling: a positron emission tomography study with [11C]-(+)-propyl-hexahydro-naphtho-oxazin and [11C] raclopride. Addiction 108:953–963
6. Campbell-Meiklejohn D, Wakeley J, Herbert V, Cook J, Scollo P, Ray MK, Selvaraj S, Passingham RE, Cowen P, Rogers RD (2011) Serotonin and dopamine play complementary roles in gambling to recover losses. Neuropsychopharmacology 36:402–410
7. Clark L, Stokes PR, Wu K, Michalczuk R, Benecke A, Watson BJ, Egerton A, Piccini P, Nutt DJ, Bowden-Jones H, Lingford-Hughes AR (2012) Striatal dopamine D(2)/D(3) receptor binding in pathological gambling is correlated with mood-related impulsivity. NeuroImage 63:40–46
8. Cowan JA, Devine CM (2013) Diet and body composition outcomes of an environmental and educational intervention among men in treatment for substance addiction. J Nutr Educ Behav 45:154–158
9. Crabbe JC (2002) Genetic contributions to addiction. Annu Rev Psychol 53:435–462
10. Crean J, Richards JB, de Wit H (2002) Effect of tryptophan depletion on impulsive behavior in men with or without a family history of alcoholism. Behav Brain Res 136:349–357
11. Dong G, Hu Y, Lin X, Lu Q (2013) What makes internet addicts continue playing online even when faced by severe negative consequences? Possible explanations from an fMRI study. Biol Psychol 94:282–289
12. Fauth-Buhler M, Mann K, Potenza MN (2016) Pathological gambling: a review of the neurobiological evidence relevant for its classification as an addictive disorder. Addict Biol *22(4)*:885–897
13. Fils-Aime ML, Eckardt MJ, George DT, Brown GL, Mefford I, Linnoila M (1996) Early-onset alcoholics have lower cerebrospinal fluid 5-hydroxyindoleacetic acid levels than late-onset alcoholics. Arch Gen Psychiatry 53:211–216
14. George MS, Aston-Jones G (2010) Noninvasive techniques for probing neurocircuitry and treating illness: vagus nerve stimulation (VNS), transcranial magnetic stimulation (TMS) and transcranial direct current stimulation (tDCS). Neuropsychopharmacology 35:301–316
15. Goldstein RZ, Volkow ND (2002) Drug addiction and its underlying neurobiological basis: neuroimaging evidence for the involvement of the frontal cortex. Am J Psychiatry 159:1642–1652
16. Goldstein RZ, Volkow ND (2011) Dysfunction of the prefrontal cortex in addiction: neuroimaging findings and clinical implications. Nat Rev Neurosci 12:652–669
17. Grant JE, Brewer JA, Potenza MN (2006) The neurobiology of substance and behavioral addictions. CNS Spectr 11:924–930
18. Heinz A, Siessmeier T, Wrase J, Hermann D, Klein S, Grusser SM, Flor H, Braus DF, Buchholz HG, Grunder G, Schreckenberger M, Smolka MN, Rosch F, Mann K, Bartenstein P (2004) Correlation between dopamine D(2) receptors in the ventral striatum and central processing of alcohol cues and craving. Am J Psychiatry 161:1783–1789

19. Hietala J, West C, Syvalahti E, Nagren K, Lehikoinen P, Sonninen P, Ruotsalainen U (1994) Striatal D2 dopamine receptor binding characteristics in vivo in patients with alcohol dependence. Psychopharmacology 116:285–290
20. Imperatori C, Fabbricatore M, Vumbaca V, Innamorati M, Contardi A, Farina B (2016) Food addiction: definition, measurement and prevalence in healthy subjects and in patients with eating disorders. Riv Psichiatr 51:60–65
21. Kalivas PW (2001) Drug addiction: to the cortex and beyond! Am J Psychiatry 158:349–350
22. Kilts CD, Gross RE, Ely TD, Drexler KP (2004) The neural correlates of cue-induced craving in cocaine-dependent women. Am J Psychiatry 161:233–241
23. Knapp CM, Tozier L, Pak A, Ciraulo DA, Kornetsky C (2009) Deep brain stimulation of the nucleus accumbens reduces ethanol consumption in rats. Pharmacol Biochem Behav 92:474–479
24. Kuo MF, Paulus W, Nitsche MA (2014) Therapeutic effects of non-invasive brain stimulation with direct currents (tDCS) in neuropsychiatric diseases. NeuroImage 85(Pt 3):948–960
25. Lee B, London ED, Poldrack RA, Farahi J, Nacca A, Monterosso JR, Mumford JA, Bokarius AV, Dahlbom M, Mukherjee J, Bilder RM, Brody AL, Mandelkern MA (2009) Striatal dopamine d2/d3 receptor availability is reduced in methamphetamine dependence and is linked to impulsivity. J Neurosci 29:14734–14740
26. LeMarquand DG, Benkelfat C, Pihl RO, Palmour RM, Young SN (1999) Behavioral disinhibition induced by tryptophan depletion in nonalcoholic young men with multigenerational family histories of paternal alcoholism. Am J Psychiatry 156:1771–1779
27. Li M, Tian J, Zhang R, Qiu Y, Wen X, Ma X, Wang J, Xu Y, Jiang G, Huang R (2014) Abnormal cortical thickness in heroin-dependent individuals. NeuroImage 88:295–307
28. Liebetanz D, Nitsche MA, Tergau F, Paulus W (2002) Pharmacological approach to the mechanisms of transcranial DC-stimulation-induced after-effects of human motor cortex excitability. Brain 125:2238–2247
29. Lim KO, Choi SJ, Pomara N, Wolkin A, Rotrosen JP (2002) Reduced frontal white matter integrity in cocaine dependence: a controlled diffusion tensor imaging study. Biol Psychiatry 51:890–895
30. Lim S, Ha J, Choi SW, Kang SG, Shin YC (2012) Association study on pathological gambling and polymorphisms of dopamine D1, D2, D3, and D4 receptor genes in a Korean population. J Gambl Stud/Co-Sponsored National Counc Probl Gambl Inst Study of Gambl Commer Gaming 28:481–491
31. Lin F, Zhou Y, Du Y, Qin L, Zhao Z, Xu J, Lei H (2012) Abnormal white matter integrity in adolescents with internet addiction disorder: a tract-based spatial statistics study. PLoS One 7:e30253
32. Lutz PE, Kieffer BL (2013) The multiple facets of opioid receptor function: implications for addiction. Curr Opin Neurobiol 23:473–479
33. Martinez D, Gil R, Slifstein M, Hwang DR, Huang Y, Perez A, Kegeles L, Talbot P, Evans S, Krystal J, Laruelle M, Abi-Dargham A (2005) Alcohol dependence is associated with blunted dopamine transmission in the ventral striatum. Biol Psychiatry 58:779–786
34. Marzbani H, Marateb HR, Mansourian M (2016) Neurofeedback: a comprehensive review on system design, methodology and clinical applications. Basic Clin Neurosci 7:143–158
35. Monteleone P, Zanardini R, Tortorella A, Gennarelli M, Castaldo E, Canestrelli B, Maj M (2006) The 196G/a (val66met) polymorphism of the BDNF gene is significantly associated with binge eating behavior in women with bulimia nervosa or binge eating disorder. Neurosci Lett 406:133–137
36. Nakama H, Chang L, Fein G, Shimotsu R, Jiang CS, Ernst T (2011) Methamphetamine users show greater than normal age-related cortical gray matter loss. Addiction 106:1474–1483
37. Nordin C, Eklundh T (1999) Altered CSF 5-HIAA disposition in pathologic male gamblers. CNS Spectr 4:25–33
38. O'Brien CP, Volkow N, Li TK (2006) What's in a word? Addiction versus dependence in DSM-V. Am J Psychiatry 163:764–765

39. Pecina S, Smith KS, Berridge KC (2006) Hedonic hot spots in the brain. Neuroscientist 12:500–511
40. Quester S, Romanczuk-Seiferth N (2015) Brain Imaging in Gambling Disorder. Curr Addict Rep 2:220–229
41. Randolph TG (1956) The descriptive features of food addiction; addictive eating and drinking. Q J Stud Alcohol 17:198–224
42. Ratsma JE, Van Der Stelt O, Gunning WB (2002) Neurochemical markers of alcoholism vulnerability in humans. Alcohol Alcohol 37:522–533
43. Spanagel R, Herz A, Shippenberg TS (1992) Opposing tonically active endogenous opioid systems modulate the mesolimbic dopaminergic pathway. Proc Natl Acad Sci U S A 89:2046–2050
44. Tao R, Huang X, Wang J, Zhang H, Zhang Y, Li M (2010) Proposed diagnostic criteria for internet addiction. Addiction 105:556–564
45. Upadhyay J, Maleki N, Potter J, Elman I, Rudrauf D, Knudsen J, Wallin D, Pendse G, McDonald L, Griffin M, Anderson J, Nutile L, Renshaw P, Weiss R, Becerra L, Borsook D (2010) Alterations in brain structure and functional connectivity in prescription opioid-dependent patients. Brain 133:2098–2114
46. Vanderschuren LJ, Ahmed SH (2013) Animal studies of addictive behavior. Cold Spring Harb Perspect Med 3:a011932
47. Volkow ND, Chang L, Wang GJ, Fowler JS, Ding YS, Sedler M, Logan J, Franceschi D, Gatley J, Hitzemann R, Gifford A, Wong C, Pappas N (2001) Low level of brain dopamine D2 receptors in methamphetamine abusers: association with metabolism in the orbitofrontal cortex. Am J Psychiatry 158:2015–2021
48. Volkow ND, Fowler JS, Wang GJ (2003) The addicted human brain: insights from imaging studies. J Clin Invest 111:1444–1451
49. Volkow ND, Fowler JS, Wolf AP, Schlyer D, Shiue CY, Alpert R, Dewey SL, Logan J, Bendriem B, Christman D et al (1990) Effects of chronic cocaine abuse on postsynaptic dopamine receptors. Am J Psychiatry 147:719–724
50. Volkow ND, Wang GJ, Begleiter H, Porjesz B, Fowler JS, Telang F, Wong C, Ma Y, Logan J, Goldstein R, Alexoff D, Thanos PK (2006) High levels of dopamine D2 receptors in unaffected members of alcoholic families: possible protective factors. Arch Gen Psychiatry 63:999–1008
51. Volkow ND, Wang GJ, Fowler JS, Logan J, Hitzemann R, Ding YS, Pappas N, Shea C, Piscani K (1996) Decreases in dopamine receptors but not in dopamine transporters in alcoholics. Alcohol Clin Exp Res 20:1594–1598
52. Volkow ND, Wang GJ, Fowler JS, Tomasi D, Telang F (2011) Addiction: beyond dopamine reward circuitry. Proc Natl Acad Sci U S A 108:15037–15042
53. Volkow ND, Wang GJ, Telang F, Fowler JS, Logan J, Jayne M, Ma Y, Pradhan K, Wong C (2007) Profound decreases in dopamine release in striatum in detoxified alcoholics: possible orbitofrontal involvement. J Neurosci 27:12700–12706
54. Volkow ND, Wang GJ, Tomasi D, Baler RD (2013) Obesity and addiction: neurobiological overlaps. Obes Rev 14:2–18
55. Wang GJ, Smith L, Volkow ND, Telang F, Logan J, Tomasi D, Wong CT, Hoffman W, Jayne M, Alia-Klein N, Thanos P, Fowler JS (2012) Decreased dopamine activity predicts relapse in methamphetamine abusers. Mol Psychiatry 17:918–925
56. Wang Y, Yin Y, Sun YW, Zhou Y, Chen X, Ding WN, Wang W, Li W, Xu JR, Du YS (2015) Decreased prefrontal lobe interhemispheric functional connectivity in adolescents with internet gaming disorder: a primary study using resting-state FMRI. PLoS One 10:e0118733
57. Welberg L (2011) Addiction: from mechanisms to treatment. Nat Rev Neurosci 12:621
58. Yuan K, Qin W, Wang G, Zeng F, Zhao L, Yang X, Liu P, Liu J, Sun J, von Deneen KM, Gong Q, Liu Y, Tian J (2011) Microstructure abnormalities in adolescents with internet addiction disorder. PLoS One 6:e20708

Erratum to: Cognitive-Behavioral Therapy

Hong An, Ri-Hui He, Yun-Rong Zheng, and Ran Tao

Erratum to:
Chapter 16 in: X. Zhang et al. (eds.), *Substance and Non-substance Addiction*, Advances in Experimental Medicine and Biology 1010, https://doi.org/10.1007/978-981-10-5562-1_16

The sequence of the authors' name was incorrect. The correct sequence is given below:

Hong An, Ri-Hui He, Yun-Rong Zheng, and Ran Tao

H. An • Y.-R. Zheng
English Department, Tianjin University of Technology and Education, Tianjin 300222, China

R.-H. He
RiHuiAddiction and Mental Disorders Medical Center, Guangzhou 510000, China

R. Tao (✉)
Department of Psychological Medicine, PLA Army General Hospital, Beijing 100700, China
e-mail: bjptaoran@126.com

The updated online version of the original chapter can be found under https://doi.org/10.1007/978-981-10-5562-1_16

X. Zhang et al. (eds.), *Substance and Non-substance Addiction*, Advances in Experimental Medicine and Biology 1010, DOI 10.1007/978-981-10-5562-1_18

Index

X. Zhang et al. (eds.), *Substance and Non-substance Addiction*, Advances in Experimental Medicine and Biology 1010, DOI 10.1007/978-981-10-5562-1

FSC
www.fsc.org
MIX
Papier | Fördert
gute Waldnutzung
FSC® C083411

Kolibri 360 GmbH
Ferdinand-Jühlke-Straße 7
99095 Erfurt, Deutschland
produktsicherheit@kolibri360.de